Healthcare Hazard Control and Safety Management

Healthcare Hazard Control and Safety Management

James T. Tweedy, M.S., CHCM, CHSP

Board of
Certified Healthcare
Safety Management

Bethesda, Maryland

LEWIS PUBLISHERS
Boca Raton London New York Washington, D.C.

Library of Congress Cataloging-in-Publication Data

Catalog record is available from the Library of Congress.

No claim to original U.S. Government works
International Standard Book Number 1-57444-035-7
Printed in the United States of America   3 4 5 6 7 8 9 0
Printed on acid-free paper

DEDICATION

With a father's heart of love I dedicate this book to the memory of my precious daughter, Elizabeth Cheryl Tweedy, who in her brief nineteen years touched the lives of so many for time and eternity. She taught all who knew her the true meaning of the words "love" and "friendship."

I have come to realize that the completeness of one's life cannot be measured in length of years but in the way in which we choose to live the time allotted us.

So I Pray
by
Elizabeth Cheryl Tweedy
18 January 1977 – 26 June 1996

In a world where sorrow is so common place
That undeniable pain is evident on every face
We are guided by theory
Led by a blackened light
And virtues are just a concept
Of a long forgotten time

Where happiness is misunderstood
And confusion reigns supreme
I look around and tremble
As I think of what it means

But I know who I am in You ...
Called out and commissioned by Christ
To go unto all the nations
And make the sacrifice

So I pray ...

Let me love with your love
Let me shine with your light
Let me care with your compassion, Lord
In this world black as night
Let me trust with all my heart
Let me speak only the truth
Let me teach so they might understand
What it means to know you

So I pray ...

Send me Lord today

TABLE OF CONTENTS

Board of Certified Healthcare Safety Management   xiii

Preface   xv

About the Author   xvii

1   Introduction to Hazard Control 1
- A.   Healthcare Hazard Control   1
- B.   Historical Perspectives   2
- C.   New Directions   2
- D.   Program Management   4
- Review Questions   5

2   Healthcare Safety Management 7
- A.   Healthcare Hazard Control   7
- B.   Healthcare Safety Management   8
- C.   1996 Joint Commission Environment of Care Standards   11
- D.   Safety Committees   14
- E.   Information Management   18
- F.   Risk Management   21
- G.   Quality Improvement   26
- H.   Occupational Health Program   30
- I.   Workers' Compensation   33
- J.   Human Resources Management   37
- K.   Review of Management   39
- Review Questions   41

3   Healthcare Hazard Control 43
- A.   Accidents   43
- B.   Hazards   44
- C.   Hazard Control   50
- D.   Worker Exposure   52
- E.   Personal Protective Equipment   54
- F.   Respiratory Protection   57
- G.   Emergency Wash Requirements   60
- H.   Accident Investigation   60

I.   Safety Training   65
J.   Promoting Safety   74
Review Questions   76

4 Safety Regulation **79**
A.   Federal Regulations   79
B.   Occupational Safety and Health Administration   81
C.   Environmental Protection Agency   91
D.   Food and Drug Administration   95
E.   Other Government Agencies   97
F.   Americans with Disabilities Act   100
G.   Joint Commission Standards   106
H.   Voluntary Compliance Agencies   111
I.   Professional Associations   115
Review Questions   117

5 Emergency Planning and Fire Safety **121**
A.   Emergency Planning   121
B.   External Disasters   124
C.   Internal Disasters   126
D.   Natural Disasters   130
E.   Other Emergency Situations   132
F.   Hazardous Materials   134
G.   Fire Safety   136
H.   Evacuation Procedures   139
I.   Ignitable Materials   140
J.   Compressed Gases   143
K.   Understanding Fires   145
L.   Fire Extinguishers   147
M.   NFPA 101, Life Safety Code   149
N.   NFPA 99, Healthcare Facilities   154
O.   Fire-Extinguishing Systems   154
Review Questions   155

6 General and Physical Plant Safety **159**
A.   Facility Fire Prevention   159
B.   Healthcare Ergonomics   159
C.   Office Safety   164
D.   Healthcare Lifting   167
E.   Preventing Slips and Falls   170
F.   Heat-Related Hazards   171
G.   Maintenance Department Safety   172
H.   Construction Safety   177
I.   Warning Signs   178
J.   Electrical Safety   179
K.   Confined Spaces   185
L.   Hearing Conservation   188
M.   Plant Management Areas   192
N.   Heating, Ventilation, and Air Conditioning System Operation   199
O.   Print Shop Safety   207
Review Questions   208

7   Managing Hazardous Materials **211**
A.   Characteristics of Hazardous Chemicals   211
B.   Hazard Categories   213
C.   Exposure to Hazards   216
D.   The OSHA Hazard Communication Program   218
E.   Hazardous Material Labeling   221
F.   Disinfecting Substances   223
G.   Glutaraldehyde/Formaldehyde   226
H.   Antineoplastic Drugs (Chemotherapeutic)   229
I.   Solvents   232
J.   Other Hazardous Substances   234
K.   Oxides   238
L.   Asbestos   240
M.   Miscellaneous Materials   244
N.   Joint Commission Standards   246
O.   Waste Management   246
P.   Hazardous Waste Regulation   248
Q.   Medical Wastes   250
R.   Waste Disposal   252
S.   Hazardous Waste Training   259
Review Questions   261

8   Biological Hazard Control **265**
A.   Infection Control Guidelines   265
B.   Infection Control   267
C.   Working Safely with Sharps and Needles   272
D.   Bloodborne Pathogens   274
E.   Hepatitis   277
F.   Human Immunodeficiency Virus (HIV)   280
G.   Tuberculosis   281
H.   Summary of Exposure Control Management Elements (29 CFR 1910.1030)   286
I.   Infectious Waste Management   290
Review Questions   291

9   Safety in Patient Care Areas **293**
A.   Healthcare Lifting Hazards   293
B.   Wheelchair Safety   297
C.   Preventing Slips and Falls   298
D.   Medication Safety   303
E.   General Safety in Patient Areas   306
F.   Bed Safety   307
G.   Special Nursing Concerns   309
H.   Dealing with Patient Emotions   312
I.   Helicopter Safety   315
J.   Managing Stress   316
K.   Volunteer Safety   319
L.   Home Health Services Safety   319
M.   Other Patient Care Areas   321
N.   Central Supply   323
O.   Pharmaceutical Safety   324
Review Questions   328

10 Healthcare Support Area Safety **331**
   A. Environmental Services   331
   B. Laundry Safety   336
   C. Food Service Safety   337
   D. Facility Security   344
   E. Radiation Safety   350
   F. Laser Safety   363
   G. Laboratory Safety   365
   Review Questions   382

11 Healthcare Safety Terms and Abbreviations **385**

12 Appendices **399**
   A. Sample Hazard Communication Program   399
   B. Bloodborne Pathogens Exposure Control Plan (29 CFR 1910.1030)   404
   C. Chemical Hygiene Plan   414
   D. Personal Protective Equipment   421
   E. Healthcare Safety Orientation/Training Record   423
   F. Trade Association Indoor Air Quality Guidelines   425
   G. OSHA Laboratory Training Requirements   427
   H. Sample Substance Abuse Policy   429
   I. Sample Lockout/Tagout Policy   433
   J. Sample Safety Policy Statement with Rules   438
   K. Sample OSHA HAZCOM Training Plan   440
   L. Healthcare Emergency Drill Evaluation   446
   M. Suggested Precautions for Exposure to Blood or Body Fluids   449
   N. Suggested Precautions for Respiratory Care Exposures   452
   O. Accident Investigation Report   453
   P. Ergonomic (CTD) Pain/Strain Report   455
   Q. Hearing Conservation Acknowledgment   457
   R. Safety and Health Program Assessment Worksheet   458
   S. Sample Healthcare Facility Slip, Trip, and Fall Report   461
   T. Sample Equipment Problem Reporting Form   462
   U. Sample Hazard Correction Form   463
   V. Sample Bomb Threat Checklist   464

13 Checklists **465**
   1. Nursing Services   465
   2. Basic Life Safety Considerations   467
   3. Healthcare Facility Slip, Trip, and Fall Prevention   468
   4. Food Service Safety   471
   5. Healthcare Facility Electrical Safety   474
   6. Hazard Communication Program Evaluation   477
   7. Annual Safety Management Evaluation   481
   8. Sample OSHA Compliance for Healthcare Organizations   483
   9. Personal Protective Equipment   486

14 Agencies **489**

15 Bibliography **491**

Index **495**

BOARD OF CERTIFIED HEALTHCARE SAFETY MANAGEMENT

The Board of Certified Healthcare Safety Management (BCHSM) is pleased to be co-publishers with GR/St. Lucie Press of *Healthcare Hazard Control and Safety Management,* which provides healthcare industry safety managers with a quality, in-depth source that comprehensively covers both the management and technical aspects of healthcare safety.

This volume provides information and answers to questions relating to government and voluntary standards and management techniques and will also serve as a valuable resource for seasoned practitioners, as well as providing focus and direction for candidates in preparation for the CHSP examination.

For information regarding the Certified Healthcare Safety Professional (CHSP) and the Certified Hazard Control Manager (CHCM) programs, contact the Board at 11900 Parklawn Drive, Suite 451, Rockville, MD 20852 (phone: 301-770-2540).

PREFACE

This work began more than two years ago as a joint project to develop a study manual for those preparing for the Certified Healthcare Safety Professional (CHSP) examination. As the project progressed, a decision was made by the author and the Board of Certified Healthcare Safety Management to publish the information as a hardcover book.

The author has attempted to produce a single-volume resource that addresses every major area of healthcare safety. The book is broad in scope but focuses primarily on safety in acute care hospitals. However, much of the information has application to other healthcare organizations, such as subacute facilities, surgery centers, and long-term care organizations. The author does not place emphasis on construction or engineering aspects of healthcare safety but rather attempts to present the principles of healthcare safety management and hazard control.

The text is written in an easy-to-use format that includes tables, lists, and bulleted paragraphs. Multiple-choice review questions appear at the end of each chapter. The text also includes several appendices which contain management plans, a glossary of healthcare safety terms, and healthcare-specific safety checklists.

The author acknowledges that many of the hazards found in healthcare organizations cannot be comprehensively addressed in a text of this type. Readers are encouraged to refer to other texts and government publications that address specific healthcare safety topics such as radiation safety, hazardous waste management, and infection control. Voluntary compliance organizations such the Joint Commission on Accreditation of Healthcare Organizations and National Fire Protection Association also publish safety material specifically for healthcare applications.

This book is presented with the sincere hope that it will fill a void in the growing discipline of healthcare safety management. The author trusts it will provide both the seasoned professional and the newcomer to the field with an accurate overview of healthcare safety and a study resource for the CHSP examination.

ACKNOWLEDGMENTS

I would like to thank the Board of Certified Healthcare Safety Management for co-publishing this book. The project would have never materialized without the vision, support, and encouragement of Mr. Harold M. Gordon, Executive Director of the Board.

A very special thanks goes to Dianna Peck for helping prepare the script for submission and for creating the many checklists and tables that appear in the text.

Most of all I would like to thank my wife, Marlene, my son, Aaron, and my daughter, Cheryl, for their love, support, and understanding during this lengthy project.

James Tweedy

ABOUT THE AUTHOR

James T. Tweedy is an experienced safety consultant, risk manager, and educator. He is President of TLC Services, a Helena, Alabama consulting and training company that specializes in conducting Certified Healthcare Safety Professional (CHSP) review seminars. Mr. Tweedy holds a Master of Science degree in Safety Management from Central Missouri State University and a Bachelor of Science degree in Liberal Arts from the University of the State of New York. He is a Certified Healthcare Safety Professional (CHSP), a Certified Hazard Control Manager (CHCM), and a Professional Member of the American Society of Safety Engineers.

Mr. Tweedy can be contacted at P.O. Box 213, Helena, AL 35080 or by phone at (205) 621-0464.

ABOUT THE AUTHOR

CHAPTER 1

INTRODUCTION TO HAZARD CONTROL

A. HEALTHCARE HAZARD CONTROL

Hazard control management is a proven means of controlling hazards and promoting safe behavior within an organization. The healthcare industry faces change on a daily basis through reorganizations, technological advances, and increased competition for market share. The human factors, including worker and visitor safety, must be addressed in this era of change. Organizations seeking to increase revenues, reduce losses, and control costs must have effective loss management programs. Hazard control management is a proven process that produces results by preventing accidents, reducing injury rates, and increasing organizational efficiency in the "environment of care" era. Topics addressed in the text include:

- Laws, standards, and compliance agencies
- Emergency planning and fire safety
- Managing hazardous materials and wastes
- Infection control and biohazard safety
- Safety in patient care areas
- Laboratory and radiation safety
- Safety in primary healthcare support functions

Benefits of Hazard Control Management—An effective safety management or hazard control program must go beyond meeting only minimum requirements found in regulations, codes, and standards. The truly effective program will be proactive and strive to reduce risks, prevent injuries, and protect patients, visitors, and staff. An effective hazard control program can:

- Enhance environment of care and quality assurance efforts
- Reduce insurance costs and premiums
- Control medical payments and workers' compensation costs

- Improve employee morale
- Reduce time lost due to absences
- Increase productivity and efficiency
- Improve facility attractiveness
- Provide a safe environment for workers, patients, and visitors
- Assist with compliance with safety and health regulations

B. HISTORICAL PERSPECTIVES

Healthcare safety traditionally has focused on patient welfare, safety engineering, and life safety issues. During past 20 years, many healthcare organizations made the safety program the responsibility of the risk management, quality assurance, plant engineering, or security department. During this same period, "healthcare safety or hazard control" has emerged as a discipline with special problems, challenges, and hazards. The Joint Commission on Accreditation of Healthcare Organizations and the American Hospital Association have done much to promote the importance of healthcare safety management.

C. NEW DIRECTIONS

The Joint Commission continues to place a strong emphasis on safety, as indicated in the Environment of Care standards first published in 1995. The philosophy of these new standards is similar to a "systems safety approach" which integrates safety into the environment of care philosophy for the entire organization. The fact is that health-care facilities differ in many ways, and these new standards allow flexibility in meeting a standard's intent. The Joint Commission recently dropped the requirement for organizations to have a "safety committee." This action further illustrates the Joint Commission's move to allow for organizations to develop and implement safety programs that will best meet the needs of their employees, visitors, and patients. The "management by committee" concept does provide a mechanism to manage or control the wide variety of hazards found in healthcare environments. However, in many organizations, managing by a committee did not result in an effectively managed program.

Healthcare Employment Trends—Controlling hazards, managing risks, and maintaining proactive safety management programs continue to be challenges in many healthcare settings. The discipline of healthcare hazard control continues to be overlooked despite the number of workers employed in healthcare-related jobs. Advances in medical technology and clinical treatment techniques expose workers to a variety of occupational hazards that must be controlled. The following information published by the U.S. Department of Labor illustrates the challenges facing healthcare safety professionals:

- According to the Bureau of Labor Statistics (BLS), health services (SIC 80) employed over 8 million workers in 1992. More than half that number worked in

TABLE 1.1 Occupational Incidence Rates

Number of Cases and Incidence Rates (in thousands)[a]
for Industries with 100,000 or More Injury Cases, 1990 and 1991

Industry	SIC Code[b]	Total Cases 1990	Total Cases 1991	Incidence Rate 1990	Incidence Rate 1991
1. Eating and drinking establishments	581	353.6	310.0	8.4	7.4
2. Hospitals	806	282.4	308.9	10.0	10.8
3. Grocery stores	541	246.4	238.4	12.1	11.7
4. Trucking and courier services	421	203.8	200.6	14.1	14.4
5. Nursing and personal care facilities	805	168.2	174.1	15.4	15.0

[a] Incidence rates represent the number of injuries per 100 full-time workers and were calculated as: (N/EH) × 200,000.
[b] Standard Industrial Classification Manual, 1987 edition.

Adapted from information published by the U.S. Department of Labor, Bureau of Labor Statistics (1993).

hospitals and personal care facilities. The outlook is for continued employment growth in healthcare-related services during the next 10 years.

- According to the BLS, hospitals employ more than 3.5 million workers who support more than 900,000 beds in over 5000 facilities. Long-term and personal care nursing account for another 1.5 million workers. A workforce of more than 5 million remains a challenge to protect.

Private hospitals reported more than 200,000 occupational injuries in 1993. Long-term and personal nursing care facilities reported 18.6 work-related injuries per 100 employees in 1992. This rate is more than double the rate of 8.9 reported for all of general industry, which includes manufacturing, mining, and construction data. Tables 1.1 and 1.2 illustrate the injury experiences of healthcare facilities as compared to other service-type industries.

TABLE 1.2 Injury Comparison Data

	Hospital	Private industry
Incident rate of occupational injury and illness cases per 100 full-time workers	10.6	8.8
Incidence rate of lost workday cases per 100 full-time hospital workers	4.2	4.1
Incidence rate of lost workdays per 100 full-time hospital workers	81.6	84.0
Average number of employees in hospital industry	3.7 million	

Adapted from information published by the U.S. Department of Labor, Bureau of Labor Statistics (1990).

D. PROGRAM MANAGEMENT

Top healthcare administrators and executives need to evaluate the organizational structure and effectiveness of their safety management programs. A well-organized safety committee can play an important role in helping control healthcare hazards, but the role should shift from management to advisory. Healthcare safety programs should be coordinated by a qualified healthcare safety specialist with the authority and responsibility to manage the program on a daily basis.

Management Functions Related to Safety—Several management functions play important roles in healthcare hazard control management. An effective risk management program can provide valuable information to enhance safety effectiveness. Continuous quality improvement programs can also identify problems and make suggestions to improve safety, especially in patient care functions. Safety management recognizes the role that human behavior plays in preventing accidents and reducing losses. Human behavior is concerned with a number of issues, such as education, training, and motivation. Human resource management and education programs can be instrumental in contributing to the success of the safety program by hiring and training qualified personnel.

Joint Commission Environment of Care Standards—The Joint Commission places strong emphasis on safety management and has included safety in its Environment of Care standards. These standards address seven areas considered crucial to safety within an organization:

1. Safety management
2. Security management
3. Hazardous materials and waste management
4. Emergency preparedness
5. Life safety
6. Medical equipment management
7. Utility systems management

Occupational Safety and Health Administration (OSHA) Standards—OSHA regulates employee safety under its General Industry Standards found in Title 29 CFR Part 1910. Some of the key OSHA standards that apply to healthcare facilities are as follows:

1. *Hazard Communication Standard (29 CFR 1910.1200)*—This standard requires that all workers be made aware of all hazardous substances that they encounter or work with in the workplace.
2. *Bloodborne Pathogens Standard (29 CFR 1910.1030)*—This relatively new standard requires facilities to develop an exposure control plan and train all workers that could be potentially exposed to pathogens found in blood or body fluids.
3. *Hazard-Specific Standards*—OSHA has also published many hazard-specific standards that cover hazards such as ethylene oxide, formaldehyde, benzene, asbestos, lead, ionizing radiation, and noise.

4. ***OSHA Tuberculosis Regulation***—OSHA enforces the Centers for Disease Control and Prevention (CDC) guidelines under the general duty clause of the OSH Act. OSHA recently published a proposed TB standard that could become a standard late in 1998.

OSHA Safety Management Guidelines—OSHA recently published some non-mandatory guidelines that healthcare organizations can use to improve worker safety. Healthcare facilities should consider the following elements essential to any effective program:

* ***Management Commitment***—Top management must demonstrate total commitment to the program. An effective program considers worker safety and health as a fundamental responsibility of the organization.
* ***Worksite Analysis***—Management must ensure that effective worksite hazard surveys are conducted. Hazard information must be accurately analyzed to better permit the organization to anticipate and prevent occurrences.
* ***Hazard Prevention and Control***—Organizations should stress accident prevention and safe work practices to all employees. Actions should be taken to control hazards by effectively designing work areas or the job task itself. When it is not feasible to eliminate hazardous conditions, the organization must implement measures to protect individuals from unsafe conditions and/or unhealthy exposures.
* ***Employee Training***—Training is the key to success. The type and amount of training depend on the type, size, and complexity of the organization. Training also is based on the nature of hazards, risks, or potential exposures.

SUMMARY

Hazard control management will continue to be a key function as organizations meet the challenges of the changing healthcare environment. Organizations will continue to look for ways to reduce costs, meet new demands, and maintain the quality of care. Healthcare hazard control as a safety management discipline can help organizations meet the challenges by ensuring the safety of patients, visitors, and employees.

REVIEW QUESTIONS

1 Healthcare safety programs have historically focused on:
   a. patient welfare
   b. safety engineering
   c. fire safety issues
   d. all of the above

2   The Joint Commission Environment of Care standards emphasize:
   a.   integrating safety into the operating philosophy of the organization
   b.   stricter standards with no flexibility in application
   c.   the need to develop a program to meet organizational requirements
   d.   a and c

3   The Bureau of Labor Statistics projects that the total number of healthcare workers
   will _____ over the next 10 years.
   a.   decrease slightly
   b.   increase steadily
   c.   not change significantly
   d.   follow cyclical patterns

4   Which of the following concerning OSHA tuberculosis regulation is true?
   a.   OSHA does not regulate tuberculosis exposure in healthcare facilities
   b.   the OSHA bloodborne pathogens standard also covers tuberculosis exposures
   c.   OSHA currently uses CDC guidelines to regulate exposure under its general
        duty clause
   d.   none of the above is true

5   Which of the following is/are *not* true about the illness and injury incidence rates
   experienced by healthcare organizations?
   a.   nursing facilities had a higher rate than hospitals in 1991
   b.   long-term nursing facilities had a rate twice that of general industry in 1992
   c.   in both 1990 and 1991, nursing and personal care facilities experienced rates
        lower than the trucking industry
   d.   a and b

Answers: 1–d   2–d   3–b   4–c   5–c

CHAPTER 2

HEALTHCARE SAFETY MANAGEMENT

A. HEALTHCARE HAZARD CONTROL

Most healthcare organizations, especially acute care and subacute care facilities, have established safety management programs based on the requirements of the Joint Commission on Accreditation of Healthcare Organizations (JCAHO). However, many facilities do not have safety programs that adequately address the hazards, exposures, and risks found in their organizations. A written hazard control management program must establish clear objectives, provide an action plan, and stress the importance of communicating safety issues to all members. An effective program must go beyond meeting the minimum requirements of regulations, codes, and standards; be proactive; and strive to reduce risks, prevent accidents, and reduce human injury.

Key Program Elements—Healthcare hazard control management emphasizes:

- Comprehensive safety orientation program for new employees
- Published accident, injury, and incident reporting procedures
- Documentation and follow-up on all safety-related problems
- Medical device safety, recall, and reporting procedures
- Information collection and evaluation system
- Coordinated quality improvement and risk management programs
- Reporting system to keep top management advised
- Hazard surveillance and written safety programs for all departments
- Recurring safety training provided to all workers

Basic Hazard Control Terms—

- *Risk*—A chance loss will occur under the right conditions
- *Hazard*—A condition or practice with potential for loss under the right circumstances
- *Safety*—Human actions to control, reduce, or prevent accidental loss while still accomplishing the task

- *Loss Control Management*—Applying management principles to control losses resulti.g from the risk(s) associated with task completion.

B. HEALTHCARE SAFETY MANAGEMENT

Several management functions play important roles in healthcare hazard control management. An effective risk management program can provide valuable information to enhance safety effectiveness. Continuous quality improvement programs can also identify problems and make suggestions to improve safety, especially in patient care functions. Safety management recognizes the role that human behavior plays in preventing accidents and reducing losses. Human behavior is concerned with a number of issues, including education, training, and motivation. Table 2.1 lists key safety management program elements.

Safety Involvement by Management—Senior management personnel must provide leadership and demonstrate commitment to the program by:

TABLE 2.1   Recommended Safety Management Program Elements

Administrative Support	• Appoint a hazard control coordinator, employee health director, and other responsible personnel • Allocate time for surveys, committee meetings, and education • Budget funds to evaluate and monitor hazards, implement controls, and conduct necessary health examinations
Hazard Identification	• Conduct regular walk-through inspections • Obtain Material Safety Data Sheets, regulatory standards, and other information on potential hazards • Maintain hazardous chemicals listing of materials that are used or stored in each department
Hazard Evaluation	• Conduct safety inspections and industrial hygiene surveys and determine needs for hazard controls • Conduct medical evaluations for exposed workers • Establish medical surveillance programs to monitor employee health
Training	• Develop an orientation program for new personnel and conduct recurring training for workers based on job assignments and hazard exposures
Control	• Select and implement controls and medical surveillance programs
Program Review	• Review safety inspection reports, industrial hygiene monitoring surveys, and medical surveillance results to determine patterns and measure effectiveness of the program • Update the program as new materials or procedures are introduced or as new hazards are identified
Recordkeeping	• Maintain records of surveys, evaluations, corrective actions, and worker medical examinations; records must be maintained in accordance with local, state, and federal regulations

- Publishing a safety policy that expresses a commitment to safety
- Establishing realistic program goals and expectations
- Providing planning support to ensure goal accomplishment
- Communicating the importance of the program to all employees
- Assigning responsibilities and authority to qualified personnel
- Holding all employees accountable through a disciplinary policy

Worker Safety Involvement—Administrators, department heads, and supervisors should encourage workers to stay actively involved by:

- Appointing employees to positions in the safety program
- Placing hourly workers on safety committees
- Requiring workers to report accidents and injuries immediately
- Providing a quick response to concerns about safety
- Assessing and correcting problems and hazardous conditions
- Training and educating workers on a recurring basis
- Promoting safety on a continuous basis
- Developing an effective preventive maintenance program

Safety Action Plans—Specific action plans must:

- Establish a target date for completion
- Determine amount of monetary investment
- Identify accountable individuals
- Provide guidance for putting together the support team

Safety Program Success—Safety programs developed to fit the needs of an organization can succeed if properly managed. There are several reasons why safety programs produce results for an organization. Successful safety programs:

- Stress results-oriented activities based on defined goals
- Investigate and analyze causal factors that produce loss
- Develop a management action plan in addition to publishing policies
- Establish measurement criteria to assess program effectiveness
- Publish contingency plans to deal with potential problems

Written Safety Program Elements—When developing a formal hazard control program, an organization should:

- Consider the program structure that would best serve the facility and ensure program effectiveness
- Assign responsibilities and delegate authority to a hazard control manager
- Establish lines of communication within the safety management function
- Develop, promote, and properly fund a comprehensive safety orientation, education, and training program
- Promote safety issues such as accident prevention, workers' compensation cost containment, and the importance of patient and visitor safety
- Implement written procedures on incident reporting, accident investigation requirements, and program evaluation methods

- Ensure that the health program monitors employee health issues and provides quality medical treatment as required
- Establish a security program that will ensure personal safety and minimize property losses of all patients, workers, contractors, and visitors
- Develop emergency plans to address internal and external contingencies

Safety Program Assessment—Healthcare administrators and executives need to evaluate the organizational structure and effectiveness of their safety management programs. A well-organized safety committee can play an important role in helping to control healthcare hazards, but the role should shift from management to advisory. Healthcare safety programs should be coordinated by a qualified healthcare safety specialist with the authority and responsibility to manage the program on a daily basis.

Written Program Development Steps—There are several key steps that should be followed when developing or revising a written healthcare safety management program:

Step 1. Initial Assessment
- Review statistics, claims, trends, and severity/frequency rates
- Evaluate effectiveness and scope of current loss control practices
- Talk with workers to get their feelings and reactions
- Identify any other problem areas that impact loss control efforts

Step 2. Assigning Authority/Responsibility
- Appoint a safety officer, director, or coordinator to manage the program
- Assign responsibilities and delegate authority to ensure program success
- Publicize management's commitment to the program
- Ensure all employees understand their responsibilities

Step 3. Publishing the Organizational Safety Policy Statement
- Communicate that the facility seeks ways to reduce injuries and accidents
- Emphasize a commitment to maintain a safe and healthful workplace
- Outline the main points to be emphasized in the program

Step 4. Establishing Documentation and Training Procedures
- Determine recordkeeping requirements for injury and accident reports
- Obtain all required regulations, publications, and standards
- Publish written safety policies, job procedures, and safety regulations
- Establish documentation requirements for all employee training sessions
- Develop concise and complete job descriptions for all employees

Step 5. Developing Evaluation Guidelines
- Determine how frequently the program will be evaluated for effectiveness
- Involve department heads by requiring periodic self-inspections
- Safety officer should evaluate each department on a regular basis

Step 6. Establishing a Safety Committee
- Oversees the safety management program
- Takes a proactive role in promoting and overseeing safety activities
- Has the authority to cross departmental boundaries

- Takes all actions necessary to accomplish program objectives
- Structured to meet the needs of the organization

Note—Steps may not always be sequential.

Safety Program Failure—There are many factors that affect safety management program effectiveness. Safety management programs must be developed to address the needs of the organization. Many well-designed programs fail because organizations:

- Focus on activities instead of organizational and behavioral elements
- Fail to communicate goals and objectives to all workers
- Do not investigate to determine root causes for losses
- Adopt a "one-size-fits-all" safety program which ensures failure
- Do not develop clearly defined goals and objectives

Factors Contributing to Losses—Losses generally result from a number of causal factors. Loss research should include a detailed analysis of accident data, turnover rates, absenteeism information, and quality assurance data. Causal factors can generally be classified in one of three categories:

- *Organizational Factors*—Relate to the management system, knowledge levels, or current safety policies
- *Motivational Factors*—Causal factors that can be attributed to the attitude, morale, and dedication of workers at all organizational levels
- *Operational Factors*—Result from physical hazards, occupational hazards, and health exposures encountered during work

C. JOINT COMMISSION ENVIRONMENT OF CARE STANDARDS

Environment of Care standards focus on providing a safe and functional environment for patients, staff members, visitors, and contractor personnel. The Environment of Care standards do not address designated structures or methods to manage the environment of care. However, the following should be stressed:

- Ways to reduce and control hazards/risks in the environment
- Implementing activities to prevent accidents and injuries
- Maintaining safe conditions for patients, visitors, and staff

Safety-Related Standards—Healthcare organizations are expected to design a safe and accessible environment consistent with their goals, mission, and laws/regulations. When providing a safe environment, the following management plans should be developed and implemented:

- Safety
- Security
- Hazardous materials
- Emergency planning

- Life safety
- Medical equipment
- Utility systems

Safety Management Plans—A safety management plan addresses how the organization will maintain a hazard-free environment and manage human activities to reduce the chance of injuries or loss. The plan must address the following activities:

- Grounds and equipment maintenance
- Conducting a proactive risk assessment program
- Continuous review of safety issues that arise
- Accident/incident reporting and investigation procedures
- The hazard surveillance program
- Identifying personnel to intervene in situations that pose immediate threat to life or health
- Training and education programs designed to meet the facility's needs
- Performance standards for safety knowledge, worker participation, inspection activities, incident reporting effectiveness, safety policy review (every three years), preventive maintenance effectiveness, and procedures to evaluate the safety management program at least annually

1. *Safety Management Program*—A documented management plan must include provisions for establishing and maintaining a safety management plan based on monitoring and evaluation of the facility's experiences, applicable laws and regulations, and operating practices. The plan must also:
   - Provide a physical environment free of hazards and outline activities that will reduce risk of human injury
   - Ensure the safe maintenance of all grounds, facilities, and equipment
   - Provide clearly identified and accessible emergency service areas
   - Establish a risk assessment program to evaluate safety
   - Provide for the appointment of a safety officer by the chief executive officer who is qualified by education or experience to monitor the organization's safety program
   - Establish accident investigation procedures for all incidents involving personal injury or property damage
   - Require departmental safety programs that are practiced and enforced
   - Promote a continuous hazard surveillance program, including response procedures for product safety recalls
   - Require that safety-related information is used in orientation and continuing education of all employees
   - Provide authority for the safety officer to intervene in situations that pose an immediate threat to life or health
   - Require an annual evaluation of all program objectives, performance, and effectiveness of the documented safety plan

2. *Safety Orientation and Education*—The following safety issues must be addressed:

- General safety processes
- Departmental safety requirements
- Specific procedures for job-related hazards
- Use of safety-related information in orientation and continuing education sessions

3. *Safety Management Performance Standards*—The Joint Commission standards require organizational assessment of:
   - Staff knowledge and skills required by the safety program
   - Monitoring and inspection activities
   - Emergency and accident reporting procedures
   - Inspection, preventive maintenance, and testing of safety equipment

4. *Safety Surveys*—Surveys must be conducted to identify environmental deficiencies, hazards, and unsafe practices. The Joint Commission requires semi-annual surveys in patient areas and a yearly survey in non-patient areas. Surveys should:
   - Focus on physical and environmental hazards
   - Emphasize human behavior and compliance with safety policies
   - Identify hazards and make recommendations for corrective actions

5. *Safety Program Improvement*—The Joint Commission suggests that healthcare facilities follow four basic steps to facilitate safety improvement:
   - Identify safety-related problems and hazards
   - Analyze hazards, unsafe conditions, and problem areas
   - Develop and implement corrective actions
   - Measure and evaluate program success on a regular basis

Organizational Safety Officer—Many healthcare organizations use the term safety director or officer. Regardless of the term used, hazard control management efforts should be directed by a qualified person. The Joint Commission recommends that personnel assigned to this position possess formal safety education and/or job-related experience. Table 2.2 lists some duties of the safety officer. Some important qualifications also include:

- Knowledge of management principles and organizational dynamics
- Effective communication to all levels within the organization
- Understanding of employee monitoring and safety evaluation requirements
- Ability to define safety issues, investigate accidents, and prioritize safety committee activities

Joint Commission Safety Officer Appointment Guidelines—

- Safety officers of a facility with less than 125 beds should devote 25% of their time to safety management issues
- Safety officers of organizations with 125 to 250 beds should devote 50% of their time to safety management
- Organizations with more than 250 beds should employ a full-time safety officer, director, or manager

TABLE 2.2   Listing of Safety Officer Duties

- Monitor accidents and maintain complete injury records
- Investigate and analyze each mishap to identify trends and problem areas
- Develop procedures to track and monitor individual employee injuries
- Be knowledgeable of workers' compensation law and procedures
- Be knowledgeable of Joint Commission standards, OSHA standards, other federal/state regulations, and state regulatory requirements
- Ensure that life safety and emergency drills are held as required and properly documented
- Keep administration informed of all work-related injury matters
- Establish and monitor required hazard communication and exposure control programs
- Identify hazardous/infectious waste and monitor disposal activities
- Conduct a periodic visit to each department to identify problems and weaknesses
- Ensure that all personnel receive a detailed initial safety orientation, monthly recurring in-service safety education, and specialized safety training as required; monitor safety training program effectiveness and maintain comprehensive training records

D. SAFETY COMMITTEES

The 1996 Joint Commission Environment of Care standards dropped the requirement for organizations to have a safety committee. However, a well-organized safety committee can be beneficial to safety program effectiveness.

1. *Committee Membership*—The committee should have non-supervisory personnel appointed as members. Representation from several healthcare functions should be considered:
   - Administration and executive management
   - Professional medical staff
   - Nursing services
   - Patient affairs
   - Risk and quality management
   - Legal counsel
   - Maintenance and engineering
   - Environmental services and housekeeping
   - Infection control
   - Employee health
   - Laboratory and radiation services
   - Pharmacy and central supply
   - Food and nutrition services

2. *Committee Responsibilities*—Committee responsibilities will vary depending on the organization. Some basic responsibilities include:
   - Reviewing progress reports of committee functions
   - Monitoring the activities of all safety subcommittees
   - Assigning special projects to individuals or hospital departments
   - Monitoring patient, visitor, and internal variance summary report

- Evaluating safety education and employee health reports
- Ensuring compliance with Occupational Safety and Health Administration (OSHA), National Fire Protection Act (NFPA), and Environmental Protection Agency (EPA) standards
- Promoting safety throughout the organization

3. *Committee Functions*—The safety committee should periodically review and evaluate the following:
   - Employee safety and health programs
   - Workers' compensation cost containment programs
   - Patient, visitor, and contractor safety
   - Employee training and safety education programs
   - Emergency preparedness plans and drills
   - Employee health management
   - Departmental safety program effectiveness
   - Hazardous materials and waste management programs
   - Security reports
   - Fire drills, inspections, and prevention activities
   - Equipment safety, utility management, and product recall
   - JCAHO, OSHA, EPA, NFPA, state, and local regulatory compliance

4. *Evaluating Program Effectiveness*—The committee should consider the following when evaluating effectiveness of the program:
   - Degree of compliance with published policies and procedures
   - Individual departments' adherence to their own safety policies
   - Actions taken to correct safety-related deficiencies or hazards
   - Communication with department heads about corrective actions
   - Levels of risk found in each patient care area
   - Review of program effectiveness, problems, priorities, and mission

5. *Committee Authority*—The safety committee must be organized under the direction, control, and authority of top hospital administration or the organization safety officer. The scope of authority involves staff support, guidance, coordination, training, recommendations, and actions to assure effectiveness of the program. The safety committee can be given authority to:
   - Cross operational and departmental boundaries
   - Recommend actions and purchase equipment
   - Delegate operational responsibilities when necessary
   - Monitor remedial actions of operations, departments, and employees
   - Make key recommendations to administrator on safety-related issues

   Note—Executive management retains authority and responsibility for ensuring program effectiveness, including regulatory compliance. Management must provide supporting systems, authority, financial resources, and information sources to achieve all objectives.

Safety Committee Organization—The committee membership should include representatives from several disciplines and individuals with knowledge of organizational goals, methods, and practices. Some important considerations are as follows.

1. *Quorum*—It is recommend that a quorum of three-fourths of the members must be present to hold a meeting. The chairman should decide whether a meeting will be held or rescheduled, depending on the importance of the issues scheduled to be addressed.

2. *Chairperson Responsibilities*—
   - Preside at all meetings of the safety committee
   - Plan the agenda of the meeting based upon recommendations from the committee membership, hospital departments, report reviews, regulatory and administrative directives, and safety inspections
   - Coordinate all hospital safety programs and activities with the assistance of the safety committee and education department
   - Maintain communication and advise the hospital administrator of safety problems, programs, infractions, and actions
   - Submit a copy of the meeting minutes and other information to the administrator, board of directors, and the quality assurance committee

3. *Meeting Minutes*—Record of minutes should include:
   - Date and time of meeting
   - Names of persons attending
   - Names of persons absent
   - Corrections to minutes of last meeting
   - Review of items from previous meeting
   - Review of variance summary reports
   - Review of safety reports and corrective actions
   - Actions of subcommittees
   - Announcements, meeting information, and upcoming events
   - Time of adjournment and signatures of chairman and recorder

4. *Recorder Duties*—
   - Keep the minutes of all meetings
   - Arrange for alternate meeting sites when necessary
   - Send notices, agendas, and minutes to members
   - Prepare committee correspondence as required

5. *Safety Committee Meetings*—
   - Meet as necessary but at least every other month
   - Special meetings called by the chairman as necessary
   - Sufficient number of members in attendance to have a quorum

Safety Committee Involvement in Organizational Programs—An effective safety committee should be involved in a number of programs that affect the safety and health efforts throughout the facility. The safety committee should work hard to ensure that a promotional program is maintained that focuses on:

- Improving job knowledge as a deterrent to unsafe practices and personal accidents through annual mandatory in-service training
- Using available media such as bulletin boards, newsletters, and films to promote safety issues

- Promoting the responsibility of good housekeeping practices
- Educating employees about OSHA and other regulatory requirements

Healthcare Safety Management Components—Listed below are some programs that should be monitored by the safety officer and committee.

1. *Departmental Safety Programs*—Each department should develop a safety program. Results of departmental surveys should be communicated to the safety officer and safety committee. Department heads should periodically attend a safety committee meeting to assess specific safety rules and policies.

2. *Electrical and Equipment Safety*—The safety officer and committee should coordinate and monitor the electrical safety program with the maintenance/engineering department to ensure that:
   - Proper procedures are used to purchase electrical devices and patient care equipment
   - All equipment is tested before being placed into service in patient care areas
   - Preventive maintenance and user inspections are accomplished using approved instruments
   - All inspections, testing, maintenance, and repairs are documented
   - A training program with appropriate levels of involvement of the entire staff has been developed
   - Suspected accidents involving electrical devices are evaluated
   - Product failures or recalls are investigated and evaluated by the facility or manufacturer

3. *Emergency Planning*—The safety officer and committee should be involved in the planning, writing, revision, and evaluation of the fire plan, internal disaster plan, and external disaster plan.

4. *Fire Prevention Program*—The safety officer, plant engineer, and safety committee should coordinate the fire prevention and inspection program. The program should include the following:
   - Quarterly fire drills for all shifts
   - Written report of fire drill evaluations
   - Drill critiques, with emphasis on assessing training needs
   - Quarterly fire inspection of each fire zone
   - Procedures to assess equipment and test alarms

5. *Hazardous Materials Management*—The safety officer and committee should monitor the effectiveness of the program to ensure that it includes:
   - Inventory and categorization of hazardous substances to include chemicals, antineoplastics, medical waste, and radioactive materials
   - Identification of degree of hazard of daily use and exceptional use by injury classification (NFPA labels and target organ information)
   - Storage, labeling, handling, and disposal procedures for each substance
   - Accident/exposures of substances identified as potentially injurious by creating a hazardous materials listing

   • Education and training as necessary to meet HAZCOM and emergency response requirements

6. *Occupational Health Program*—The safety committee and safety officer should monitor the effectiveness of the employee health program. This service should be organized as a separate hospital function to include:
   • Pre-employment health screening and services
   • Employee education for health promotion
   • Employee illness analysis and reporting
   • Quarterly reports to the committee

7. *Radiation Safety*—The committee should work with the radiation committee and radiation safety officer to ensure that all individuals who work with or in the vicinity of radioactive material have sufficient training and experience to enable them to perform their duties safely and in accordance with Nuclear Regulatory Commission regulations and the conditions of the license.

8. *Security Management Program*—The committee should support the security and workplace violence programs that protect patients, visitors, employees, property, equipment, and valuables. The committee should ensure that security assessments and evaluations are conducted periodically.

9. *Training and Education*—The organization should place a strong emphasis on training program effectiveness. Training in the following subjects should be provided as required:
   • General safety policies and procedures
   • Electrical safety policies and procedures
   • Fire prevention, safety, and fire plan response
   • Internal disaster and evacuation plan
   • External disaster, severe weather, and hazardous substance release
   • Variance reporting policies
   • Patient lifting and transfer techniques and proper body mechanics
   • Hazardous materials training to address use, storage, and disposal
   • Fire extinguisher training offered annually to all employees

E. INFORMATION MANAGEMENT

Information management is a function that is focused on meeting the information needs of healthcare organizations. Information managers attempt to obtain, manage, and use information to enhance organizational performance. Good information management techniques must be implemented regardless of the technology used.

1. *Joint Commission Requirements*—The program should address:
   • Identification of the organization's information needs
   • Structural design of the information management system
   • Definition and capture of data and information
   • Data analysis and transformation of data into a usable form

- Transmission and reporting of information and data
- Assimilation and use of available information

2. *Organizational Leadership*—Organizational leaders must ensure that roles and responsibilities are clearly defined. Staff members at all levels must be educated and trained in how to manage and use the information.

3. *Program Objectives*—The following are some basic objectives of an effective information management program:
   - Provide timely and easy access to organizational information
   - Improve the accuracy of informational data
   - Maintain proper security while providing ease of access for users
   - Design collection processes that improve program efficiency
   - Provide opportunities for collaboration and information sharing

Information Collection and Evaluation System (ICES)—The Joint Commission requires facilities to have an ICES. An effective ICES allows for the effective management of information and is crucial to the safety management and improvement process.

1. *General ICES Considerations*—
   - The ICES contributes valuable information for measurement of performance, including management of the environment of care.
   - Information collected by the system comes from a number of sources, including hazard surveillance studies, incident report data, periodic assessment studies, infection control information, and reports from engineering and maintenance departments.
   - The ICES must be evaluated continually to ensure that safety-related information is being collected, evaluated, and disseminated to appropriate departments and functions. Each facility must determine and develop performance indicators and methods to measure compliance.
   - Communication and the exchange of information are important to the success of any program. Committees, departments, and functions must strive to avoid competition.
   - Those involved in risk management, quality improvement, safety, and infection control programs must ensure that open lines of communication are maintained.

2. *ICES Performance Indicators*—
   - Safety deficiencies reported during the period
   - Causal factors that contributed to accidents
   - Hazards abated during a time period
   - Hazards discovered during next evaluation period

3. *Safety Management Report*—The following are some suggested areas to be addressed in the report:
   - General information on patient, visitor, and worker incidents
   - Security information, including incidents and problem areas
   - Information on training and in-service education programs

- Status of policy reviews currently being conducted
- Emergency and disaster preparedness drills or plan activation
- Status of the hazardous materials and waste management program

4. *Information Collection and Evaluation System*—The ICES gathers information from a number of sources, including:
   - Accident reports
   - Equipment management reports
   - Utilities management reports
   - Life safety
   - Infection control incidents
   - Information on quality improvement programs or evaluations
   - Risk management concerns or topics of interest
   - Hazard surveillance information and results of surveys
   - Departmental inspection results

Quarterly Report—A quarterly report should be published by the safety officer or safety committee to address organizational compliance with safety rules, regulations, and standards. The report should reflect departmental or program performance. A sample report format is presented in Table 2.3. Some other considerations include:

- The report should help focus on important issues, costly trends, or hazardous conditions.
- The report helps administration evaluate safety program effectiveness while documenting problems and accomplishments.
- The report provides the organization with a way to communicate safety and other topics to standing committees and top management personnel.
- The report should be distributed to the governing body, chief executive officer, risk manager, and quality assurance coordinator.
- The report format, contents, and scope will vary depending on the organization served.
- The report should include a single summary page to provide managers with a brief overview of trends, problem areas, or accomplishments.

1. *Report Indicators*—The report should contain indicators and thresholds that provide managers with meaningful and useful information. Some indicators address department-specific issues while others affect the entire healthcare organization. Examples of organizational indicators include:
   - Safety education participation rates
   - Orientation program effectiveness
   - Surveys of staff knowledge or safety performance
   - Laboratory inspection results
   - Hazard surveillance program results
   - Life safety discrepancies
   - Disaster drill participation and effectiveness
   - Hazardous and medical waste information
   - Workers' compensation injury rates

TABLE 2.3 Safety Management Report Format

This report represents a summary of the activity which took place in the facility's safety management program. The report is broken down by program areas and includes information such as status reports, reportable occurrences, issues, or activities which relate to safety program effectiveness.

- General Safety Information (including patient, visitor, student, employee, and department safety):

- Security Issues:

- Safety Training and Education:

- Policy Review:

- Emergency Preparedness:

- Hazardous Materials/Wastes:

- Information Collection and Evaluation System:

    1. Accident Information:

    2. Equipment Management:

    3. Utilities Management:

    4. Life Safety Management:

    5. Other (including infection control, quality assurance, and risk management):

- Miscellaneous Information:

- Hazard Surveillance (department inspections):

    Departments Inspected This Period   *Issues/Problems Identified*

    _____   _____

    _____   _____

    _____   _____

Period of Report: _____

---

2. ***Departmental Indicators***—Indicators required for specific departments should be developed to address employee injuries, Joint Commission requirements, and compliance with OSHA standards. Indicators must be expressed in objective terms so that comparison of data can be made. Indicators must provide managers with an incentive to ensure results are achieved.

F. RISK MANAGEMENT

Risk management describes the department or function that attempts to control financial losses and liabilities of the healthcare organization. Most losses result from the activities of the clinical and patient care departments. Healthcare risk managers normally focus on incidents involving patients, employees, and problem areas. This section covers some fundamentals of risk management, including a look at obtaining, using, and managing risk information.

Risk Management Functions—

- The risk management function should monitor initiatives, regulations, and standards that are under consideration by a facility committee. Examples are OSHA violations, workers' compensation issues, Joint Commission recommendations, and state or local compliance issues.
- Loss control actions should be evaluated in light of humanitarian issues, legal–corporate objectives, and cost containment interests. These items should be reviewed and summarized quarterly and annually using the risk management, quality assurance, and safety committee integration mechanism.
- Employee, visitor, and patient incidents/accidents should be analyzed and reported through the risk management function and reported to the safety committee.
- The safety committee or officer should initiate measures for accident reduction and safety improvements based on information received from the risk management, infection control, employee health, and quality administration areas.
- The safety committee should also review hygiene assessments, actions, and recommendations based on accident occurrence, work duty assessment, employee suggestions, and safety and infection control data.

Risk Management Committee—The size and needs of the healthcare organization will dictate the size or scope of the committee.

1. *General Responsibilities*—
   - Assist the risk manager by reviewing specific incidents
   - Evaluate each situation and make recommendations to reduce the chances of further losses
   - Help develop and implement changes in policies and procedures

2. *Committee Membership*—The committee should include key organizational members and representation from several departments, including:
   - Risk manager
   - Hospital administrator
   - Chief financial officer
   - In-house legal counsel
   - Nursing services director
   - Quality assurance coordinator
   - Safety officer
   - Facilities director
   - Infection control coordinator
   - Medical director or chief of staff
   - Member of governing board or trustees
   - Representation from high-risk areas such as surgery and pharmacy

- Patient affairs representative
- Others deemed necessary

American Society of Healthcare Risk Management (ASHRM)—ASHRM is dedicated to promoting the discipline of risk management within healthcare organizations. ASHRM encourages organizations to:

- Develop systems that can identify, evaluate, and control exposures
- Employ a qualified risk manager
- Develop methods and procedures for sharing risk information
- Create a risk management continuing education program
- Obtain formal commitment to the program from top management

Gathering Risk Information—Some risk management information systems do not have the capability to accommodate risk and hazard control information. A claims-oriented system concentrates on transmitting information about the consequence of loss, whereas a risk control system also addresses the prevention of loss. Regardless of the risk control system, it is crucial that the following information be obtained:

- Time of day of the accident/incident
- Age and condition of the person involved
- Length of employment and time on the job
- Job training and education
- Environmental factors such as lighting or noise levels

1. *Risk Information Sources*—Risk managers must rely on a number of information sources to effectively manage risk. Many organizations maintain a database of information referred to as Risk Management Information System (RMIS). RMIS data come from sources such as variance summaries, requests for medical records, safety committee actions, security reports, legal actions, patient complaints, and incident reports.

2. *Information from Outside Reports*—Risk managers must also depend on reports from regulatory agencies and the Joint Commission to evaluate situations that need additional action. Some items to be reported include:
   - Missed diagnosis or misdiagnosis that results in patient injury or death
   - Surgery-related incidents such as performing incorrect procedures
   - Treatment or procedures resulting in adverse reactions
   - Blood-related incidents such as giving wrong blood to a patient
   - Administering the wrong medication or the wrong dosage
   - All incidents that involve falls of patients or visitors
   - Any incident resulting in injury to a patient or a visitor

3. *Informal Reporting*—Informal reporting systems allow the risk manger to learn of incidents through telephone conversations, the organizational grapevine, or by talking to someone who witnessed the event.
   - Other hospital departments can provide the risk manager with information regarding patient concerns over bills or complaints of inadequate care.
   - An effective risk manager should also stay informed about legislative initia-

tives, changes in statutory laws and regulatory requirements, and the impact of jurisdictional decisions.

Risk Coordination—The risk manager is dependent on other departments and functions for information. An effective risk manager must establish good working relationships and be effective in communicating with others in the organization. The hospital administrator provides a link to the governing board. The administrator may also be involved in settlement decisions of large claims. Other offices contributing include:

- Medical records office can provide information about requests for medical records which can be a signal of pending litigation
- Infection control nurse can provide valuable information about infections which could lead to a rise in liability claims
- Quality assurance coordinator can assist the risk manager with information on clinical incidents and quality improvement programs
- Human resources and/or employee health department can provide workers' compensation injury and claims data
- Medical director can be very helpful in educating physicians about the importance of risk management
- Patient affairs and accounts offices can keep the risk manager informed of patient complaints and areas of dissatisfaction
- Department heads can provide technical and medical information that can help the risk manager identify and evaluate risk exposures

Role of the Risk Manager—Many healthcare organizations require the risk management department to review all incident reports, prepare injury claims, monitor employee medical treatments, and manage the workers' compensation program. The risk manager must work with the safety and quality improvement departments to ensure that problem areas are addressed and corrective actions implemented. Small healthcare organizations may require the risk manager to serve in another capacity, such as safety director or controller. The size, scope, and responsibility of the risk management function depend on a number of factors, including organizational structure and type of risk financing. Many hospitals manage their own risks by being self-insured or belonging to self-insurance associations/trusts. The risk manager attempts to identify, evaluate, and control exposures that could result in financial loss to the organization.

Comparing Accident Data and Costs—A good way to evaluate the risk management program is to compare accident data and loss costs in a number of clearly measurable variables. These variables include:

- Physical exposures, including square footage of property, personnel units, production figures, and labor hours
- Financial considerations (payroll and gross revenues)
- If accident frequency and loss costs viewed as a function of physical exposure and financial considerations are down over a several-year period, then the risk management program is doing quite well

- If accident frequency and loss costs are steady or not declining, there is a problem that needs to be addressed

Determining Loss Trends without Financial Impact Information—Some organizations make the mistake of comparing costs of past claims with claims from more recent years. These comparisons reveal faulty data. This can be remedied by using accident costs and injury rates to extrapolate a cost-per-injury rate. This figure can be multiplied by the number of injuries in any given year to arrive at a financial impact projection. The results do not account for inflation but can be used to evaluate the effectiveness of the risk management program.

Patient-Related Risks—Most risk managers are primarily concerned with issues that are patient related. The risk manager should be very concerned about emergency and disaster planning activities. Understanding these issues will allow the risk manager to better focus risk financing efforts to ensure that losses are minimized should a situation arise. Other concerns include:

- Hazardous materials and medical waste management.
- Risks can vary by facility but could include issues related to contractors, volunteers, and even medical students.
- Personal safety and security issues also need to be monitored by the risk manager.
- Joint Commission standards address the importance of healthcare facilities maintaining a proper environment of care. In many cases of litigation, it has been difficult to educate judges and juries about the differences between the institution and the physicians who have privileges.

Relationship-Based Risks—The unique relationships within healthcare organizations requires that the risk manager emphasize the following issues:

- Medical staff peer reviews
- Medical staff quality assurance activities
- Credential verification, appointment, and privileging procedures
- Medical staff disciplinary processes
- Issues related to due process, anti-trust, and restraint of trade

Employee Risk Concerns—Employee issues create concerns for the risk manager, with the following being potentially serious:

- Discrimination claims based on recruitment, hiring, and promotion practices
- Americans with Disabilities Act complaints
- Claims filed with the Equal Employment Opportunity Commission
- Workers' compensation disability actions
- Wrongful termination suits
- OSHA violations

Ergonomic Risks—Healthcare organizations continue to deal with the high costs and incidence of back injuries. Senior management must be presented with detailed information that proposed ergonomic improvements will reduce injury costs, increase quality of care, and improve worker morale. Healthcare organizations that have implemented ergonomic solutions to reduce lifting exposures have been highly successful in reducing the costs associated with back-related injuries.

Risk Management Activity Categories—

1. *Pre-Loss Activities—*
   * **Elimination**—Identify, minimize, and/or eliminate if possible activities or exposures that could result in the possibility of a loss
   * **Prevention**—Activities that allow the organization to integrate knowledge and technical advances in an environment that protects the human
   * **Safety**—A conduct or behavior that is the last defense in stopping a loss

2. *Post-Loss Activities—*
   * **Controls**—Policies and procedures implemented to minimize the severity of a loss
   * **Claims Management**—Activities and decisions aimed at reducing or minimizing loss severity and enhancing recovery from loss events
   * **Funding**—The mechanism(s) selected to pay for losses incurred

G. QUALITY IMPROVEMENT

Recently, the Joint Commission placed increased emphasis on quality assurance. The 1995 JCAHO standards introduced the theme "Quality of Care." This approach means the Joint Commission is stressing that accreditation standards should focus on an integrated approach, with total quality management as the foundation. All functions of a healthcare organization should be designed to be systematic, objective, continuing, and integrated. Any quality improvement program should strive to:

* Provide quality patient care, maintain accurate and complete clinical records, and ensure a safe environment for patients, visitors, and staff
* Ensure all staff members adhere to the highest professional standards
* Establish processes to monitor, evaluate, and improve the environment of care activities
* Document quality assurance activities that relate to safety issues
* Report safety issues that relate to medical staff to the appropriate committee
* Coordinate quality efforts with safety, security, risk management, and patient care departments
* Ensure that the quality assurance coordinator communicates and shares information with the risk manager and infection control director (for example, nosocomial infections can impact the quality of care and become a financial risk to the facility)

Quality and Risk Management—Quality improvement and risk management are different functions with some similarities. As described earlier, risk management involves managing financial risks, whereas quality assurance is simply meeting or exceeding preestablished objectives. The three basic components of quality management are:

- *Establishing Standards*—Develop appropriate and realistic standards, qualification, or structures.
- *Identifying Deviations*—Create methods or processes to identify deviations from the standards.
- *Correcting Actions*—Implement a system that can correct deviations by minimizing recurrence and improving outcomes.

Quality and the Joint Commission—Quality improvement activities in most healthcare organizations are guided by Joint Commission standards. Historically, the Joint Commission has had a major impact on the evolution of healthcare quality assurance. The Joint Commission focuses on the following quality issues:

- Competence of people providing care
- Accuracy and completeness of clinical information
- Safety of patients, staff, and visitors
- Compliance with established standards of medical care
- Organized process to evaluate and monitor care provided to patients

Process and Performance Improvement—Quality management over the years has focused on care processes, with the objective of analyzing and improving the steps in each process.

- Performance improvement combines the traditional quality methods into a process that is more than just quality assurance.
- Improving methods and processes requires leadership, goal-driven designs of new products or methods, a system to measure effectiveness, and an assessment system based on performance data.
- Performance improvement strives to meet external and internal customer needs while encouraging changes in the total organizational structure.
- Performance improvement places an emphasis on objective evaluations and measurements. As explained previously, the 1995 Joint Commission standards shifted from quality improvement to performance improvement, which focuses on individual performance rather than care processes.

Keys to Implementing Quality Improvement in Healthcare Safety—

- Develop a quality strategy and communicate it to all employees
- Identify external and internal customer needs
- Develop a spirit of ownership
- Integrate quality principles into safety training sessions
- Encourage teamwork among key program players
- Implement a quality improvement suggestion system
- Track and respond to all complaints
- Conduct comprehensive safety audits and assessments
- Recognize employees who contribute to the quality improvement program
- Develop quality improvement teams to focus on specific problems

Improving Organizational Performance (1996 JCAHO Standards)—The Joint Commission standards on improving organizational performance continue to stress the objective of improving the health and welfare of patients. A continuous improvement program addresses all functions that affect patient care, costs of improving care, and perceptions of the quality of services. The process should address the four common elements that historically have been known as the improvement cycle. The necessary elements are:

- Design a process or procedure
- Measure performance
- Assess or evaluate performance and/or compliance
- Improve the process or procedure

The new standards require facilities to have a planned organization-wide system which ensures that performance is measured, assessed, and improved as required. All new processes must be properly designed, and data collection is required for processes/results relating to patient welfare and facility functions. Finally, the organization must use a systematic data collection process and make improvements using a systematic approach. Many of the standards can only be achieved through establishing long-term goals. The Joint Commission realizes that full implementation will vary by facility. Another important aspect of any organizational improvement effort is that all activities should be collaborative and interdisciplinary. The design of new process must meet the needs of the organization and be clinically sound. All processes must establish baseline expectations. To be effective, the organization must ensure that data collection activities provide information to help establish baselines, identify areas that need improvement, and evaluate whether or not new changes meet established objectives. The organization must also assess any data collected by making internal comparisons, comparing data with current standards or practice guidelines, and referencing data from other healthcare organizations. Regulations, standards, and legal ramifications should also be considered. Finally, the organization must ensure that performance or process design is improved by taking the appropriate actions.

Evaluating Performance—The Joint Commission *Environment of Care Safety Handbook* outlines nine elements of performance:

- Efficacy of treatment relative to patient's condition
- Appropriateness of test, procedure, or service
- Availability of needed test or procedure
- Effectiveness of tests, treatments, and procedures
- Timeliness of tests, service, treatment, or procedures
- Safety of patient, staff, and others involved in services provided
- Efficiency with which services are provided
- Continuity of services with respect to other services
- Respect and caring shown during tests, treatment, or procedures

Improvement Process Elements—An improvement process should have the following elements:

- Means of achieving clearly defined goals or objectives
- Objectives that can be expressed in a function or process
- Methods for measuring performance of function or process

- Use of outside information to assist the assessment process
- Assessment should identify and prioritize improvement opportunities
- Create, test, and implement new improved processes or designs

Organizational Considerations—The improvement process raises three issues that should be considered by the organization:

- *External Relationships*—These are relationships to the external environment and include such things as community healthcare issues; demands of employers; concerns of regulators, consumers, and accrediting agencies; laws and regulations; and the pressures of increased competition.
- *Internal Characteristics*—Internal considerations must consider professional knowledge, clinical expertise, modern technology, technical competence, organizational structure, and leadership.
- *Evaluation System*—Organizations must implement a system for designing, measuring, evaluating, and improving key processes within the organization and must consider a cycle for improving performance.

Continuous Improvement Process (CIP)—This process is a systematic method used to plan, prioritize, and implement improvement efforts using data. In reality, CIP is a systems approach. The seven steps in the process are:

1. *Identify Improvement Opportunity*—Any process for improvement must be based on plans, priorities, and data. Those involved must understand how the improvement supports the goals of the organization.
2. *Evaluate the Process*—Select a challenge or problem and then establish a goal or target for improvement. This step allows members to focus on the improvement idea identified in Step 1.
3. *Analyze*—This step allows the improvement process to focus on the real problem rather than symptoms.
4. *Take Action*—Plan and implement corrective actions. This step involves developing a plan that will address eliminating causes or improving processes.
5. *Evaluate Results*—Verify if the actions taken have addressed the problem or improved the process.
6. *Standardize Solutions*—Ensure the improvement effort becomes part of the daily operation of the organization.
7. *Future Planning*—This final step allows members to review accomplishments, seek ways to continually improve the process, and to re-evaluate the whole process.

The JCAHO Improvement Cycle—The Joint Commission's improvement cycle can be described as follows:

- Design
- Teach
- Implement
- Assess and/or measure
- Improve the process/design

Applying Quality Improvement Principles to Safety—Organizations can apply the principles of quality improvement to safety program management in a number of ways.

1. *Develop a Policy*—The organization should publish a safety policy that outlines the purpose of the program and goals to be achieved. This statement of purpose and goals should be communicated to all members of the organization. Everyone must participate if the program is going to be successful.

2. *Comprehend the Importance of Inspections*—All levels within the organization must comprehend the purpose of assessments for improving processes and reducing costs.

3. *Constantly Improve the System*—All safety and management personnel should endorse the philosophy of continuous improvement.

4. *Teach Leadership*—Managers and supervisors must be engaged in the real job of management, which is to lead others to accomplish things in a more efficient way. Management must provide staff members with the tools and time to pursue improvement ideas.

5. *Promote Trust and Innovation*—Provide a climate where workers can identify problems and make suggestions without being fearful. This type of environment creates an atmosphere of innovation which can help improve organizational performance. Organizational or common goals need to be discussed. Provide workers with opportunities to work on common goals. Encourage coordination and communication between departments.

6. *Eliminate Meaningless Slogans*—Safety should be expressed in clearly defined work practices, rules, or job procedures.

7. *Learn and Implement Improvement Methods*—Quotas hinder quality efforts because the emphasis is on quantity. Safety hinges on doing a job correctly. People are usually motivated to do a quality job unless situations exist that hinder job pride. Workers need to be properly trained, provided with the right equipment, and have the opportunity to communicate with others.

8. *Encourage Self-Improvement*—Performance improvement will never be realized by all workers unless training and personal development opportunities are provided.

9. *Facilitate Change*—Managers and supervisors must have the courage to change their management styles for the improvement of the organization. Leaders must not just accept change but promote it within the organization.

H. OCCUPATIONAL HEALTH PROGRAM

Employee Health Program—A key concern in managing risks within a healthcare organization is employee safety and health. The employee or occupational health program normally is a function within the safety or risk management department. Many large organizations have created a health service department to manage the employee health program and monitor health assessments. The primary goal of the program is to protect the health of employees. A written policy should be developed that addresses the organization's philosophy and goals. An effective program should have the following components:

- Post-offer medical examinations and screening of medical history
- Periodic health assessments, screening, and examinations
- Health, safety, and infection control education
- Care for work-related injuries and illnesses
- Health, wellness, and off-the-job safety counseling/training
- Medical surveillance and immunization programs
- Evaluations and training in the use of personal protective equipment

Treatment of Occupational Injuries—Workplace injuries should be treated immediately. The organization should establish and publicize medical treatment protocols. An initial post-injury assessment should be accomplished as soon as possible by the employee health department. If the injury is a musculoskeletal disorder, the post-injury assessment should include the following:

- Job and personal health history
- Review of the treatment examination and test results
- Visual assessment, palpation determination, and range of motion
- Treating healthcare provider should make a specific diagnosis
- A written recovery management plan should be completed

The plan must include information about treatment, rehabilitation, and return to work. The plan should also address restrictions to be followed during the recovery period. The employee health department should monitor the worker's recovery and provide assistance when needed, which could include case management. The written recovery plan ensures that all parties are aware of the worker's condition and the steps recommended to promote recovery.

Health Assessments—The main objective of health assessments is to establish baseline data and to prevent disease by identifying early changes in health status. The health assessment program should be designed to evaluate the effects of potential worker exposures based on information from the worksite analysis and the requirements of applicable OSHA standards. Components of a health assessment program might include employee pre-placement evaluations, periodic post-exposure assessments, and exit assessments. Policies and procedures should also be established for a health assessment in the event of the employee's separation from the hospital. The purpose of the assessment is to establish the current health status upon separation. Items that need to be considered at this time are future plans for follow-up assessments and testing and for treatment of any ongoing occupational illness or injury.

Pre-Placement Assessment—The purpose of pre-placement assessment is to develop a baseline for medical surveillance and to determine whether a new employee is capable of performing essential job functions. Organizations must comply with the provisions of the Americans with Disabilities Act (ADA). ADA requirements are covered more completely in Chapter 4. The pre-placement assessment may include the following:

- A medical history including reproductive health history if exposure to known or potential reproductive toxins is likely
- An occupational health history
- A review of immunization record
- Physical exam of systems pertinent to job performance
- Baseline medical surveillance, such as pulmonary function tests if a respirator will be required or audiometric testing if noise exposure is a hazard

Periodic Assessment—This assessment and its frequency will depend on the dose and type of the exposure(s), as well as the characteristics of the employee, such as age. Some OSHA regulations require that a physical evaluation be performed when employees are involved in the use of certain personal protective equipment such as respirators or are exposed to certain hazards such as ethylene oxide or formaldehyde. Assessments may include an update of the occupational and medical histories, biological monitoring, and medical surveillance.

Post-Exposure Assessment—Post-exposure assessment should be performed following an exposure incident. The purpose of post-exposure assessment is to determine the extent of exposure, provide treatment, and develop measures to prevent recurrence. Policies and procedures should be developed to respond to exposure incidents. There should be an established policy and procedure to provide post-exposure evaluation and treatment following a needlestick injury as required by the OSHA bloodborne pathogens standard (29 CFR 1910.1030).

Return-to-Work Assessment—A return-to-work assessment should be conducted for employees who have been absent from work due to injury or illness. The purpose of this assessment is to ensure that the employee is well enough to return to work in some capacity and to ensure that the employee does not pose a threat to the safety and health of patients or other workers. This assessment is useful to determine if restrictions are needed before the employee is permitted to return to duty.

Modified Duty Categories—When evaluating workers for return to work, the following categories should be considered:

- *Sedentary Work (S)*—Exerting up to 10 pounds of force occasionally and/or a negligible amount of force frequently to lift, carry, push, pull, or otherwise move objects including the human body. Sedentary work involves sitting most of the time, but may involve walking or standing for brief periods. Jobs are sedentary if walking and standing are required only occasionally and all other sedentary criteria are met.
- *Light Work (L)*—Exerting up to 20 pounds of force occasionally, and/or up to 10 pounds of force frequently, and/or a negligible amount of force constantly to move objects including the human body. Physical demand requirements exceed those of sedentary work.
- *Medium Work (M)*—Exerting 20 to 50 pounds of force occasionally, and/or 10 to 25 pounds of force frequently, and/or greater than negligible up to 10 pounds

of force constantly to move objects. Physical demand requirements exceed those for light work.

- *Heavy Work (H)*—Exerting 50 to 100 pounds of force occasionally, and/or 25 to 50 pounds of force frequently, and/or 10 to 20 pounds of force constantly to move objects. Physical demand requirements exceed those for medium work.
- *Very Heavy Work (V)*—Exerting in excess of 100 pounds of force occasionally, and/or in excess of 50 pounds of force frequently, and/or in excess of 20 pounds of force constantly to move objects. Physical demands exceed those for heavy work.

Rehabilitation—This involves facilitating the employee's recovery to a pre-injury or pre-illness state. Occupational health personnel may be involved in the rehabilitation of workers who experience illnesses or injuries regardless of whether or not they are work related.

Case Management—Case management is an important strategy that should be utilized with employees who are out of work because of injury or illness. The goal of case management is to work with the employee to facilitate a complete and timely recovery. The occupational health professional who is performing case management for the organization must have an understanding of insurance laws, medical care, and hospital policies.

Informed Consent—Consent forms should be given to workers to explain the purpose of procedures and their possible side effects. These forms should be read, signed, dated, and maintained in the employee's medical file.

- Employees have a right to refuse to participate in the occupational health program.
- Privacy becomes an issue when someone other than the employee seeks access to the information generated through health assessment procedures.
- Employee personal health information is confidential and cannot be shared with management, supervisors, or co-workers without consent.
- The healthcare professional can only provide information about employee work restrictions as necessary to ensure the health of the employee and his or her co-workers.
- Employees have a right to know the results of their health assessment and medical surveillance. These medical testing results should be provided directly and confidentially to individual workers.

I. WORKERS' COMPENSATION

All states and U.S. territories have enacted statutes that protect and compensate workers injured on the job. Some laws are limited to physical injuries and others only compensate for specific injuries. Many statutes protect employees from discharge or other retaliatory actions for filing a workers' compensation claim. Most workplace injuries are the result of unsafe actions and not unsafe conditions. The majority of claims are filed by a fraction of the workforce. During the past 20 years, occupational diseases have become compensable in most workers' compensation systems if the condition can be related to work exposures. Occupational diseases can result from exposure to hazardous substances including radiation and can be attributed to exposures to heat, vibration, and repetitive motion work. Such claims have caused workers' compensation costs to soar tremendously in some states. Workers' compensation is viewed by many workers as a way to get fast and easy money. Healthcare organizations should have an effective cost containment program that includes hazard control programs, effective employer–employee communication, managed medical care, and effective claims handling procedures.

Purpose of Workers' Compensation—An effective workers' compensation system should:

- Provide adequate, equitable, prompt, and sure income and medical benefits to work-related injury victims or income to their dependents, regardless of fault
- Provide a single remedy and reduce court delays, costs, and workloads arising from personal injury litigation
- Relieve public and private charities of financial drains incident to uncompensated industrial accidents
- Reduce payments to lawyers and expert witnesses
- Encourage maximum employer interest in safety and rehabilitation through an appropriate experience rating mechanism

Workers' Compensation Coverage—Workers' compensation coverage depends on two considerations:

1. The injury or illness must have resulted from employment. In some states, injuries resulting from horseplay are not covered.
2. The injury or illness must have occurred during employment periods considering time, place, and circumstances.

Types of Coverage—Two types of coverage are provided to workers:

- *Employee Benefits*—Workers' compensation provides benefits to the injured worker, including medical coverage and wages during periods of disability.
- *Employer Protection*—Employers' liability protects employers from litigation based on work-related injury resulting from employer negligence.

Workers' Compensation Rates—The costs associated with providing coverage are paid by the employer in the form of purchasing commercial insurance, establishing a self-insurance program, becoming a member of a workers' compensation self-insurance association fund, or being placed in a state-controlled fund. Basic rates paid by employers are normally determined by the state or the National Council of Compensation Insurance (NCCI), an independent rating organization. The actual rates would be based on a number of other factors, with the state insurance department approving rates. Other factors considered include:

- The organization or fund quoting the coverage
- Standard industrial classification code(s) of the employer
- Payroll amount for the workforce covered
- Experience rating—Losses experienced by a company can result in a premium reduction or a surcharge being added to the premium. Experience ratings can be influential for large companies which can expect to pay for past losses in future premium projections. Controlling losses can result in a lower experience rating and lower premiums. Organizations such as NCCI use three years of payroll and loss information when calculating the rating. The current and most recent policy period is not used in the calculation. An experience rating of 1.0 is considered average, while an experience rating above 1.0 could result in an additional premium. A rate below 1.0 could result in a premium discount.

Workers' Compensation Cost Containment—Containing costs and maintaining employee welfare must be top concerns of management. Injured employees are entitled to benefits, but companies do not have to tolerate false or fraudulent claims. A successful program will have a plan of action and will ensure that each individual case is treated fairly and consistently. An cost containment program should focus on several goals:

- *Prevention Activities*—Injury and accident prevention should be the main objective of any cost containment program. Employers must be committed to establishing a comprehensive occupational safety and health program.
- *Timely Reporting of Injuries*—Employers must report work-related injuries on a state-approved First Report of Injury Form within a specified time as determined by statute. The claims-handling agency normally reports for the employer to state agencies. Timely reporting can reduce costs and allow the claims adjuster to better manage the claim.
- *Effective Claims Management*—Timely and professional claims handling can be crucial in controlling costs. Workers' compensation cases must be managed and not just paid. Some companies are using managed care and preferred provider networks to control medical payment costs.
- *Injury Reporting*—Formal accident and injury reporting procedures that require prompt reporting by injured workers should be established. Supervisors must ensure that the "first report" of injury is sent to the claims adjusting agency immediately.
- *Establish Medical Treatment Protocols*—Employers should develop medical treatment protocols and communicate requirements to all workers.
- *Accident Investigations*—Each accident should be investigated immediately and corrective measures implemented.
- *Post-Injury Substance Abuse Testing*—Positive results or impairment contributing to injury can affect worker compensation.

Disability Categories—There are several categories of disability under workers' compensation:

- **Permanent Total Disability**—A permanent disability rating occurs when an injured employee is not capable of doing work of any type. Most states award a benefit of two-thirds of average wages within a low and high range. The amount is paid for life or for a maximum number of weeks as allowed by law.
- **Temporary Total Disability**—This type of disability involves an employee who is totally unable to work for a temporary period of time but is expected to recover. Benefits normally begin after a brief waiting period and payments are normally determined using the average wage. The payments are made for the period during which the worker is recovering.
- **Permanent Partial Disability**—Workers determined to have permanent injuries that make them partially disabled and unable to perform their normal jobs can receive benefits according to a fixed schedule or be compensated for lost wages that can no longer be earned due to their disability.
- **Temporary Partial Disability**—This refers to a worker who is recovering from an injury but is still capable of performing a modified duty job. The benefit pays the difference between the previous and current jobs.
- **Death Benefits**—These are sometimes called survivor benefits and are paid to a spouse and/or dependent children. Partial benefits can often be paid to other family members if the deceased did not have dependents.
- **Special Injury Funds**—Some states have established a secondary injury fund to help an employer pay benefits to an employee with a pre-existing condition who is injured again, resulting in total disability. The current employer pays for the recent injury, with the additional disability benefits coming from the subsequent injury fund.

Limited Duty Status—A well-organized return-to-work/modified duty program cannot only save but also deter fraudulent claims by returning the employee to work. Any modified duty job must be assigned to meet the physical capabilities of the injured worker.

- The treating physician must evaluate the injured worker and determine the specific physical limitations. A worker should never be placed in a modified duty job based on the statement "Return to Light Duty."
- Human resources should develop job descriptions that are expressed in physical capacity terms. Medical personnel can use these descriptions to evaluate an injured worker's return-to-work status.
- When returning a worker to a modified duty status, part-time status and/or placement throughout the company should be considered if the worker cannot return to his or her department. The worker's supervisor must be involved in the placement process to reduce the chance of re-injury. The worker must be oriented in the new job and understand the modified duty requirements.
- Some disadvantages of modified duty include the possibility of re-injury, work designated may not be light duty, poor match of skills with the job, and the unknown length of the modified duty assignment.
- Some advantages of modified duty include a return to financial productivity, encouragement to contribute while recovering from injury, and relief of stress

associated with being off work. A modified duty job should be evaluated and determined to meet the physical capabilities of the individual.

J. HUMAN RESOURCES MANAGEMENT

The human resources department within a healthcare organization plays an important role in quality of care and safety management issues. The Joint Commission publishes human resources standards that emphasize the importance of assessing individual competence, including work experience, education, credentials, qualifications, and abilities. The human resources department supports the organization by providing manpower in the areas needed to accomplish its mission and taking care of the people who provide care to the patients. An important function of most human resources departments is the orientation and training provided to new hires as they enter the healthcare organization. The human resources department also collects data regarding staff competence patterns and trends related to staff educational needs. In many organizations, the human resources department is involved in a number of other employee issues such as workers' compensation, healthcare insurance, substance abuse testing, and employee health programs.

Chemical Dependency—Healthcare workers at all levels are at risk of becoming dependent on alcohol and other drugs. The availability of drugs and the stress related to working in healthcare environments make these workers very vulnerable. Most substance abuse is thought to occur from using street drugs such as crack, cocaine, and marijuana. However, addiction to prescription drugs and over-the-counter products such as diet pills can be just as devastating. Healthcare administrators must realize that chemical dependency can and does exit in the organization. Abuse of drugs not only affects the employee or staff member but also affects the quality of care provided to patients. Healthcare organizations should strongly consider adopting a formal substance abuse policy. The written program should be communicated to all employees and include a substance abuse testing provision. The policy should also ensure that an employee assistance program (EAP) is implemented. An effective program will pay great dividends by helping to ensure quality of care standards are maintained.

Drug-Free Workplace—The Drug-Free Workplace Act of 1988 requires federal contractors and grantees to maintain a drug-free workplace. The act has limited applicability and does not apply in private workplaces unless government contracts of more than $25,000 are involved. The act does cover some specific requirements that could help any facility concerned about substance abuse. The basic requirements of the act are as follows:

- Publish a substance abuse policy statement which notifies workers that the unlawful manufacture, distribution, possession, and/or use of a controlled substance is not allowed in the workplace.
- Employees must be informed that action will be taken against those who violate the policy.

- Create a drug-free information program that informs workers about the effects of drug abuse, counseling, the scope of the EAP, and the penalties for abusing drugs.

Substance Abuse Testing—Testing employees for illegal drug use can be accomplished through several programs, including:

- Pre-employment screening
- For cause with reasonable suspicion by employer
- Post-accident testing during medical treatment
- Periodic screening at pre-set intervals
- Random testing whereby anyone can be tested without prior notice

Drugs Detected by Screening—The five categories of drugs normally detected by testing are:

- Amphetamines (speed and uppers)
- Cannabinoids (marijuana)
- Cocaine (coke, crack)
- Opiates (morphine, codeine, heroin)
- Phencyclidine (PCP)

Chain of Custody—This refers to the method of tracking a single specimen through the entire testing process. Each specimen is accompanied by a form that provides a written record and is signed by all who had access. The chain ensures an accurate result and provides integrity to the program. Organizations with drug or substance abuse programs must ensure that the policy meets constitutional due process to include:

- Providing workers and job applicants prior notice about any drug testing requirements
- Obtaining written voluntary consent prior to administering the test
- Only implementing workplace testing of employees on the basis of objective criteria or on reasonable suspicion of abuse
- Ensuring the chain of custody and that all specimens are tested by an independent laboratory
- Review of the organizational drug testing program and policy by legal counsel before implementation and periodically thereafter

Losses Attributed to Substance Abuse—Losses incurred by employers and organizations as a result of substance abuse include:

- Measurable losses such as absenteeism, overtime pay, tardiness, sick leave abuse, health insurance claims, and disability payments
- Hidden costs are tougher to recognize; nevertheless, they are expensive to employers and include friction among workers, waste, damage to equipment, diverted supervisory time, and poor workplace decisions
- Legal claims such as workers' compensation payments, employee complaints, disciplinary actions, security issues, and drug dealing on the job

Supervisory Involvement—Supervisors play a key role in an effective substance abuse program. Supervisors need training to know what to do and what not to do. The supervisor should document worker behavior and signs of substance abuse such as:

- Increased absenteeism
- Poor decision making
- Ineffectiveness on the job
- Poor quantity and/or quality of production
- High accident rate
- Resentment by co-workers who pick up the "slack"
- Poor morale in the department

1. *Handling a Suspected Substance Abuser*—
   - Supervisors must be thoroughly familiar with the policy regarding substance abuse. Normally, the EAP or the personnel department has established procedures to handle these situations.
   - Confront the employee with documented information and only discuss work-related problems.
   - Explain in detail and in writing what the employee must do to improve.
   - If no improvement is noted, conduct a follow-up meeting and inform the employee that an evaluation is desired.
   - If the facility has a treatment program, explain the requirements that must be met to remain employed after treatment.

2. *Documenting Performance*—Supervisors must be objective and fair when documenting an employee's performance. Performance files should contain the following information:
   - Date and time of incident
   - Who was involved
   - What happened and where
   - What caused the incident and results
   - Witnesses

3. *Profile of a Worker on Drugs*—
   - Late three times more often than other workers
   - Uses three times more sick leave than others
   - Five times more likely to file a workers' compensation claim
   - Involved in accidents four times more often than other employees

K. REVIEW OF MANAGEMENT

Unsafe conditions, injuries, accidents, property damage, and other losses can be attributed to ineffective management. Hazard control, loss prevention, and safety-related activities must be managed to produce results. Safety must also consider the human element.

Review of Basic Management Principles—

- *Planning*—Pre-determine the best course of action considering available information and resources.
- *Organizing*—Arrange work or tasks so that they can be accomplished in the most efficient manner.
- *Directing*—Provide the necessary guidance to others during job accomplishment.
- *Controlling*—Measure and regulate the performance of work being accomplished to achieve desired outcome(s).
- *Coordinating*—Communicate facts and decisions to ensure that all parties with an interest have had the opportunity to provide input.
- *Leading*—Creating an atmosphere that allows people to accomplish their work. Some components of leadership include developing, inspiring, and communicating to get optimum results from others. Things are managed; people are led.

Theory X and Theory Y—This classic theory of management first proposed by Douglas McGregor can play an important role in safety program organization and success. Theory X or the traditional safety approach involves rules and procedures established and enforced by management. Theory Y, on the other hand, promotes worker participation. A good example of a Theory Y approach would be to include non-supervisory employees on the facility safety committee. Safety managers today should look for ways to motivate, educate, and encourage workers to work safely.

Worker Motivation—Many managers do not understand the factors that motivate workers. Workers can be motivated by being included on the safety committee or allowed to help develop a safety policy. People are motivated and satisfied by different things:

- *Hygiene Factors*—These factors can lead to increased job satisfaction and can include such things as organizational rules, departmental policies, interpersonal relationships, open lines of communication, and even job status.
- *Motivational Factors*—These are things that motivate people to develop and grow as productive employees. Opportunities for achievement, increased responsibility, recognition, and personal growth can be very motivational.

Basic Organizational Theory—Healthcare safety professionals must understand some basic dynamics about organizations. A healthcare facility organized to provide an environment of care for patients can create an environment in many departments where personal growth and job satisfaction are lacking. The attitude or motivation of the workers may not be promoted by the organizational structure of the facility. The high turnover rates in many healthcare organizations is a good example of the problem. Healthcare safety professionals should consider the following concepts when evaluating safety program effectiveness:

- *Line Organization*—This creates a hierarchy with a chain of command. Workers in many departments rely on guidance from the next in command.

Healthcare organizations are a little more complicated because of the professional medical staff, which does not adhere to the formal bureaucratic structure of the organization.

- *Span of Control*—The span of control reduces the freedom and independence of the worker.
- *Unity of Command*—Workers are normally responsible to one supervisor or leader. The superior and subordinate roles are strongly emphasized.
- *Departmental Specialization*—Healthcare facilities can have many departments that accomplish specialized tasks. Jobs and tasks within departments can be broken down further. This creates a lack of growth and can damage self-esteem, which can affect motivation and even safety.

SUMMARY

Safety management in healthcare organizations should be viewed as a comprehensive program that relies on several functional areas for information and assistance. Successful hazard control programs must have support from top managers and stress safety for patients, visitors, contractors, and workers. This chapter provided an overview of management functions related to safety, including risk management, quality improvement, information management, human resources management, and safety committee recommendations. Information on Joint Commission safety requirements was also presented, including the new Environment of Care standards.

REVIEW QUESTIONS

1 Loss control management is *best* defined as:
   a. controlling hazards through engineering designs
   b. applying management principles to the loss control process
   c. taking actions to reduce risks within the organization
   d. all of the above

2 The main reason why many safety management programs fail is:
   a. organizational emphasis is placed on activity-oriented elements
   b. management fails to clearly define goals
   c. managers ignore basic management techniques
   d. a and c

3 Which of the following is *not true* concerning the Joint Commission's 1996 Environment of Care standards for hospitals?
   a. organizations are to maintain a physical environment free of hazards
   b. facilities must outline activities to reduce human injury
   c. healthcare organizations must establish a working safety committee
   d. an infection control committee is no longer required

4   The hospital information management system is crucial to the safety management program because:
    a.   it collects important organizational data
    b.   it helps identify problem areas and trends related to safety
    c.   it provides an opportunity for collaboration and information sharing
    d.   all of the above

5   The primary function of the risk management department is to:
    a.   monitor workers' compensation and insurance expenses
    b.   support the safety management function by providing key loss data
    c.   control financial losses and liabilities of the organization
    d.   ensure organizational compliance with all regulatory requirements

6   Which of the following is the best indicator of the effectiveness of a risk management program?
    a.   analysis of insurance premiums and loss ratios
    b.   auditing internal risk management statistics
    c.   relying on technical reports prepared by outside consultants
    d.   comparing accident/incident data with costs and other losses

7   Quality improvement combines traditional total quality management methods into a process that also emphasizes:
    a.   meeting external and internal customer needs
    b.   using objective evaluation and measurement instruments
    c.   encouraging changes in the total organizational structure
    d.   all of the above

8   Creating an atmosphere that motivates people to do their jobs is an example of:
    a.   directing
    b.   leading
    c.   controlling
    d.   coordinating

9   To be effective, a safety action plan must:
    a.   establish a target date for completion
    b.   determine budget required to support the plan
    c.   provide guidance in establishing the safety support team
    d.   all of the above

10   Senior managers can best communicate safety issues to workers at every organizational level by:
    a.   formally adopting and publishing an organizational safety policy
    b.   establishing realistic safety goals
    c.   assigning responsibility and authority to manage
    d.   all of the above

Answers: 1–b   2–d   3–c   4–d   5–c   6–d   7–d   8–b   9–d   10–d

CHAPTER 3

HEALTHCARE HAZARD CONTROL

A. ACCIDENTS

This chapter covers the key concepts in the hazard control process, including accident investigation, personal protective equipment, and safety training. Healthcare loss control and injury prevention involves controlling hazards and training people to work safely. The previous chapter focused on safety-related healthcare management functions involved in the loss control process. This chapter discusses and presents information on the hazard identification and control process. The chapter also briefly addresses human behavior elements in accident and injury prevention. As illustrated in Chapter 1, healthcare organizations traditionally experience accident frequency and severity rates well above national rates. It is crucial that top management personnel understand that hazard control:

- Does not cost but pays great dividends to the organization by improving morale and promoting safe job accomplishment
- Directly affects the profit margin of the organization by helping reduce insurance premiums, workers' compensation losses, and litigation costs
- Directly affects organizational goals and quality of patient care services

Accidents—An accident has been defined as any unplanned event that interferes with job or task accomplishment. The event could cause measurable loss such as injury or property damage. An accident can also be classified as "near miss" and have no measurable loss. An accident occurs when two or more hazards come together, come close, or interact. A lack of understanding about the nature of accidents can hinder accident investigation and hazard control efforts.

Accident Myths—The following are some common misconceptions about accidents:

- An accident can be explained by determining a single cause.
- Accidents can be prevented by placing blame on the person responsible.

- Measurable loss such as serious injury or property damage must occur for an event to be classified an accident.
- Accidents are just chance events that happen as the result of random variables.
- Accidents are unavoidable and unforeseeable events that are going to happen regardless of attempts to prevent them.
- Accidents are acts of God or nature.

Accident Prevention—Identifying and understanding accident causal factors including unsafe human behaviors can be beneficial when evaluating accidents and implementing preventive measures. The following prevention principles should be taken into consideration:

- Accident prevention programs must place strong emphasis on identifying, evaluating, and correcting hazards and hazardous conditions.
- Accident prevention must also address human behavior, which is the most unpredictable aspect of the accident prevention program.
- Accident prevention programs must be organized, planned, and directed to achieve desired results.
- Identifying the causal factors responsible for the accident is important in accident prevention.
- Preventing accidents and controlling hazards include some type of process innovation, machine safeguarding, personal protective equipment, training, and administrative procedure.
- Monitoring systems can also assess the effectiveness of hazard-reducing controls and the accident prevention program.

B. HAZARDS

Physical hazards can be classified as liquids, solids, or gases existing in the work environment. When a hazard contacts or comes close to another hazard, the result can cause death, injury, or property damage. Hazard control is more than safety engineering and/or correcting physical/environmental hazards. Hazards can be identified, analyzed, and controlled by using management principles. Loss control management personnel can better understand hazards and hazardous situations by periodically conducting hazard publication reviews. This review process involves researching current healthcare, safety, environmental, and management publications to obtain current information on potential healthcare-related hazards. Many publications provide detailed information on effective control strategies. This review enables safety management personnel to identify, analyze, and control similar hazards in their facilities. The various categories of healthcare hazards are provided in Table 3.1.

Managers can also obtain hazard information by reviewing technical and regulatory standards, conducting hazard surveys, and evaluating data to identify trends/problem areas. Worksite analysis identifies existing and potential hazards within the facility.

Importance of Identifying Hazards—Properly classifying and identifying hazards enables safety management personnel to determine potential consequences and the probability of occurrence. Table 3.2 lists healthcare hazards by location.

TABLE 3.1   Categories of Healthcare Hazards

Hazard Categories	Definition	Examples
Biological	Infectious/biological agents, such as bacteria, viruses, fungi, or parasites, that may be transmitted by contact with infected patients or contaminated body secretions/fluids.	Human immunodeficiency virus, hepatitis B, tuberculosis
Ergonomic	Strives to fit the job to the worker as opposed to the traditional method of fitting the worker to the job. Ergonomics is the study of human behavioral and biological characteristics for the appropriate design of the living and working environment.	Lifting, standing for long periods of time, poor lighting, video display terminal usage, keyboard repetition
Chemical	Various forms of chemicals that are potentially toxic or irritating to the body system, including medications, solutions, and gases.	Ethylene oxide, formaldehyde, glutaraldehyde, disinfectants, waste anesthetic gases, cytotoxic agents, pentamidine, ribavirin, solvents
Psychological	Factors and situations encountered or associated with one's job or work environment that create or potentiate stress, emotional strain, and/or other interpersonal problems.	Stress, shiftwork, overtime, personal problems
Physical	Agents within the work environment that can cause tissue trauma.	Radiation, lasers, noise, electricity, extreme temperatures

- Hazard surveys enable management to quickly understand, evaluate, assign priorities, and make decisions regarding hazard control.
- Hazard inspectors should describe unsafe conditions, environmental hazards, broken equipment, or deviations from accepted safety practices.
- Hazard surveys or safety inspections should focus on the human factors as well as situational work factors such as facilities, tools, equipment, and materials.
- Environmental factors such as noise, vibration, temperature extremes, and illumination should also be examined.
- An effective hazard control management system can reduce the likelihood of accidents occurring in the workplace.
- Hazard control management attempts to prevent accidents and does not consider accidents as chance occurrences or acts of God.
- Hazard control and prevention efforts place a strong emphasis on managing many interfacing components and maintaining losses within acceptable limits.

Worksite Hazard Analysis—An effective worksite analysis should:

- Identify employees at risk of exposure and evaluate the effectiveness of control measures
- Establish a baseline to be used in a continuous or ongoing process

TABLE 3.2   Healthcare Facility Hazards by Location

Location	Hazard
Central supply	Ethylene oxide, infection, broken equipment (cuts), soaps, detergents, steam, flammable gases, lifting, noise, asbestos insulation, mercury
Dialysis units	Infection, formaldehyde
Dental service	Mercury, ethylene oxide, anesthetic gases, ionizing radiation, infection
Food service	Wet floors, sharp equipment, noise, soaps, detergents, disinfectants, ammonia, chlorine, solvents, drain cleaners, oven cleaners, caustic solutions, pesticides, microwave ovens, steam lines, ovens, heat, electrical hazards, lifting
Pathology	Infectious diseases, formaldehyde, glutaraldehyde, flammable substances, freon, solvents, phenols
Pharmacy	Pharmaceuticals, antineoplastic agents, mercury, slips, falls
Print shop	Inks, solvents, noise, fire
Radiology	Radiation, infectious diseases, lifting, pushing, pulling
Visitors	Falls, slips, elevators/stairs, fire, assault
Environmental services	Soaps, detergents, cleaners, solvents, disinfectants, glutaraldehyde, infection, needle punctures, wastes (chemical, radioactive, infectious), electrical hazards, lifting, climbing, slips, falls
Lab	Infectious diseases, toxic chemicals, benzene, ethylene oxide, formaldehyde, solvents, flammable and explosive agents, carcinogens, teratogens, mutagens, cryogenic hazards, wastes (chemical, radioactive, infectious), radiation
Laundry	Wet floors, lifting, noise, heat, burns, infection, needle punctures, detergents, soaps, bleaches, solvents, wastes (chemical and radioactive)
Maintenance and engineering	Electrical hazards, tools, machinery, noise, welding fumes, asbestos, flammable liquids, solvents, mercury, pesticides, cleaners, ammonia, carbon monoxide, ethylene oxide, freon, paints, adhesives, water treatment chemicals, sewage, heat stress, cold stress (refrigeration units), falls, lifting, climbing, strains, sprains
Nursing services	Lifting, pushing, pulling, slips, falls, standing for long periods, infectious diseases, needle punctures, toxic substances, chemotherapeutic agents, radiation, radioactive patients, electrical hazards
Nuclear medicine	Radionuclides, infection, x-irradiation
Office and data processing	Video display terminals, air quality, ergonomic/body mechanics, chemicals, ozone
Operating rooms	Anesthetics, antiseptics, methyl methacrylate, compressed gases, sterilizing gases, infection, electrical hazards, sharp instruments, lifting
Security	Infectious agents, assaults, slips, falls

Adapted from U.S. Department of Health and Human Services, National Institute for Occupational Safety and Health. Guidelines for Protecting the Safety and Health of Health Care Workers. Government Printing Office, Washington, D.C., September 1988, pp. A3-1, A3-2.

- Identity the specific characteristics of hazards and the work environment to be evaluated
- Evaluate hazard severity, effects on workers, personal protective equipment effectiveness, and environmental factors
- Be done after the walk-through inspection to assess the hazards of specific areas, processes, and/or types of work
- Be an orderly process for evaluating hazards that are most likely to occur or have the severest consequences
- Involve selecting the processes to be analyzed, studying and recording each step, identifying existing/potential hazards, and recommending changes to eliminate or reduce the hazards
- Make recommendations such as substitution of a less hazardous chemical, facility alterations, equipment and materials selections, or redesign of job tasks

Records Review—When conducting a worksite analysis, a review of all available records can reveal some very helpful information. Listed below are some records that should be reviewed:

- Monitoring and surveillance records
- Accident, incident, and near-miss reports
- Occupational Safety and Health Administration (OSHA) injury and illness logs for past five years
- Workers' compensation loss runs for past four or five years
- Required OSHA plans and inspection reports
- Joint Commission on Accreditation of Healthcare Organizations survey results
- Environmental surveys and regulatory visits
- Self-inspection results and hazard surveys
- Insurance surveys and underwriting reports
- Safety committee minutes and quarterly reports
- Employee health program records and data analyses
- Emergency planning documents and drill evaluation forms
- Workers' compensation records and return-to-work information
- Written safety, health, and risk management programs
- Federal, state, and local regulatory/accrediting reports
- Organizational policy and procedures manual
- Infection control reports and information
- Quality improvement information/processes affecting safety

Hazard Surveys—Some facilities use a multi-disciplinary team comprised of personnel with clinical or technical expertise to assist in maintaining an effective program. The team, led by a health or safety professional, should have expertise in several specialties, including policy development. The team leader must be able to plan, develop, and maintain an effective program. The team could include the following professionals, depending upon the facility and/or situation being evaluated:

- Industrial hygiene/toxicology
- Occupational medicine

- Ergonomics
- Physical or occupational therapy
- Infection control/epidemiology
- Safety and security management

Hazard Surveillance—Surveillance is collecting, obtaining, and using employee data or information to determine trends, problems, and risks associated with hazards. There are two primary types of surveillance:

- *Passive Surveillance*—This involves utilizing existing data to describe past trends. Documentation that is collected through recordkeeping provides data for an analysis of trends. The availability of, and access to, these records will depend on hospital policy and legal limitations such as access to employee medical records. The person accessing and reviewing these records must be cognizant of the limitations of access to such information.
- *Active Surveillance*—This involves collecting data that are not currently documented to describe current trends and identify problem areas. The data can be obtained from sources such as questionnaires, screening, or surveys. Survey results can be used with other surveillance information to determine whether a problem exists.

Hazard Audits—Comprehensive hazard or safety audits can be conducted by a special team, the hazard control management staff, or an outside source. Evaluating hazards includes gaining information from the following sources:

- Accident reports can assist in identifying broken equipment, unsafe operations, or hazardous work areas. Reports can also provide insight about the human elements in accident prevention and hazard control.
- Joint Commission surveys can provide valuable information on hazards, problem areas, and safety deficiencies. These surveys should be reviewed each time a hazard evaluation effort is underway.
- Inspections from regulatory agencies such as OSHA, the Environmental Protection Agency (EPA), the Nuclear Regulatory Commission (NRC), state licensure boards, county health departments, and local fire marshal reports provide excellent information about hazards.
- Insurance companies are concerned about hazardous conditions and many conduct periodic loss control surveys to determine insurability risks.
- Workers' compensation injury data provide hazard control managers with information about accidents that resulted in occupational injuries or illnesses.
- Material Safety Data Sheets received from manufacturers and distributors of hazardous materials should be thoroughly reviewed upon receipt.

Evaluating Hazards—The following should be considered when evaluating hazards:

- Types of hazards found in the facility
- Problems created by the hazardous condition
- Contribution of the hazard to accidents or injuries
- Employee attitudes and compliance with policies

- Hazards that could create risks for the facility
- Ways to reduce or eliminate major hazards first
- Risks posed by the hazard
- Likelihood a hazard will contribute to loss in some way

Analyzing Hazards—Determine what exposures or unsafe conditions are present in a given operation, task, or equipment. When conducting the analysis:

- Seek to identify hazards that developed as the result of the task or job operation.
- Break down a job to determine the sequence of task accomplishment. This also allows for hazards to be identified.
- Look at all factors, including environmental conditions.
- Identify unsafe work practices and exposure to chemical or physical hazards.
- Attempt to identify tasks and/or equipment that should be modified to reduce human exposures.

Job Hazard Analysis—Job hazard analysis (JHA) is a method for identifying, removing, or minimizing workplace hazards. This process can be referred to as a "hazard hunt." The analysis can be performed for all jobs in a department or workplace. Normally, a supervisor conducts the analysis with input from a worker who does the job.

1. *General JHA Information*—
   - The JHA can be an investigative tool when an accident occurs.
   - Supervisors often determine whether a worker failed to follow recommended procedures or whether the analysis missed identifying a hazard.
   - Most jobs, tasks, and procedures can be broken down into steps.
   - When analyzing a job for hazards, list each step in the order of occurrence as the job is accomplished.

2. *JHA Process*—
   Step 1. Break Down the Job—Examine each step in the process for hazards or unsafe conditions that could develop during job accomplishments

   Step 2. Identifying Hazards—Consider the following:
   - Clothing or jewelry worn by the worker that creates a hazard
   - Fixed hazards or objects such as sharp corners
   - Point of operation hazards or moving machine parts
   - Unusual or dangerous positions assumed by the worker
   - Repetitive motions and lifting requirements
   - Environmental hazards such as noise, heat, and chemicals

   Step 3. Evaluating Hazards—This step concerns causal factors:
   - Worker wearing proper protective equipment
   - Work areas and hazardous operations properly guarded
   - Lockout procedures properly followed
   - Work flow properly organized
   - Adequate ventilation to disperse harmful substances
   - Sources of noise, radiation, and heat
   - Adequate job safety training and supervision

Step 4. Implementing Safe Job Procedures—Attempt to develop safer job procedures based on the analyses of the hazards. This step requires coordination with other management personnel. It also requires that the new procedures and recommendations be approved and communicated to workers. Management must ensure that workers have the opportunity to provide suggestions.

Step 5. Revising the JHA—A JHA must be reviewed and updated periodically to remain effective. When a job is changed or a process modified, the JHA should be reviewed; a new hazard may have been introduced or an old hazard may have re-emerged.

Using Hazard Information—The hazard evaluation process determines the seriousness of the exposure and also seeks to develop measures to control, reduce, or eliminate the hazard. Evaluation criteria for many hazards are found in OSHA, EPA, NFPA, ANSI, or other published standards. Sometimes a hazard must be evaluated by a professional such as an industrial hygienist, a healthcare safety professional, or a certified hazard control manager. The National Institute of Occupational Safety and Health (NIOSH) and the OSHA Consulting Service agencies can assist facilities in evaluating workplace hazards.

Probability decisions can be difficult and should be based on historical information and empirical calculations. When an evaluation reveals potential hazards and inadequate controls, the hazard or hazardous condition must be re-evaluated carefully to determine how the situation could better be controlled. Other important factors to consider include:

- The frequency of monitoring depends on extent of exposure, severity of the effects, complexity of the work process, protective measures, and environmental factors such as temperature or humidity.
- OSHA 29 CFR 1910 requires many hazardous situations to be monitored on a periodic or annual basis. When not mandated by a standard, the healthcare facility must decide when hazardous processes will be evaluated.
- A thorough workplace assessment can include a short survey form or informal conference with workers to try to identify hazards not recognized during the walk-through survey.
- Ask workers about health problems that developed while working the job. Inquire about any noticeable health problems of other workers. Give the workers an opportunity to share their concerns about the safety and health risk associated with the job.

C. HAZARD CONTROL

Controlling Hazards—Once existing and potential hazards have been identified, and existing control measures have been evaluated, the next step is to assess the need for prevention, control, and/or personal protective equipment.

Policies and procedures should be written to describe the use of appropriate methods of control, such as engineering, work practice, and administrative controls, and appropriate personal protective equipment. These methods are sometimes organized into a "hierarchy of controls" to indicate that some methods of control are preferred over others. Once controls are installed, they must be checked periodically to ensure they are working correctly and protecting the workers adequately.

1. *Control Priorities*—Facilities must establish priorities for controlling hazards.
   • Identified safety hazards should be corrected immediately, and a training program should be developed to address topics such as correct lifting, patient transfer techniques, and electrical safety.
   • Workers must be informed of all hazards in their work areas and trained to avoid or control these hazards.
   • Control methods for environmental hazards include engineering, work practices, personal protective equipment, and administrative controls.

2. *Hazard Warning Systems*—Facilities must also implement warning systems where necessary to:
   • Provide immediate warning of all dangers
   • Describe known acute or chronic health effects of exposure to the hazard
   • Describe hazards and exposures that might result in traumatic injuries
   • Indicate actions for preventing or reducing hazard exposures
   • Provide instruction for reducing/minimizing illness or injury
   • Identify actions to be taken in case of illness or injury
   • Describe procedures used in emergency situations
   • Identify the population at risk so information can be provided

Engineering Controls—The preferred method for controlling hazards is the use of technological means to isolate or remove hazards from the workplace. Examples of engineering controls include the use of systems to prevent needlesticks and the use of a scavenging system in the operating room to prevent exposure to waste anesthetic gases. A good equipment maintenance program can also keep engineering control systems working as intended to prevent hazards. It is important, therefore, that the safety and health program provide for the maintenance of the facility and equipment.

Work Practice Controls—Work practices are another preferred control method that reduces the likelihood of exposure to occupational hazards by altering the manner in which a task is performed. Examples of work practice controls are prohibiting the recapping of needles by a two-handed technique and prohibiting mouth pipetting.

Administrative Controls—These controls reduce or eliminate worker exposure by changing the duration, frequency, and/or severity of exposure. Where traditional toxic air exposures may exist and cannot be reduced to permissible exposure limits through engineering or work practice controls, administrative controls may be used as an alternative means to reduce worker exposure. Examples of administrative controls include rotating employees to jobs free of the specific hazard, adjusting work schedules, and providing adequate staffing when the work output is increased.

D. WORKER EXPOSURE

Exposure—The amount of a hazardous substance to which an employee is exposed during a workshift. Exposure levels must be measured while work is occurring or a task is being accomplished. OSHA sets exposure limits for the amount of a specific air contaminant and the duration of employee exposure to that substance. OSHA standards use permissible exposure limits to describe this exposure, measured as time-weighted averages, short-term exposure limits, or ceiling limits.

Categories of Exposure—

- *Permissible Exposure Limit (PEL)*—The maximum allowed exposure for workers based on exposure during a time-weighted average (TWA, 8-hour workday or 40-hour work week)
- *Short-Term Exposure Limit (STEL)*—The maximum exposure to a hazardous substance allowed at one time (normally measured in a single 15-minute period)
- *Ceiling Limit*—The airborne concentration that cannot be exceeded at any time during the workday

Toxicity of Exposure—The toxicity of a material is not identical to its potential as a health hazard. Toxicity is the capacity of a material to produce injury or harm. The amount of exposure depends on three primary factors:

- Length of exposure
- Toxicity of hazardous substance
- Individual susceptibility

Effects of Exposure—

- *Acute Effects*—These usually involve short-term high concentrations that can cause irritation, illness, or death.
- *Chronic Effects*—These usually involve continued exposure to a toxic substance over a period of time. When a substance is absorbed more rapidly than the body can eliminate it, the chemical begins to accumulate in the body.

Exceeding Exposure Limits—If the monitoring results show a contaminant level that exceeds an exposure limit, some action or control must be implemented. It is important to recognize that adverse effects are still possible at exposure levels lower than those indicated by OSHA.

- If an adverse reaction occurs, the employee should be evaluated and treated by an occupational healthcare professional. The findings must be documented and recorded on OSHA Form 200.
- Based upon individual susceptibility and sensitivity to a particular hazard, it may be necessary to remove an employee from the work area where exposure occurs.
- Recommendations about occupational exposure limits are also available from agencies such as the NIOSH and the American Conference of Governmental

Industrial Hygienists (ACGIH). Recommendations from these organizations are useful in developing a comprehensive safety and health program.

Monitoring—The observation and measurement to determine levels of exposure to a specific substance in a worksite. Some OSHA standards require that employees be notified in writing of the results of environmental monitoring. Various types of monitoring are listed below.

1. *Area Monitoring*—This is done by measuring the contaminants in the air at the work area. This type of sampling provides information about the amount and type of exposures found in a given area. Area sampling is an important technique to determine the need to develop, implement, or improve control measures. An example of area sampling is the use of a sound level meter to monitor the noise level in a workplace.

2. *Personal Monitoring*—A measurement of air contaminants in the worker's breathing zone. Personal samples determine exposure for individuals by sampling for specific substances in the worker's immediate vicinity for durations corresponding to the exposure limit being evaluated. To ensure accuracy, the sampling must be representative of employee exposure; therefore, measurements should be taken during a typical workday. An example of personal sampling is the OSHA requirement to monitor the breathing zone of employees exposed to ethylene oxide to determine the level of exposure.

3. *Biological Monitoring*—This is the measurement of a chemical, its metabolite, or a non-adverse biochemical effect in a person to assess exposure. It is mandated by a number of OSHA standards.
   - Typically, specimens are blood, urine, or exhaled air.
   - It provides a measure of the quantity of a hazard absorbed by an individual regardless of the route of exposure.
   - A well-known example of biological monitoring is measuring blood lead levels to screen for evidence of exposure to lead.

4. *Medical Monitoring*—Medical procedures exist to evaluate exposures such as lead levels in the blood or threshold shifts in hearing.
   - A medical monitoring program should be designed for each department based on the hazards and exposures.
   - Tests should be designed to monitor for specific exposures.
   - General examinations do not target specific hazards. Also consider adverse health effects or other risks associated with the tests.

5. *Specific Monitoring*—Tests for each job category should be incorporated into the monitoring program. The program may also test for conditions that are not job related but important to general good health.

Medical Evaluations—Medical evaluations may be required by standards or by the nature of the hazards to which the worker is exposed. Organizations are encouraged to maintain an occupational medical history to assist with evaluation of long-term effects of exposure. Organizations must comply with the provisions of the Americans with

Disabilities Act (ADA) when gathering information about workers. OSHA standards also require periodic evaluations or medical surveillance of workers exposed to some hazards.

Employee Confidentiality Procedures—To ensure document consent and confidentiality, workers should read, sign, and date consent forms designed to alert them to side effects of certain immunizations. This is especially true of the hepatitis B vaccine given under the OSHA bloodborne pathogens standard. The results of all tests should be provided directly and confidentially to each worker. While confidentiality is vital, there may some exceptions, such as hearing threshold shifts. In such cases, some management personnel need to be aware of the problem. The ADA greatly impacted what an employer can ask a potential employee regarding medical history and current health. These areas cannot be addressed until after a job offer has been made. Once a job offer has been made, a pre-placement physical exam is crucial to establish baseline data and to ensure the worker can do the job.

Screening Workers—The Centers for Disease Control and Prevention and the American Hospital Association have developed guidelines for screening new hospital workers. Long-term care workers are normally screened using guidelines established by the Health Care Financing Administration and state licensure agencies. Worker safety is monitored by OSHA or by an OSHA-approved state-run agency.

- NIOSH conducts research and recommends new and improved standards to OSHA.
- ACGIH is the professional association that recommends limits for airborne contaminants, called threshold limit values (TLVs).
- Recordkeeping tracks the safety of workers and provides documentation for future evaluations. Effective records can assist healthcare hazard control personnel in identifying problems and evaluating the effectiveness of the safety and health program.
- Specific OSHA standards many times require detailed recordkeeping. Some examples are hearing conservation, asbestos exposures, and bloodborne pathogens recordkeeping requirements.

E. PERSONAL PROTECTIVE EQUIPMENT

Personal protective equipment (PPE) should be used when hazardous exposure cannot be controlled through one of the previously discussed control methods. OSHA recently issued new PPE standards which complement manufacturers' requirements. Many PPE standards refer to the specifications published by the American National Standards Institute (ANSI) and the American Society for Testing and Materials (ASTM). The purpose of personal protective equipment is to shield or isolate individuals from physical, chemical, and biological hazards. According to OSHA standards, PPE shall be provided, used, and maintained in a sanitary and reliable condition for all hazards of processes or environment, chemicals, radiological hazards, or mechanical irritants capable of causing injury or illness through absorption, inhalation, or physical contact. Anyone encountering hazardous conditions must be protected against the potential hazards. PPE should never be used as a substitute for engineering, work practice, or administrative controls. All PPE must be used in conjunction with other hazard control methods. The management element in any PPE program is the evaluation of equipment needed to protect against the hazards.

1. ***PPE Hazard Assessment***—OSHA 29 CFR 1910.132 requires the employer to conduct a hazard survey or assessment of each work area or department. The survey must be documented and show the date of assessment, workplace evaluations, and the name of the certifying official. An effective survey should consider:
   - Impact and penetration hazards
   - Compression or rollover activities
   - Hazardous chemicals and toxic materials
   - Heat extremes and associated hazards
   - Harmful fumes, mists, or dust
   Note—After the survey analysis, the employer must select the proper PPE to protect against the hazards.

2. ***PPE Training***—Training must include the following elements:
   - When and under what conditions PPE must be worn
   - The type of PPE necessary
   - How to put on, adjust, wear, and remove PPE properly
   - Training on any limitations of the PPE
   - Procedures for maintaining, storing, and disposing of PPE

3. ***Employee Requirements***—The new standard requires employees to demonstrate an understanding of all requirements. Employers must also have procedures in effect to identify and remove from service all defective PPE. All PPE must meet ANSI standards or be equally effective. No combination of protective equipment is able to protect against all hazards. PPE should be used in conjunction with other protective methods. The use of PPE can cause hazards, such as heat stress, physical stress, impaired vision, and reduced mobility. Using PPE improperly is worse than using no protection at all. Without any protection, the worker knows he or she is vulnerable and, perhaps, takes precautions. The two objectives of any PPE program should be to protect the wearer from hazards and to prevent injury. To accomplish these goals a program should also include medical monitoring and environmental surveillance.

4. ***Program Review***—The program should be reviewed annually and should consider the following elements:
   - Number of hours worked
   - Accident and illness data
   - Levels of exposure to hazard(s)
   - Effectiveness of training
   - Program documentation and costs
   - Recommendation(s) for program improvement/modification

Head Protection (29 CFR 1910.135)—Protective helmets provide some safety from impact and penetration from falling and flying objects. Some helmets also provide protection from limited electric shock and burns. Helmets must meet the requirements and specifications established in the American National Standard Safety Requirements for Industrial Head Protection, Z89.1-1986, if purchased after July 5, 1994. The revised standard changed the testing requirements to give a more accurate reflection of a helmet's protective abilities. The wearer should be able to identify the type of helmet by looking inside the shell for the manufacturer and ANSI designation and class. Classes include:

- *Class A*—General service, limited voltage protection
- *Class B*—Utility service, high-voltage helmets
- *Class C*—Special service, no voltage protection

Eye and Face Protection (29 CFR 1910.133)—Eye and face protective equipment is required by OSHA where there is a reasonable probability of preventable injury when such equipment is used. The following minimum requirements must be met:

- Protectors must provide adequate protection against the particular hazards for which they are designed.
- They must be comfortable when worn under designated conditions.
- They must fit snugly without interfering with movement or vision.
- Protectors must be durable, capable of being disinfected, easy to clean, and maintained in good repair.
- Employers must provide a protector suitable for the type of work to be performed and employees must use the protectors.
- Eye protection must be provided where machines or operations have hazard(s) of flying objects, glare, liquid, injurious radiation, or a combination of these.
- When selecting protectors, consideration should be given to the kind and degree of hazard. Design, construction, testing, and use of eye and face protection must be in accordance with ANSI Z87.1-1989, American National Standard Practice for Occupational and Educational Eye and Face Protection.

General Torso Protection—Hazards such as heat, splashes from hot metals and liquids, impacts, cuts, acids, and radiation can threaten the body. A variety of protective clothing is available including vests, jackets, aprons, coveralls, and full-body suits.

- Wool and specially treated cotton are two natural fibers which are fire resistant and comfortable because they adapt well to changing workplace temperatures.
- Duck is a closely woven cotton fabric that is good for light-duty protective clothing. It can protect against cuts and bruises on jobs where employees handle heavy, sharp, or rough material.
- Heat-reflecting clothing such as leather is often used to guard against dry heat and flame.
- Rubberized materials, neoprene, and plastics give protection against some acids and chemicals. Disposable clothing is particularly important for protection from dusty materials or materials that can splash.

Foot Protection (29 CFR 1910.136)—Protection of feet/legs from falling or rolling objects, sharp objects, molten metal, hot surfaces, and wet, slippery surfaces requires the use of appropriate foot guards, safety shoes, or boots and leggings.

- Most workers in selected occupations who suffered impact injuries to the feet were not wearing protective footwear.
- The typical foot injury is caused by objects falling less than 4 feet and the median weight is about 60 pounds.
- Most were injured while performing their normal job activities. Shoes should be sturdy and have an impact-resistant toe. Safety shoes come in a variety of styles and materials including leather, rubber boots, oxford, and athletic styles.
- Safety footwear is classified according to its ability to meet minimum requirements for both compression and impact tests. The requirements and testing procedures may be found in ANSI Z41-1991, American National Standard for Personal Protection–Protective Footwear.

Arm and Hand Protection—There is a wide assortment of gloves, hand pads, sleeves, and wristlets to protect against a variety of hazards. Employers need to determine what type of hand protection is needed by studying the degree of dexterity required, duration of exposure, frequency of task(s), types of hazardous exposure, and the physical stresses that will be applied. Purchasers of gloves should ensure that gloves meet appropriate hazard test standards. The protective device should always fit the task or job. Rubber protective equipment for electrical workers must conform to ANSI or ASTM standards.

F. RESPIRATORY PROTECTION

When engineering controls and/or ventilation do not provide protection against air contaminants, workers must be required to wear respiratory protection. As discussed earlier, some of the most common hazards are lack of oxygen and harmful dust, fogs, smokes, mists, fumes, gases, vapors, or sprays. Respirators prevent the entry of harmful substances into the lungs during breathing. The prevention of atmospheric contamination at the worksite generally should be accomplished as far as feasible by engineering control measures, such as enclosing or confining the contaminant-producing operation, exhausting the contaminant, or substituting with less toxic materials. The user should be aware that respirators have their limitations and are not substitutes for effective engineering controls. OSHA recently issued a new standard that brings requirements up to date with technological advances. Several new definitions are included in the newly proposed standard, including hazardous exposure level, adequate warning properties, and oxygen-deficient atmosphere. NIOSH will be responsible for certifying or approving respirators. The new standard is designated 29 CFR 1910.139.

1. *Written Program Requirements*—The written program content is essentially the same as the current standard and requires:

   - Written standard operating procedures to be established
   - Respirator selection to be based on worker exposure hazards
   - Users to be trained in proper use and limitations
   - Respirators to be assigned to individuals
   - Respirators to be regularly cleaned and disinfected

- Respirators to be stored in a convenient, clean, and sanitary location
- Respirators used routinely to be inspected during cleaning
- Appropriate surveillance of work area conditions
- Inspections to determine the continued effectiveness of the program
- Approved respirators to be used when available
- A qualified person to manage and administer the program

Note—Workers should not be assigned tasks requiring use of respirators unless they are physically able to perform the work and use the equipment.

2. ***User Requirements***—Users must be familiar with these procedures as well as with available respirators and their limitations. The procedures should contain all information needed to ensure proper respiratory protection of a specific group of workers against a specific hazard or several particular hazards. Guidelines must be established for medical surveillance of workers, including pre-employment physical examinations to eliminate those physically or psychologically unfit to wear respirators and periodic physical examinations conducted to review the overall effectiveness of the respirator program on the basis of physiological factors.

   - A much simplified document may serve the small user who has only a few workers, but it must cover the same subjects. Records should be kept on the issuance and use of each respirator and should include the date of initial issue, the dates of re-issue, and a listing of repairs.
   - Choosing the right equipment involves several steps, including determining what the hazard is and its extent, choosing equipment that is certified for the function, and assuring that the device is performing the intended function.

3. ***Respirator Selection***—Employers must provide equipment at no cost. If elasto-meric respirators are used, the employer must provide three sizes for each type of facepiece from at least two manufacturers. Selection criteria in the proposed standard include:
   - Nature of hazard and chemical/physical properties of the contaminant
   - Adverse health effects and relevant PEL or recommended exposure limit
   - Workplace sampling results
   - Nature of work operations/processes and time period of wear
   - Work activities, stress factors, and fit testing
   - Warning properties and physical characteristics
   - Functional capabilities and limitations of respirators

4. ***Medical Evaluations***—Employers would be required to obtain a written opinion from a licensed physician on each employee required to wear a respirator for five hours of more during the work week. The physician must determine if the worker is physically fit to wear a respirator.

5. ***Respirator Fit Testing (29 CFR 1910.134, Appendix A)***—Fit-testing procedures would be used when evaluating a proper fit, including both qualitative and quantitative guidelines.

- Quantitative fit testing uses an instrument to measure the challenge agent inside and outside the respirator. It is more exact than the qualitative fit test.
- Qualitative fit testing is a pass/fail test that relies on the wearer's ability to detect any of three challenge agents: banana oil (isoamyl acetate), saccharin solution aerosol, and irritant fume. For banana oil, the respirator must have an organic vapor filter. Saccharin requires a particulate filter.

6. *Training and Program Evaluation*—All substance-specific standards will be updated to conform to the requirements of 29 CFR 1910.139. Employers must review the protection program at least annually and conduct random inspections to ensure procedures are being followed and are required to conduct initial training prior to the time the worker would wear the respirator. Annual retraining would be required. Topics for which training must be provided include:
   - Nature of hazards encountered
   - Limitations and capabilities of the respirator
   - Inspection procedures
   - How to check for proper seal and fit
   - Maintenance and storage requirements
   - Dealing with emergency situations
   - Information about the written program

Respirator Types—

- *Air-Purifying Respirators*—These half-mask or full-face respirators come in disposable and reusable varieties with chemical or mechanical cartridges to protect against fumes, vapors, dusts, mists, and gases.
- *Gas Masks*—These provide protection against higher concentrations of contaminants than air-purifying respirators.
- *Powered Air-Purifying Respirators*—These respirators use a blower to pass the contaminated air through a filter. The purified air is then delivered into the mask or hood. These cannot be used in oxygen-deficient environments.
- *Self-Contained Breathing Apparatus*—These supplied-air respirators provide the highest level of protection by providing an air supply that is carried by the user.
- *Emergency Escape Breathing Apparatus*—This type of respirators provides 5, 10, or 15 minutes of oxygen for emergency escape from dangerous environments.
- *Air-Line Respirators*—These respirators have an air hose connected to a central supply source.

Cartridge Types—All respirator cartridges come in a color that designates the type of contaminant it protects against:

- *Orange*—Mists, dusts, and fumes
- *Purple*—Radioactive materials
- *Yellow*—Acid gases and organic vapors
- *White*—Acid gas
- *Black*—Organic vapors

- **Green**—Ammonia gas
- **Olive**—Other gases and vapors

Note—Protection for workers exposed to tuberculosis is addressed in Chapter 8.

G. EMERGENCY WASH REQUIREMENTS

Another method of reducing or preventing injuries is the use of emergency showers and eyewash stations. OSHA requires suitable facilities to be available in the work area for quick drenching of the eyes and body when workers may be exposed to injurious corrosive materials. Refer to OSHA Standard 29 CFR 1910.151. Other OSHA requirements address specific industries or operations such as open surface tanks, handling anhydrous ammonia, and hazardous material operations.

ANSI Standard—Published in 1990, ANSI Standard Z-358.1 helps employers select and install emergency equipment to meet OSHA requirements. The best solution is to have plumbed showers and eyewash stations. However, OSHA standards can be met by installing "self-contained" units, which contain flushing liquids that require refilling or replacement after use.

General Requirements—General requirements for emergency units include:

- Valves must activate in 1 second or less.
- Valves must be installed 100 feet or 10 seconds from the hazard.
- Valve must be located in a lighted area and identified with a sign.
- Employees must be trained in equipment use.
- Plumbed units must be activated weekly.
- Self-contained units must be maintained according to manufacturers' specifications.

H. ACCIDENT INVESTIGATION

Healthcare facilities should have or develop a standardized system for reporting all incidents. Timely and accurately reporting of accidents will allow safety personnel to understand patterns, discover contributing causal factors, identify risks, and reduce loss by preventing or reducing mishaps. Forms and procedures should be easy to complete and provide for confidential reporting. The same office should receive, analyze, and document all accident information. Workers should be trained in proper reporting of accidents and injuries. Accident investigation is especially important to determine direct causes, uncover contributing causes, prevent similar accidents from occurring, document facts, provide information on costs, and promote safety. Accident investigations should focus on:

- Causal factors contributing to the event
- Why and how the hazard control system failed
- Documenting which safety policies and regulations were violated
- Determining whether defective machinery contributed to the accident

- Cataloging the environmental factors that contributed to the event
- Looking for problems that indirectly contributed to the accident

1. *Purpose of Accident Investigation*—The purpose of an investigation is to locate and define procedural errors and hazards that contributed to the incident. An investigation focuses on primary and contributing causal factors. Causal factors generally fall into one of three categories:
   - Hazardous mechanical or physical conditions such as unsafe equipment or improper lifting
   - Violation of safety procedures such as knowingly using unsafe equipment or operating without authority
   - Personal factors such as willfully neglecting safety rules, lack of knowledge or skill, or failure to understand the proper procedure

2. *Determining Causal Factors*—A thorough investigation can uncover contributing factors such as sloppy housekeeping, failure to follow procedures, inadequate maintenance, poor supervision, or faulty equipment. Accident investigations can help identify actions and improvements that could help prevent similar accidents. Table 3.3 catalogs accident causal factors.

3. *Fact Finding*—Human error contributes to most accidents in some manner, but simply placing blame may allow other hazards to go undiscovered and uncontrolled.
   - When conducting an accident investigation, seek to look beyond the obvious and try to uncover causal factors, determine the true loss potential of the occurrence, and develop practical recommendations to prevent recurrence.
   - A major weakness of many accident investigations is the failure to establish and consider all human, situational, and environmental factors.

4. *Documenting Findings*—Properly conducted and documented investigations can serve as the permanent record. It may become necessary to reconstruct an accident situation long after the occurrence for legal purposes.
   - Accident investigations provide information on direct and indirect costs of accidents.
   - The investigation also demonstrates the organization's interest in worker safety and health.
   - When investigating an accident, never assume that the person involved in the accident was at fault.

5. *Management Involvement*—Management should emphasize the importance of timely reporting, investigation, and analysis by:
   - Promptly reviewing and analyzing reports
   - Conducting follow-up actions when required
   - Assisting in coordinating and developing preventive procedure
   - Making constructive, purposeful, and timely recommendations
   - Reviewing all major incidents within 48 hours

The Accident Investigation Process—The elements in the investigative process include:

TABLE 3.3 Accident Causal Factors

Causal Factors	Identification of Factors	Possible Corrective Actions
Environmental: unsafe procedure or process	Hazardous process; management failed to make adequate plans for safety	Job analysis; formulation of safe procedure
Defective through overuse	Buildings, machines, or equipment that have become rough, slippery, sharp-edged, worn, cracked, broken, or otherwise defective through use or abuse	Inspection; proper maintenance
Improperly guarded	Work areas, machines, or equipment that are unguarded or inadequately guarded	Inspection; checking plans, blueprints, purchase orders, contracts, and materials for safety; provide guards for existing hazards
Defective design	Failure to provide for safety in the design, construction, and installation of buildings, machinery, and equipment; too large, too small, not strong enough	Source of supply must be reliable; checking plans, blueprints, purchase orders, contracts, and materials for safety; provide guards for existing hazards
Unsafe dress or apparel	Management's failure to provide or specify the use of goggles, respirators, safety shoes, hard hats, and other articles of safe dress or apparel	Provide safe dress, apparel, personal protective equipment if management could reasonably be expected to provide; specify the use or non-use of certain dress, apparel, or protective equipment on certain jobs
Unsafe housekeeping	No suitable layout or equipment as necessary for good housekeeping—shelves, boxes, bins, aisle markers, etc.	Provide suitable layout and equipment necessary for good housekeeping
Improper ventilation	Poorly ventilated or not ventilated at all	Improve the ventilation
Improper illumination	Poorly illuminated or no illumination at all	Improve the illumination
Behavioristic: lack of knowledge or skill	Unaware of safe practice; unpracticed; unskilled; not properly instructed or trained	Job training
Improper attitude	Worker was properly trained and instructed but failed to follow instructions because he or she was willful, reckless, absent-minded, excitable, or angry	Supervisor; discipline; personnel work
Physical deficiencies	Worker has poor eyesight, defective hearing, heart trouble, hernia, etc.	Pre-placement physical examination; periodic physical examination; proper placement of employees; identification of workers with temporary bodily defects

- Determining the persons involved in the accident, their status or job description, and the supervisor responsible for the area or job task
- Documenting exact date, time, location of, and witnesses to the event
- Identifying all objects, equipment, hazards, or other causal factors that could have contributed to the accident
- Obtaining a detailed description of what happened, including as many witness statements as possible
- Evaluating only after all available evidence and information have been collected
- Determining appropriate corrective actions or countermeasures

1. *Formal Investigations*—Safety professionals should ensure a formal investigation is conducted when preliminary reports indicate inconsistencies with written policies, safety rules, or organizational procedures. Investigations should also be conducted if the event:
   - Fits an established pattern of time, place, hazard, or person
   - Resulted in personal injury of employee, patient, or visitor
   - Involved possible faulty equipment
   - Involved workers who failed to use personal protective equipment
   - Demonstrated a possible training deficiency
   - Had inadequate staffing
   - Involved personal factors such as tension or domestic problems

2. *Interviewing Witnesses*—An important element in the accident investigation process is effectively interviewing witnesses and documenting their observations of the event.
   - Interview one person at a time as soon as possible after the event.
   - If possible, interview eyewitnesses and those involved first.
   - Talk with anyone who may contribute useful information.
   - Conduct interviews at the accident site if possible.
   - Do not be confrontational and always focus on the facts, not fault.
   - Ask for suggestions on how to prevent future events.
   - Take notes accurately to ensure all facts are properly documented.
   - Be courteous and have a positive attitude.

3. *Using Video Recorders*—Large organizations should consider using video recorders to conduct accident investigations. Videotaping an accident scene can be beneficial in analyzing hazards and determining causal factors. Accident videos can be used in training sessions. Videos can also provide excellent documentation for future reference.

Reviewing Accident Investigation Reports—Reviewing accident information can be instrumental in reducing losses. The accident information review process is critical because:

- It serves as a bridge between accidents and safety training and education.
- Many workers' compensation adjusting agencies can provide excellent statistics that greatly enable healthcare facility risk management personnel to better manage the safety and health program.

- OSHA logs can provide important information when developing accident review procedures, but the program requires total support from department managers and supervisors.
- Training and hazard surveillance programs can be changed or modified based on information provided by the accident review process.
- Controlling hazards in the healthcare environment is a challenging process that requires organization and coordination.

Summary of Accident Causal Factors—Many things contribute to accidents; listed below are some of the more common causal factors:

1. *Poor Supervision*—
   - Lack of proper instructions
   - Job and/or safety rules not enforced
   - Inadequate PPE
   - Correct tools or equipment not provided
   - Inadequate inspection of equipment or jobs
   - Poor planning or improper job procedures
   - Rushing the worker

2. *Worker Job Practices*—
   - Using shortcuts and/or working too fast
   - Not using proper tools or equipment
   - Incorrect, or failure to use, protective equipment
   - Horseplay
   - Disregard of safety rules
   - Inattention or inexperience
   - Physically or mentally impaired
   - Improper body motion
   - Action of fellow worker
   - Improper personal clothing

3. *Unsafe Materials, Tools, or Equipment*—
   - Ineffective machine guard
   - Unguarded equipment
   - Defective materials or tools
   - Improper or poor equipment design

4. *Unsafe Conditions*—
   - Poor lighting
   - Poor ventilation
   - Crowded work area
   - Poor storage or piling
   - Inadequate exits
   - Poor housekeeping practices
   - Unsafe environmental conditions such as slippery floors

I. SAFETY TRAINING

Importance—One of the greatest hazard control techniques is simply educating people to follow safe work practices. Safety training, new employee orientation, and continuing education sessions are crucial to the effectiveness of healthcare safety management programs. Safety training programs must strive to build understanding, positively affect worker attitudes, and improve safety knowledge. Training must facilitate the transfer of knowledge and skills to normal work activities. Many training programs are not being evaluated to determine their effectiveness. Healthcare organizations should follow sound educational principles. A program that merely takes attendance cannot be effective. The safety training function could very easily be the weak link in many healthcare safety management programs. Most organizations do not want to dedicate sufficient time, allocate sufficient resources, or require workers to attend lengthy training sessions. The around-the-clock operation of healthcare facilities also makes training second- and third-shift employees difficult. Professional and medical staff normally receive in-service education as required. However, in many instances, personnel in support departments do not receive adequate safety training. Safety training and education must be a priority because the human element is a causal factor in the majority of accidents.

Healthcare Training Topics—Specific topics will vary with the type and size of the facility. Training is provided in most healthcare facilities for the topics listed below:

- Fire and life safety
- Handling hazardous chemicals
- Accident reporting and investigation
- Use of respirators and PPE
- Biological hazards (bloodborne pathogens)
- Laboratory safety
- Radiation control
- Safe lifting and back care
- Lockout/tagout
- Equipment safety
- Internal emergencies
- Slip, trip, and fall prevention
- Electrical safety
- Waste handling
- External disasters
- Formaldehyde exposure

Training Program Development—Training should focus on good practices which can be learned through repetition. Policies should always translate into actual procedures. Workers usually support the program when they see management's concern or the importance of the training.

Participation strengthens the program while enhancing learning. Safety training must cover the theory and purpose of each safety measure. An overview of healthcare safety training is provided in Table 3.4.

TABLE 3.4   Overview of Healthcare Safety Training Requirements

Orientation Training
- General safety
- Cost center safety
- Fire safety
- Specific job-related hazards
- Accident reporting
- Hazardous materials and wastes program
- Hazard communication program
- Emergency preparedness program
- Security program
- Infection control, bloodborne pathogens (exposed workers)

Annual Training
- General safety
- Fire safety
- Emergency preparedness
- Hazardous material and waste management program
- Departmental safety
- Specific job-related hazards (ethylene oxide, formaldehyde, etc.)
- Changes implemented by the safety committee or senior administration
- Bloodborne pathogens exposure

Ongoing Training
- Fire safety (via drills)
- Emergency preparedness program (via drills and/or implementations)

As-Needed Training
- Employees involved in multiple incidents
- Hazardous materials and wastes program
- Emergency preparedness program

1. *Address Real Situations*—Employees are most likely to resent a new safety program when no one has bothered to explain its purpose. Instructors should always emphasize the risks that threaten to cause bodily harm and the facility's ability to operate. If the hiring process recruits qualified new workers, orientation will heighten their existing safety skills. Objectives should be developed immediately for supervisors and other key staff personnel. Developing an effective safety management program will take time, and developing the training will also require time and attention.

2. *Program Development Considerations*—The following should be taken into consideration when developing a program:
   - Assess what needs exist and what may be projected, and determine the facility's training capabilities.
   - Assign training responsibilities, and encourage professional development of those involved in safety.
   - Remember that safety is simply part of job training, and effective training can maintain enthusiasm among workers.

- Safety bulletins, posters, and films supplement a training program but cannot substitute for a good program.
- Employee safety meetings are effective; off-the-job safety topics can also be presented.
- Publish a training policy statement that does not conflict with other facility policies or programs.

3. *Training Objectives*—All training and education programs should have measurable, observable, realistic goals or objectives.
   - The objectives should state exactly what the employee should be able to do or know.
   - The information collection and evaluation system (ICES) can be an important tool for monitoring and evaluating each element of the training program.

Instructional Techniques—Training programs must use a variety of techniques to ensure that workers are learning the necessary information. Some training experts believe that using appropriate videos or other visual aids greatly increases student retention. Hands-on practice and guided discussions or informal lectures can greatly increase retention levels. Listed below are some training techniques that healthcare training specialists should consider to improve the quality of training:

- *Pre-Test*—A pre-test allows evaluation of the student's knowledge level. It can motivate students to look for the concepts or principles on the test.
- *Informal Discussion or Lecture*—The informal session coupled with a question-and-answer session allows student participation.
- *Demonstration*—The hands-on technique allows for application of knowledge and demonstration of required skills.
- *Simulated Drills/Exercises*—Some training topics can only be effectively trained when students participate in hands-on situations.
- *Interactive Software Programs*—The use of computers allows the student to control the flow of information during the training session.
- *Information Sheets/Handouts*—Students can keep the material for future reference.
- *Visual Aids*—The proper use of overhead transparencies, marker boards, or blackboards to communicate visual clues can enhance learning.
- *Formal Test*—Testing at the end of the sessions can reinforce key points while documenting training effectiveness and student learning.

Know the Audience—One of the important elements in the preparation process is to know the audience. Knowing as much as possible about the students will allow the training to be presented in a more effective manner. Some things to know about the audience include:

- Type of workers
- Average age or experience level
- Educational level and/or literacy level
- Job skills and interests
- Whether all students need the training

Training Preparation Checklist—

- Review regulatory/accreditation requirements for the subject being presented.
- Review lesson plans, books, videos, and all training aids.
- Know the number of students to be trained, and make sure the training area is adequate for the session.
- Coordinate all training to be done on the job or in a work area with the supervisor or department head.
- Check all equipment before beginning the session.
- Gather all materials needed for the session, including test booklets, answer sheets, and student handouts.
- Arrive at the training site before the students and ensure things are in order.

Determining Training Needs—Facilities should also use a variety of ways to determine training needs. Regulatory and voluntary agencies require safety training and education to ensure quality care and/or worker protection. Organizations must place a strong emphasis on this important human element in the safety equation.

- Training must publicize the importance of safety while reinforcing and providing information on a specific subject.
- Training must focus on hazardous conditions and emphasize ways workers can reduce their risk of exposure.
- Healthcare safety training should be coordinated by the safety department or be managed by an education department.
- The safety director along with the safety committee should determine training needs, monitor sessions, and evaluate training effectiveness.
- Safety training responsibilities should be clearly communicated to all management personnel and department heads.

Training Methods—The Joint Commission requires that employees receive annual training or continuing education classes. Healthcare facilities can use a variety of ways to train and reinforce the knowledge of their workers. The program begins with a strong orientation program and continues with ongoing sessions. Healthcare organizations use a number of training methods, including:

- Safety posters, flyers, bulletins, newsletters, self-study programs, classroom presentations, on-the-job training sessions, professional seminars, safety training fairs, and computer-assisted programs.
- Some facilities delegate a number of training responsibilities to the individual departments, while others employ a full-time educational coordinator.
- Large or specialized departments in some organizations, such as laboratories, conduct most of their own training programs.
- The safety department or educational coordinator must be active in monitoring the program regardless of the system.
- Training program reviews should take place when new standards are issued, but at least annually.
- Information provided by accident reviews and hazard surveillance programs should be analyzed to help evaluate training program effectiveness.

- Testing and surveying employees also provide important information on program effectiveness and worker needs.

Adult Training Techniques—Effective training allows all employees, supervisors, and managers to actively participate in the health and safety program. Participation encourages commitment to the program and the desire to solve safety problems. Management must communicate clearly with all employees its commitment to provide a safe and healthful work environment. Employees must be made aware of all potential hazards to which they may be exposed. This can be done in a number of ways. Some important suggestions are listed below:

- Use adult education and training techniques to facilitate learning.
- Some occupational safety and health standards require employees to demonstrate competency in the topic being trained.
- Present material in understandable language and a format that encourages learning.
- Evaluate training presentation to ensure learning goals are mastered by the students.
- Document all training and keep good records on training required by a safety and health standard.
- OSHA requires records to be made available for review during compliance inspections.

Joint Commission Safety Education and Orientation Requirements—The Joint Commission requires organizations to provide safety training during orientation and on a recurring basis. Safety education should receive the same priority as other professional continuing education. The Joint Commission has expanded the scope of healthcare safety to include more than traditional safety concerns.

1. *New Employee Orientation*—A staff orientation program should provide initial job training and information, including an assessment of a person's ability to perform specified job responsibilities.
   - Organizations should maintain individual employee records that document both facility-wide and departmental orientation training.
   - New worker orientation should address emergency procedures, accident reporting, hazard identification, security measures, smoking regulations, and equipment management.

2. *Continuing Education*—Continuing or ongoing education should cover on-the-job training and refresher sessions to ensure employees remain current in worker-related issues, including safety topics. Changes covered might include updated technology procedures, new government regulatory requirements, and improved standards of practice. The Joint Commission requires that at least 91% of all workers participate in an in-service education or training program once in the previous 12 months. Some safety regulations may require training in some subjects at different intervals (e.g., all workers exposed to bloodborne pathogens must receive annual refresher training). Joint Commission standards also require

organizations to collect data on a regular basis concerning staff competence and to identify/respond to worker training needs.

3. ***Specialized Training***—Specialized training normally applies to specific hazards, safety problems, or exposures.
   - This training could include bloodborne pathogens exposure training, use of PPE, or procedures for working with patient care equipment.
   - Not all training requirements are covered under Joint Commission standards. Hazard control managers must be aware of the need for other training requirements including OSHA, NRC, and EPA requirements.

Training Evaluation and Documentation—Documentation of all safety education programs is needed to ensure that the programs will continue if there is a change in personnel. Documentation also provides data that will assist managers in assessing the overall effectiveness of individual programs. A review of the safety training can help to determine how frequently certain programs should be held. The evaluation of these programs is also important in determining the quality of the education programs, particularly the relevance of programs to daily activities.

Training Frequency—The Joint Commission provides guidance on new employee orientation and recurring training sessions. However, some regulatory agencies such as OSHA mandate training in specific standards. Employees exposed to certain levels of ethylene oxide must receive annual training. Facilities should also provide training when:

- New equipment, materials, or processes are introduced
- Procedures are revised or updated
- Information indicates employee performance must be improved
- A new standard becomes effective
- New hazards or exposures are identified

Note—The Joint Commission does not prescribe a set frequency schedule, but most programs should be conducted annually.

OSHA Healthcare Training Requirements—OSHA has requirements that may differ from Joint Commission requirements. More than 100 OSHA standards require employers to provide workers some type of training. Listed below are some training requirements for standards applicable to healthcare organizations:

- ***Emergency Action and Fire Prevention (1910.38)***—Training required upon job assignment, when transferred to new job, or when emergency procedures change.
- ***Hearing Conservation (1910.95)***—Training for all employees exposed above the action level of 85 dB for an eight-hour TWA. Annual retraining is required for all workers in the hearing conservation program. Employers must train employees in the use and care of hearing protectors.
- ***Ionizing Radiation (1910.96)***—Training must be provided to all individuals working in or frequenting areas where radiation exposure could occur. Training should cover safe work practices and precautions to minimize exposure.

- *Liquid Propane Gas Handling (1910.110)*—Workers involved in installation, removal, operation, and maintenance must be trained in each function.
- *Hazardous Waste Operation and Emergency Response (1910.120)*—All workers involved in handling hazardous waste must be trained to a level required by their job. Scope of training varies depending on exposures and responsibilities. Emergency responders are also trained to meet their expected level of response. The standard requires annual refresher training.
- *Respiratory Protection (1910.134)*—Workers required to wear respirators must be trained in their proper use and limitations. Users must receive fitting instructions to include demonstration and practice. Retraining should be done annually.
- *Confined Spaces (1910.146)*—Training must be provided for all employees involved in working with confined spaces. Training must be completed before being assigned duties involving confined spaces.
- *Lockout of Hazard Energy Sources (1910.147)*—Workers authorized to conduct lockout/tagout actions must be trained on lockout procedures to include training on isolating specific energy sources and/or machines. Other workers in the area must receive awareness-type training. Training must also be conducted on procedures for removing lockout devices.
- *Laundry Machine Operating Rules (1910.264)*—Laundry workers must be instructed on the hazards of their work and be trained in safe work practices including training on preventing exposure to bloodborne pathogens.
- *Electrical Safe Work Practices (1910.332)*—Workers exposed to electrical circuits must be trained on specific job hazards.
- *Lead (1910.1025)*—Employers must inform employees of any potential lead exposures using Appendices A and B of the standard. Initial and annual training must be conducted if exposure exceeds action level or if skin or eye irritations are possible.
- *Benzene (1910.1028)*—Employers must provide training and information to employees assigned to work areas where benzene is present. Annual training must be accomplished for all employees exposed at or above the action level.
- *Bloodborne Pathogens (1910.1030)*—Exposed or potentially exposed workers must receive training upon initial assignment. Retraining must be accomplished within 12 months of initial training. Training must be documented and records maintained for three years.
- *Ethylene Oxide (1910.1047)*—Employers must provide initial training to employees assigned to areas with potential ethylene oxide exposures. Annual training is required for all employees potentially exposed to ethylene oxide.
- *Formaldehyde (1910.1048)*—A training program must be established for workers exposed to 0.1 ppm or more. Initial training must be provided at the time of assignment, with annual retraining required for all potentially exposed employees.
- *Hazard Communication (1910.1200)*—Workers exposed to hazardous materials must receive initial training in program requirements upon assignment. Training must be accomplished when new hazardous substances are introduced into the workplace.

- *Lab Safety (1910.1450)*—Workers must be trained on the hazardous substances found in their work areas. Training must be accomplished when new hazards are introduced into the work area.
- *Asbestos (1910.1001)*—Employers must conduct training for all workers exposed to asbestos fibers at or above the action level during initial assignment. Retraining must be conducted annually.

Training Budget—Budget considerations are always crucial to the program. Top management should consider and communicate the following:

- Training is the foundation of the safety program.
- Facilities must provide certain regulatory training.
- Costs of accidents and injuries must be determined.
- Training helps identify, eliminate, or control hazards.
- Training allows for personal development and increased efficiency.

Supervisor Responsibilities—The supervisor's role in establishing and enforcing safety policies should be supported with training.

1. *Basic Responsibilities*—Supervisory training includes an explanation of supervisory responsibilities:
   - Analyze work areas to identify unrecognized potential hazards
   - Maintain PPE and ensure its proper use
   - Provide training on potential occupational hazards
   - Ensure workers know protective measures to follow
   - Reinforce employee training through continual performance feedback
   - Enforce compliance with safety rules and practices

2. *Training Provided by Supervisors*—Supervisors should discuss the items below with workers during job safety training:
   - Hazardous operations and tasks found in the workplace
   - Department safety procedures and rules
   - Hazardous chemicals used in the department
   - Standards and regulations that apply to job safety
   - PPE and how, when, and where to use it
   - Location and use of emergency and fire protection equipment
   - Emergency procedures that apply to job and workplace
   - How to report unsafe equipment, conditions, or procedures
   - Accident, incident, and injury reporting procedures
   - Location of written safety programs and information

Evaluating Training Effectiveness—Training should be presented so that its organization and meaning are clear to employees. Supervisors should provide overviews of the material to be learned, relate information to the job, and reinforce the key points of the information covered.

- Training should have a method for measuring effectiveness. An evaluation plan should be developed when the course objectives and content are developed.

- Evaluation will help employers or supervisors determine the level of education achieved and whether an employee's performance has improved on the job.
- Evaluation methods include student opinion surveys, supervisor's observations, and workplace improvements.
- One way to differentiate between employees who have priority needs for training and those who do not is to identify employee populations that are at higher levels of risk.
- The nature of the work will provide an indication that such groups should receive priority for information on occupational safety and health risks.
- One way to identify employee populations at high levels of occupational risk is to pinpoint hazardous occupations.
- Another way to identify employee populations at high levels of risk is to examine the incidence of accidents and injuries.
- If employees in certain occupational categories are experiencing higher accident and injury rates than other employees, training may be one way to reduce that rate.

Understanding Human Behavior—Working safely depends on a worker's current needs, present situation, and past experiences.

1. *Preventing Unsafe Behavior*—Many variables, such as work environment and job requirements, can affect unsafe behavior.
   - Supervisors must understand that the needs of people can link organizational goals to individual goals.
   - Workers often become frustrated when their goals cannot be achieved.
   - Reaction to frustration can be disruptive to safety, morale, and job performance of others.
   - Understanding factors that contribute to unsafe job performance is key to changing safety-related behaviors.

2. *Encouraging Safe Behavior*—Healthcare organizations should:
   - Provide frequent and effective training
   - Document unsafe behavior and poor attitudes
   - Provide a positive work environment
   - Recognize and reward satisfactory performance
   - Remove obstacles that prevent satisfactory performance
   - Look for easier methods to do the job
   - Listen to the opinions and concerns of workers

3. *Establishing and Enforcing Safety Rules*—All rules, policies, and regulations should be published and communicated. All employees should know what is expected. This can be done in several ways:
   - Include safety policies and rules in the employee handbook
   - Provide a copy of the rules and regulations to each worker
   - Have employees sign a safe work agreement
   - Explain the rules and regulations in regular meetings
   - Post important rules and regulations in a visible location
   - Enforce safety and job rules fairly and firmly

- Issue verbal warnings for the first or minor infractions
- Provide a written warning for serious or repeated infractions
- Ensure workers know consequences of failing to follow procedures
- Deal with flagrant job or safety violations immediately

Safety Meetings—Many organizations fail to use one of the most effective tools available for promoting safety. Safety meetings are overlooked but should be an important part of any healthcare facility safety program. Safety meetings keep the lines of communication open and can reinforce training concepts. Effectively held meetings provide the following benefits:

- Promote safety awareness in the workplace
- Encourage worker involvement
- Motivate workers to practice safety on the job
- Pinpoint problem areas and solicit corrective solutions
- Introduce workers to new rules, procedures, and equipment
- Provide information on causes and types of accidents

J. PROMOTING SAFETY

Healthcare organizations must seek ways to promote interest in safety. Top administration personnel must communicate with key personnel and supervisors on a regular basis. Management must also demonstrate its commitment by actively supporting and promoting a program that involves all personnel. The most important considerations are helping workers develop safe work habits and providing a safe and healthful work environment. Some ways to promote safety include:

- Focus attention on accident causal factors
- Provide effective safety training to all employees
- Give workers an opportunity to participate in the program
- Encourage communication between workers and management
- Stress the facility's commitment to visitor and patient safety
- Require employees to practice working safely

Promotional Activities—Management personnel should select activities that will yield results. Budget constraints can affect the safety promotion program. The size and type of the organization affects the choice of activities or materials necessary for maintaining interest. Second- and third-shift employees must be included because their support is vital to the overall accident prevention effort. An effective safety promotion program demonstrates management's commitment, reminds employees to work safely, and, in the healthcare setting, contributes to quality care. Awards and other incentives can help promote the program, but they must be meaningful.

Safety Posters—Safety posters cannot compensate for inadequate safety management, broken equipment, unsafe job procedures, poor supervision, or ineffective training. Workers prefer viewing posters over reading procedures.

- Posters should be simple, clever, and colorful; should be designed to stimulate worker interaction; and should demonstrate management's interest in the welfare of workers.
- Safety posters should never be demeaning to the viewer's intelligence, race, or sex. Posters should be placed in well-lighted locations where employees congregate, such as lunchrooms, washrooms, entrances, and loading points. Many facilities rotate posters at regular intervals.

Safety Bulletins and Newsletters—One of the most effective tools an organization can use to promote safety is the safety bulletin or informational newsletter. A well-designed and informative bulletin that is employee oriented will pay immediate dividends.

- A balanced publication will provide information and recognize employee accomplishments. A publication that is written sincerely using simple sentences will be effective.
- Stories should be personalized whenever possible.
- Look for ways to grab the reader's attention without relying on cute gimmicks with no substance.
- Ask for suggestions, articles, feedback, and comments from readers.
- Develop some features or subjects that will appear on a regular basis.
- Look for ways to sell and promote the safety message. An effective newsletter that helps prevent one injury is well worth the effort and cost.

Off-Duty Safety and Health Programs—Management should communicate its concern for employee health and safety off the job. Healthcare organizations have access to professionals who could contribute to the development of a safety, health, and wellness program for workers.

1. *Off-the-Job Safety*—When incorporating off-the-job safety topics into the safety program, strive to meet the special needs of the facility and its employees. Some suggested topics include:
   - Seasonal hazards
   - Traffic safety
   - Stress management
   - Diet and exercise
   - Home safety
   - Health risks
   - Alcohol and drugs
   - Smoking cessation

2. *Wellness Programs*—Healthcare organization have the resources to establish effective wellness programs. Organizations are now realizing that wellness programs can save money and keep people working. Healthcare organizations should be leaders in providing wellness and health information to workers because of the resources available and the specialists that could contribute to the program. An effective wellness program should:
   - Be established as a part of the employee health program and be adequately funded to provide quality service to all workers
   - Include a newsletter or bulletin to promote the program and keep workers informed on new topics or programs
   - Distribute educational materials, including informational booklets and videos

- Schedule formal presentations, risk assessment clinics, and health-related workshops
- Develop preventive activities such as aerobic classes, athletic teams, and smoking cessation programs
- Look for ways to show support for the program by rewarding participation

SUMMARY

Understanding accidents, analyzing hazards, and implementing controls will reduce injuries. Top management should receive information from the safety and health committee about current problems and program effectiveness. This information can be effective in convincing managers and department heads that changes are needed. In today's regulatory environment, management has no choice as to whether hazard control is necessary. Increased operating costs coupled with governmental and voluntary compliance agency requirements suggest that proactive hazard control activities are important to the success of a safety program. It is important to remember that increased knowledge, communication, and motivation have a positive effect on facility operational effectiveness.

REVIEW QUESTIONS

1 An accident can be defined as any unplanned event that interferes with job or task accomplishment. Which of the following statements concerning accidents is true?
   a.   an accident must always produce some type of loss
   b.   an accident can best explained by determining the single cause
   c.   an accident is a chance or random event
   d.   an accident investigation should attempt to determine causal factors

2 Identifying hazards can best be accomplished through:
   a.   conducting comprehensive worksite analyses/surveys
   b.   reviewing hazard control publications and journals
   c.   analyzing accident and injury data from previous years
   d.   understanding regulatory standards and requirements

3 The primary goal of any hazard evaluation should be to:
   a.   identify all categories of hazards within a facility
   b.   assess employee attitudes and compliance with safety rules
   c.   determine risk of injury or other loss potential posed by the hazard
   d.   a and b

4 Hazard probability calculations are best made when using:
   a.   historical data
   b.   empirical calculations
   c.   regulatory standards
   d.   a and b

5   The preferred method for controlling hazards is to:
    a.   reduce or eliminate the risk to workers by reducing the frequency and/or duration of exposure
    b.   establish safety rules and implement work practice controls
    c.   use technological or engineering methods to isolate or remove the hazard
    d.   implement a hazard warning system as soon as the hazard is identified

6   Which of the following terms describes the maximum allowed OSHA exposure for workers during an 8-hour workday or 40-hour work week?
    a.   short-term exposure limit
    b.   permissible exposure limit
    c.   length of exposure
    d.   ceiling limit

7   Which organization recommends limits for exposure to airborne contaminants using the term threshold limit values?
    a.   Centers for Disease Control and Prevention
    b.   Occupational Safety and Health Administration
    c.   American Conference of Governmental Industrial Hygienists
    d.   National Institute of Occupational Safety and Health
    e.   National Association of Certified Industrial Hygienists

8   Irritation, illness, or death resulting from a highly concentrated short-term exposure to a contaminant or hazardous substance is known as:
    a.   individual susceptibility
    b.   acute effects
    c.   toxic exposure
    d.   chronic effects

9   Which of the following is *not* true about the use of personal protective equipment?
    a.   OSHA-regulated facilities must conduct and document a PPE survey for all work areas
    b.   many PPE standards refer to specifications published by the American National Standards Institute
    c.   employees using PPE must demonstrate an understanding of all requirements before using such equipment
    d.   in most situations, PPE can be used as a substitute for other hazard controls

10 Which of the following organizations publishes guidance on respirator selection?
    a.   OSHA
    b.   ANSI
    c.   NIOSH
    d.   ASTM

11 A black cartridge on a respirator would protect against:
    a.   radioactive materials
    b.   organic vapors
    c.   acid gas
    d.   dusts and mists

12 When required because of hazardous materials, an emergency eyewash unit must be located no more than _____ from the hazard.
  a. 50 feet
  b. 75 feet
  c. 100 feet
  d. 150 feet

13 Healthcare safety training objectives should be:
  a. reasonable, observable, and realistic
  b. based on safety theory and regulatory compliance
  c. stated in terms that address what should be done in all situations
  d. developed by the instructor before class begins

14 Workers exposed to formaldehyde at the action level must be trained:
  a. when assigned a position where exposure occurs
  b. if periodic monitoring reveals exposure above the OSHA PEL
  c. at the time of initial assignment and annually thereafter
  d. when medical conditions warrant training

Answers: 1–d 2–a 3–c 4–d 5–c 6–b 7–c 8–b 9–d 10–c 11–b
              12–c 13–a 14–c

CHAPTER *4*

SAFETY REGULATION

A. FEDERAL REGULATIONS

Healthcare administrators, risk managers, care providers, and hazard control professionals must be aware of the numerous agencies involved in healthcare safety regulation. Regulatory and compliance issues are vital to quality of care and safety. This chapter provides a brief look at some of the laws, regulatory agencies, and voluntary compliance organizations involved in healthcare safety. Hazard control and safety control professionals must be familiar with federal publications, including the Federal Register, the Code of Federal Regulations (CFR), and the U.S. Code (USC).

Federal Register Act—In 1934, Congress recognized the need for a centralized system to publish and catalog federal regulations. The Federal Register Act became law on July 26, 1935 (44 U.S.C. Chapter 15).

1. *Uniform System*—This act established a uniform system for handling agency regulations by requiring:
   - Filing of documents with the Office of the Federal Register
   - Availability of documents for public inspection
   - Publication of documents in the Federal Register
   Note—A 1937 amendment established permanent codification through a numerical arrangement of rules in the Code of Federal Regulations.

2. *Administrative Procedure Act*—Passed in 1945 (5 U.S.C. 551), the act added the following dimensions to the Federal Register system:
   - Gave the public the right (in most instances) to participate in the rulemaking process by commenting on proposed rules
   - Required that the effective date for a regulation be not less than 30 days from the date of publication unless there is a good cause for an earlier date
   - Provided for publication of agency statements of organization and procedural rules
   Note—The primary function of the Federal Register and Code of Federal Regulations system is to communicate proposed rulemaking actions to the public.

The Federal Register—The Federal Register is published daily by the National Archives and Record Service to ensure federal regulations and legal notices are made available to the public. Federal agencies publish regulation proposals in the Federal Register and announce details of the comment period. When the regulation is finally issued, it is again published in the Federal Register. Information is referenced by volume number and page.

1. *Ways to Search for a Certain Document*—
   - **Federal Register Index**—This index is arranged by agency. The index is published monthly and cumulated for 12 months.
   - **List of CFR Sections Affected (LSA)**—This is a monthly cumulative list of CFR sections affected by rules and proposed rule documents published in the Federal Register.
   - **CFR Index and Finding Aids**—These annual indexes contain a subject/agency index with references to CFR title and parts of a specific subject or issuing agency. There are four annual issues:

December	Titles 1–16
March	Titles 17–27
June	Titles 28–41
September	Titles 42–50

   - **Parallel Table of Authorities and Rules**—This information, which is also contained in the LSA, has references to CFR titles and parts that implement specific legislative or presidential directives.

2. *Strategies for Searching the Federal Register System*—
   - Agency name, such as Department of Labor
   - Looking for subject, such as Hazard Communication
   - Type of action, such as "final rule"
   - Approximate date of document

   Note—The search technique used depends on the currency of the document sought.

The Code of Federal Regulations—Published by the National Archives and Record Services, the code is printed annually in paperback volumes. The CFR is a compilation of the general and permanent rules first published in the Federal Register by the executive departments and agencies of the federal government, divided into 50 titles representing areas subject to federal regulation. The CFR is kept up to date by the individual issues of the Federal Register which are published each working day. The pages in the Federal Register are numbered sequentially through the year beginning January 1 and ending December 31. The CFR format uses a period instead of a decimal to separate part numbers from the section number. For example, 29 CFR 1926.3 would precede 29 CFR 1926.25. A lower case "a" precedes an upper case "A," which can be confusing to unfamiliar users.

1. *Revision Schedule*—CFRs are revised annually according to the following schedule:
   - Titles 1–16 are revised as of January 1
   - Titles 17–27 are revised as of April 1

- Titles 28–41 are revised as of July 1
- Titles 42–50 are revised as of October 1

2. **Titles of Interest**—Some titles of interest to healthcare hazard control professionals include:
   - Title 10   Energy (Radiation)
   - Title 21   Food and Drugs
   - Title 29   Labor (OSHA Standards)
   - Title 40   Protection of the Environment (EPA)
   - Title 44   Emergency Management and Assistance
   - Title 49   Transportation (DOT)

   Note—The CFR and the Federal Register must be used together to determine the latest version of any given rule.

U.S. Code—The U.S. Code is the official document containing general and permanent laws. Statutory authority is cited in parentheses at the end of each section or group of sections. The cited authority may be congressional acts, joint resolutions, presidential executive orders, or reorganization plans. The publication is well indexed, with a table of contents for each of the 50 titles. The U.S. Code is a compilation of law in force from 1789 to the present. It is *prima facie* or presumed to be law and therefore does not include repealed or expired acts.

Superintendent of Documents—The Federal Register and CFR are available from the Government Printing Office at the address below:

U.S. Government Printing Office (GPO)
Washington, DC 20402-9329
Telephone (202) 512-2457

The Public Law Update Service—This service provides information on recent bills signed or vetoed by the president; the phone number is (202) 523-6642. The phone number for the National Archives and Record Service is (202) 523-5235.

B. OCCUPATIONAL SAFETY AND HEALTH ADMINISTRATION

The Williams–Steiger Act of 1970, more commonly known as the Occupational Safety and Health Act, was passed to ensure workers safe and healthful working conditions. The act also created the Occupational Safety and Health Administration (OSHA). OSHA continually reviews and redefines specific standards and practices; its basic purpose remains constant. OSHA strives to implement its mandate fully and firmly with fairness to all concerned. In all its procedures, from standards development through implementation and enforcement, OSHA guarantees employers and employees the right to be fully informed, to participate actively, and to appeal actions.

1. **OSHA**—OSHA is an agency of the Department of Labor; its director holds the title of assistant secretary. OSHA is tasked to:

- Encourage employers and employees to reduce workplace hazards and to implement new or improve existing safety and health programs
- Provide for research in occupational safety and health to develop innovative ways of dealing with safety and health problems
- Establish "separate but dependent responsibilities and rights" for employers and employees for the achievement of better safety and health conditions
- Maintain a reporting and recordkeeping system to monitor job-related injuries and illnesses
- Establish training programs to increase the number and competence of occupational safety and health personnel
- Develop mandatory job safety and health standards and enforce them effectively
- Provide for the development, analysis, evaluation, and approval of state occupational safety and health programs

2. *The OSHA General Duty Clause*—The employer is obligated under the General Duty Clause (5a1) to protect employees from recognized hazards even if there is not a standard that applies to the situation. Employers must go the extra mile in protecting workers. A strong safety emphasis can help employers comply with the intent of the clause. Steps to help meet the requirements include:
   - Review accident/injury records to identify trends or patterns
   - Promptly investigate every accident and take corrective actions
   - Conduct job hazard analyses to identify worker exposures
   - Conduct and document training sessions on a regular basis
   - Establish a safety committee with hourly employees as members
   - Conduct thorough safety evaluations on a regular basis

OSHA Standards—OSHA promulgates legally enforceable standards. Healthcare organizations are covered under the provisions of the General Industry Standards, 29 CFR 1910. OSHA standards may require compliance with certain conditions, practices, methods, or processes and that appropriate measures be taken to protect workers on the job. Employers must be familiar with standards applicable to their establishments and ensure employees use personal protective equipment as required. Employees must also comply with all safety regulations applicable to their jobs.

- Where OSHA does not specify a specific standard, employers are responsible for the general duty clause, which states that "each employer shall furnish a place of employment free from recognized hazards that are causing or likely to cause death or serious physical harm to his employees."
- States with OSHA-approved occupational safety programs must set standards that are at least as effective as the federal standards. Most states adopt standards identical to the federal standards, but in some instances states adopt stricter safety laws.
- OSHA begins the standards-setting procedures or initiates action in response to petitions from other agencies. Many times, action starts with the secretary of health and human services, the National Institute for Occupational Safety and Health, state and local governments, and even nationally recognized voluntary

standards organizations. At other times, requests may begin with the employer, labor representatives, or even workers.

- OSHA publishes its intentions to develop, propose, amend, or revoke a standard in the Federal Register as a "Notice of Proposed Rulemaking" or as an "Advance Notice of Proposed Rulemaking." The proposed notice includes the terms of the new rule and provides a specific time period (30 to 90 days) for the public to respond.
- Interested parties submitting written arguments and evidence may request a public hearing on the proposal if one has not been announced in the notice. When requested, OSHA will schedule a hearing and publish the time and place in the Federal Register. OSHA must publish in the Federal Register the final text of any standard amendment adopted and the effective date. OSHA also publishes an explanation of the standard and the reasons for implementation.
- Standards can be "horizontal" or "vertical." Most OSHA standards are horizontal, or have general application to any employer in any industry. Horizontal standards include those relating to fire protection, working surfaces, and first aid. Vertical or particular standards apply only to a specific industry. Vertical standards include those applicable to the construction industries and special industries covered under 29 CFR 1910, Subpart R. OSHA standards were originally promulgated from three sources—federal laws, proprietary standards, and consensus standards.

Types of OSHA Standards—

- *Proprietary Standards*—These are prepared by professionals within specific industries, professional societies, and associations. They are determined by a straight membership vote and not by consensus.
- *Consensus Standards*—Consensus standards developed by industry-wide developing organizations are agreed upon through consensus by industry, labor, and other representatives. Many of the current OSHA standards were promulgated from the American National Standards Institute, National Fire Protection Association standards.

OSHA Inspections—OSHA enforces its standards by conducting workplace inspections. Every establishment covered by OSHA standards can be inspected by OSHA compliance safety and health officers. Worst-case situations need attention first, and OSHA uses a system of inspection priorities. There are several types of OSHA inspections:

- *Imminent Danger*—An imminent danger is any condition where there is reasonable certainty that a danger exists which can be expected to cause death or serious physical harm immediately, or before the danger can be eliminated through normal enforcement procedures. Serious physical harm is any type of harm that could cause permanent or prolonged damage to the body or which, while not damaging the body on a prolonged basis, could cause such temporary disability as to require in-patient hospital treatment. OSHA considers that "permanent or prolonged damage" has occurred when, for example, a part of the body is crushed or severed; an arm, leg, or finger is amputated; or sight in one or both eyes is lost.

- *Workplace Fatalities*—Second priority is given to investigation of fatalities and catastrophes resulting in hospitalization of five or more employees. Such situations must be reported to OSHA by the employer within 48 hours. Investigations are made to determine if OSHA standards were violated and to avoid recurrence of similar accidents.
- *Worker Complaints*—Third priority is given to employee complaints of alleged violation of standards or of unsafe or unhealthful working conditions. The act gives each employee the right to request an OSHA inspection when the employee feels he or she is in imminent danger from a hazard or when he or she feels there is a violation of an OSHA standard that threatens physical harm. OSHA will maintain confidentiality if requested, will inform the employee of any action it takes regarding the complaint, and, if requested, will hold an informal review of any decision not to inspect. Just as in situations of imminent danger, the employee's name will be withheld from the employer, if the employee so requests.
- *Programmed Inspections*—Programmed or planned inspections are aimed at specific high-hazard industries, occupations, or health substances. Industries are selected for inspection on the basis of factors such as the death, injury, and illness incidence rates and employee exposure to toxic substances. Special emphasis may be regional or national in scope, depending on the distribution of the workplaces involved. States with their own occupational safety and health programs may use somewhat different systems to identify high-hazard industries for inspection.

Records Disclosure—Employers can voluntarily disclose the OSHA Log (Form 200) and the Supplemental Record (Form 101). Compliance officers are entitled to copy these. Other records required by OSHA, such as those relating to lead exposure, noise exposure, radiation exposure, HAZCOM training records, and bloodborne pathogens training, should only be disclosed pursuant to a specific warrant. Compliance officers should not be allowed to conduct fishing expeditions by searching through records, and courtesy copies should not be made for inspectors.

OSHA Violations and Citations—Employers should thoroughly question inspectors about areas of possible and probable violations. Exact descriptions of problems and standards allegedly violated should be provided.

- Employers should admit nothing; many (perhaps most) OSHA violations result from statements or evidence provided by the employer. Employees should never agree or argue with an inspector and should not indicate knowledge or negligence.
- Employers should ask about a feasible means of abatement for a possible violation. OSHA must recommend a feasible means of abatement or correction. The employer's defense can be enhanced if the inspector cannot answer this question.
- Employers should never agree to or propose an abatement date. Employers should realize there is very little information they are required to provide to an inspector.
- The OSHA area director determines what citations will be issued and penalties

imposed. A formal citation informs the employer and employees of the alleged violation. The citation will also have a proposed abatement date.

- Citations and notices of proposed penalties are sent by certified mail. Employers must post a copy of each citation near the violation site. The citation must remain posted for 3 days or until abatement of the condition, whichever is longer.

Types of OSHA Violations—

- *Other Than Serious Violation*—A violation issued for conditions which have a direct relationship to job safety but probably would not cause death or serious physical harm. A penalty can be assessed for each violation. Penalties can be adjusted downward, depending on the employer's good faith, previous violation history, and size of the facility.
- *Serious Violation*—A violation issued when there is substantial probability that death or serious physical harm could occur and the employer knew or should have known of the hazard. A mandatory penalty for each violation is proposed. A penalty for a serious violation can also be adjusted downward based on the employer's good faith and previous violation history, the gravity of the alleged violation, and size of business.
- *Willful Violation*—A violation issued when the employer knows that current processes or procedures constitute a violation or knows that a hazardous condition exists but has made no reasonable effort to eliminate it. Criminal conviction of a willful violation can result in a fine of up to $250,000 if employee death has occurred.
- *Repeat Violation*—A repeat of a previous violation of a standard, regulation, rule, or order. Reinspection confirms that a substantially similar violation has been found. Repeat violations can result in stiff fines. A citation under contest may not serve as the basis for a subsequent repeat citation.
- *Failure to Abate*—Failure to correct or abate a previous violation can result in a civil penalty for each day the violation continues beyond the prescribed abatement date.
- *Falsifying Records, Reports, or Applications*—Conviction can bring a criminal fine and jail sentence.
- *Violation of a Posting Requirement*—This can result in a civil penalty being assessed.

Note—Citation and penalty procedures may differ in states operating an OSHA-approved state safety and health program. Small and medium employers may get penalty discounts based on good faith efforts and size of the company.

Healthcare OSHA Violations—Healthcare facilities have a number of hazards which, if not controlled or managed properly, could result in OSHA citations. The ten most common OSHA violations are as follows:

1. Improper disposal of materials containing bloodborne pathogens
2. Deficiencies in hazard communication procedures
3. Errors found in the OSHA Log (Form 200)

4. Improper maintenance/access to employee exposure/medical records
5. Deficiencies in lockout/tagout procedures
6. Lack of proper respiratory protection
7. Formaldehyde hazards/exposures
8. Hazardous wiring methods and unsafe equipment
9. Electrical hazards
10. Insufficient means of emergency egress

OSHA Recordkeeping and Reporting—When OSHA legislation was passed, there was no centralized and systematic method for monitoring occupational safety and health problems. Statistics on job injuries and illnesses were only collected by a few states and some private organizations. National accident and injury figures often were based on estimates or unreliable projections. OSHA allowed for the establishment of consistent national procedures for tracking injury and illness statistics. Table 4.1 lists records to be retained and retention requirements. General recordkeeping guidelines are listed below:

- Employers of more than ten employees in hazardous and selected service industries must keep injury and illness records.
- Healthcare organizations covered under OSHA or a state-approved plan must also maintain records.

TABLE 4.1   Occupational Safety and Health Act Records

Records to Be Retained	Period of Retention/Filing Requirements
Log and summary of occupational injuries and illnesses • Briefly describing recordable cases of injury and illness • Extent and outcome of each accident • Summary totals for calendar year (OSHA Form 200 or equivalent)	Five years following end of year to which records relate Must post OSHA Form 200 by February 1 and must remain posted until March 1 of each year
Supplementary records • Containing more detailed information for each occurrence of injury or illness (OSHA Form 101 or equivalent)	Five years following end of year to which records relate
Employee medical records • Excluding health insurance claims if maintained separately • Certain minor first-aid records • Records of employees who worked less than one year and who received records upon termination	Duration of employment plus 30 years, unless a specific OSHA standard requires a different time period
Records of biological or environmental monitoring of exposure to hazardous materials and related analyses	30 years
Material Safety Data Sheets	

- Maintaining records permits the Bureau of Labor Statistics (BLS) to analyze injury data and define high-hazard industries.
- Some businesses in the service sector are exempt from maintaining OSHA injury and illness records.
- General merchandise establishments, food stores, hotels, repair services, recreation services, and health facilities are required to maintain records.
- Some exempt businesses may be requested to maintain records if selected to participate in the Annual Survey of Occupational Injuries and Illnesses. The BLS will notify establishments and provide the necessary forms and instructions.
- All employers, regardless of classification, must:
  a. Comply with OSHA standards
  b. Display the OSHA poster in a prominent location
  c. Report to OSHA within eight hours any accident that results in one or more fatalities or the hospitalization of three or more employees

Recording Occupational Injuries and Illnesses—

1. *Occupational Illness*—An occupational illness is any abnormal condition or disorder, other than one resulting from an occupational injury, caused by exposure to environmental factors associated with employment. Included are acute and chronic illnesses or diseases which may be caused by inhalation, absorption, ingestion, or direct contact with toxic substances or harmful agents. All occupational illnesses must be recorded regardless of severity.

2. *Occupational Injury*—This is a work-related injury such as a cut, fracture, sprain, or amputation resulting from a single exposure or event. All occupational injuries must be recorded if they result in:
   - Death of employee regardless of the time between the injury and death
   - One or more lost workdays
   - Restriction of work or motion
   - Loss of consciousness
   - Transfer to another job
   - Medical treatment (other than first aid)

3. *Location of Records*—Employers must keep injury and illness records for each establishment. An establishment is defined as a "single physical location where business is conducted or where services are performed."
   - An employer whose employees work in dispersed locations must keep records at the place where the employees report for work.
   - In some situations, employees do not report to work at the same place each day. In that case, records must be kept at the place from which employees are paid or at the base from which they operate.

OSHA Forms—Forms are maintained on a calendar year basis and are retained for five years by the employer at the establishment. They are not sent to OSHA. Forms must be available for inspection by representatives of OSHA, Department of Health and Human Services agencies, and the BLS.

- *OSHA Form 200 (Summary of Occupational Injuries and Illnesses Log)*—Each recordable occupational injury and illness must be logged on this form within six working days from the time the employer learns of it.
  1. A log prepared at a central location must be current to within 45 calendar days and be present at all times in the establishment.
  2. A substitute for OSHA Form 200 is acceptable if it is detailed, easy to read, and understandable.

- *OSHA Form 101 (Supplementary Record of Occupational Injuries and Illnesses)*—This form is prepared in cases where individuals are injured or experience an occupational illness. Form 101 contains more detail about each specific injury or illness; it must be completed within six working days from the time the employer learns of the work-related injury or illness. Other forms may be used in lieu of Form 101 if all required information is included.

- *OSHA Form 2203 (Informational Poster)*—All work sites must display the OSHA poster in a location where employees are likely to see it.

Completing OSHA Form 200—A copy of the totals and information following the fold-line on the last page of OSHA Form 200 must be posted at each establishment in a highly visible location. The form must be posted no later than February 1 and remain for 30 days.

- Establishments with no injuries or illnesses during the year must enter all zeros on the total lines and post the form.
- Employers that want to use a different recording system may apply for a recordkeeping variance. Requests must detail and justify the employer's intended procedures and must be submitted to the regional commissioner of the BLS.
- The BLS is the only agency that can grant a variance from recordkeeping requirements.

Proposed Changes to OSHA Recordkeeping Requirements and Forms—OSHA recently announced its intent to revise regulations for recordkeeping and reporting occupational injuries and illnesses. The proposed changes would also affect states that operate their own OSHA programs. In 1990, recordkeeping functions were transferred from the BLS to OSHA. The BLS did retain responsibility for conducting the Annual Survey of Occupational Injuries and Illnesses. The announced changes would be the first major revisions to Title 29 CFR Part 1904. Listed below are some of the proposed significant changes:

- For recordkeeping purposes, employees would be defined as workers the employer supervises on a day-to-day basis. This would include temporary, contract, and leased personnel.
- OSHA Form 200 would be replaced with OSHA Form 300, and OSHA Form 101 would become OSHA Form 301.
- First aid would be redefined to mean any treatment or care other than those treatments found on an OSHA-developed list. Work-related musculoskeletal dis-

orders, bloodborne pathogen exposures, and positive tuberculosis conversions would be listed on the log.

- OSHA logs must be signed by a responsible company official to attest to their accuracy and completeness. This could be the owner, an officer of the corporation, or the highest ranking company official at the facility.
- The new rule proposes to eliminate the requirement to record the days of restricted work and only count the number of days off the job. Days beyond six months would not be counted. The year-end summary would have to be posted for the entire year.
- The five-year retention period would be lowered to three years.
- The new rule proposes that illness/injury data maintained by employers may be collected periodically and used for a variety of purposes. The information could be used to schedule injury/illness surveillance programs, evaluate OSHA's enforcement and training programs, and direct OSHA program activities including inspections.

Employee Health Records—An individual health record for each employee should be kept in the employee health service department. An employee medical record details the health status of an employee.

- The record must be made or maintained by a physician, registered nurse, or other healthcare professional in accordance with 29 CFR 1910.20. Health and exposure records are maintained for the duration of employment plus 30 years.
- Some OSHA standards may require a different period of time (29 CFR 1910.20). Laboratory reports and worksheets must be kept for one year.
- Confidentiality of employee health records must be maintained to protect the employee's privacy. The employer must make these records available to the employee or an employee's designated representative.

Exposure Monitoring Records—Employees and their designated representatives have the right to examine and copy results of exposure monitoring. Only employees or their designated representatives can review individual employee records describing employee exposures. Exposure records contain the following type of information:

- Specific environmental monitoring or sampling results, collection methodology, a description of the analytical/mathematical methods used, and a summary of data relevant to interpretation of the results
- Biological monitoring results directly assessing absorption/exposure of a hazardous substance
- Material Safety Data Sheet or hazard inventory information describing chemicals and indicating where and when they were used

Reporting Injuries and Fatalities—OSHA recently amended the reporting rule for all employers covered under federal OSHA jurisdiction. States with approved plans must adopt a comparable rule. OSHA made three changes to the reporting requirements:

- Employers are now required to report work-related incidents resulting in the death of an employee or the hospitalization of three or more employees.

- Employers must verbally report incidents resulting in employee death or multiple hospitalizations within eight hours of their occurrence.
- Accidents resulting in death or inpatient hospitalization of three or more employees must be reported within 30 days of the incident.

OSHA Incidence Rates—Incidence rates are based on the exposure of 100 full-time workers using 200,000 employee-hours as the equivalent (100 employees working 40 hours per week for 50 weeks per year).

- An incidence rate can be computed for each category of cases or days lost, depending on what number is used in the numerator of the formula.
- The denominator of the formula should be the total number of hours worked by all employees during the same time period as that covered by the number of cases in the numerator.

$$\text{Incidence Rate} = \frac{\text{no. of injuries/illnesses} \times 200,000}{\text{total hours worked by all employees during period covered}}$$

Preparing for an OSHA Inspection—Most healthcare facilities direct hazard control and safety efforts to complying with Joint Commission on Accreditation of Healthcare Organizations standards. Hazard control managers and safety officers should be familiar with OSHA procedures.

1. *Ensure OSHA Compliance*—Listed below are some things that can be done to improve facility safety and ensure OSHA compliance:
   - Develop a comprehensive hazard identification program
   - Evaluate hazards and implement corrective measures
   - Ensure that supervisory personnel enforce safety rules
   - Provide and document safety training required by OSHA standards
   - Post OSHA informational posters throughout the facility
   - Hazard control managers should review OSHA standards regularly
   - Evaluate recordkeeping and documentation procedures periodically

2. *Managing OSHA Compliance Inspections*—
   - Never explain processes or the operation of any machinery
   - Do not point out problem areas to the inspector
   - Do not indicate awareness of any alleged violations
   - Refrain from discussing extenuating circumstances
   - Two facility representatives should be with the inspector(s)
   - Ensure one person takes detailed notes
   - Do not give any notes taken to the inspector
   - Arrange a time and place for worker interviews
   - Attempt to interview employees who met with the inspectors
   - Document names of employees interviewed
   - Employees are entitled to have their interviews in private
   - Management personnel are not required to cooperate
   - Use caution when making statements to the compliance officer

C. ENVIRONMENTAL PROTECTION AGENCY

Environmental Protection Agency (EPA)—The agency was created in 1970 to protect the environment, to exercise control over the release of harmful substances that could threaten public health, and to control and abate environmental pollution. Many of the EPA's rules define which substances can be hazardous to human health and/or pose a threat to the environment. The EPA or state-approved agencies provide guidance for handling hazardous materials, regulate the operation of waste disposal sites, and establish procedures for dealing with environmental incidents such as leaks or spills. The EPA also has an interest in hazardous and infectious wastes produced by healthcare facilities. It publishes informative guides to assist risk managers in understanding and complying with a number of environmental laws and regulations. Environmental laws are published in 40 CFR.

Resource Conservation and Recovery Act (RCRA)—The RCRA is one of the EPA's main statutory weapons. The act created a "cradle-to-grave" management system for current and future wastes, while the EPA authorizes cleanup of released hazardous substances. Several statutes are media specific and limit the amount of wastes introduced into the air, waterways, oceans, and drinking water. Other statutes directly limit the production, rather than the release, of chemical substances and products that may contribute to the nation's wastes.

The RCRA is unique in that its primary purpose is to protect human health and the environment from the dangers of hazardous waste. The RCRA has a regulatory focus and authorizes control over the management of wastes from the moment of generation until final disposal.

1. *Superfund Response Focus*—Whenever there has been a breakdown in the waste management system, such as a release of a hazardous substance, the statute authorizes cleanup actions. The RCRA was passed in 1976 as an amendment to the Solid Waste Disposal Act.

2. *General RCRA Goals*—
   - To protect human health and the environment
   - To reduce waste and conserve natural resources and energy
   - To reduce or eliminate the generation of hazardous waste as expeditiously as possible

3. *Waste Generator Classifications*—
   - Large quantity generators (over 1000 kg/month)
   - Small quantity generators (100 to 1000 kg/month)
   - Conditionally exempt generators (less than 100 kg/month with no more than 1 kg acutely hazardous waste)

4. *RCRA Regulations*—Regulatory requirements of the RCRA require facilities that generate waste to:
   - Identify and label all wastes
   - Notify the EPA of hazardous waste operations

- Maintain secure storage areas
- Keep records and train waste handlers
- Use permitted treatment, storage, and disposal facilities

Comprehensive Environmental Response, Compensation and Liability Act (CERCLA)—CERCLA is designed to remedy the mistakes in past hazardous waste management, whereas the RCRA is concerned with avoiding such mistakes through proper management in the present and future. The RCRA mainly regulates how wastes should be managed to avoid potential threats to human health and the environment.

CERCLA regulation comes into play primarily when mismanagement has occurred or when there has been a release or a substantial threat of a release of a hazardous substance or contaminant that presents an imminent and substantial threat to human health or the environment. CERCLA authorizes a number of government actions to remedy the conditions or the effects of a release. CERCLA, as originally enacted in 1980, authorized a five-year program by the federal government to perform the following primary tasks:

- Identify those sites where release of hazardous substances has already occurred or might occur and poses a serious threat to human health, welfare, or the environment
- Take appropriate actions to remedy those releases
- See that the parties responsible for the releases pay for the cleanup actions

Underground Storage Tanks (USTs)—The RCRA was amended in 1984 through passage of the Hazardous and Solid Waste Amendment. This action enabled the EPA to regulate USTs to better control and prevent leaks. Title I of the RCRA regulated substances including petroleum products and CERCLA-regulated substances. Tanks with hazardous wastes are not regulated under Subtitle I but are regulated under Subtitle C of the RCRA. A UST has been defined as a container that has at least 10% of its contents underground. Facilities with underground tanks should take action to ensure piping does not fail, control corrosion of tanks and piping, and prevent spills and overflows. Refer to 40 CFR Part 280 for a listing of tanks excluded from the regulation. Most regulated tanks are those that hold petroleum-based substances.

Superfund Amendments and Re-Authorization Act of 1986 (SARA)—During a five-year period, it became clear that the problem of abandoned hazardous waste sites was more extensive than originally believed. Solutions would be more complex and time consuming. SARA established new standards and schedules for site cleanup and also created new programs for informing the public of risks from hazardous substances in the community and for preparing communities for hazardous substance emergencies.

1. *SARA*—Under Public Law 99-499, SARA specified new requirements for state and local governments. It also has provisions for the private sector related to hazardous chemicals.

2. *The Emergency Planning and Community Right-to-Know Act (SARA Title III)*— Specifically requires states to establish a State Emergency Response Commis-

sion (SERC). The SERC must designate emergency planning districts within the state. The SERC appoints a Local Emergency Planning Committee (LEPC) for each district. SERC and LEPC responsibilities include implementing various emergency planning provisions of Title III and serve as points of contact for the community right-to-know reporting requirements.

3. ***Title III Provisions***—The provisions that affect the private sector are included in the following sections of the law:
   - **Section 302 (Substances and Facilities Covered and Notification)**—Requires facilities producing, storing, or using certain "extremely hazardous substances" in excess of threshold planning quantities to notify the SERC if these substances are present at the facility. (A list of 406 extremely hazardous substances with corresponding threshold planning quantities was published in the April 22, 1987 Federal Register, which is available in local libraries.) The presence of these chemicals triggers certain local emergency planning requirements under Title III.
   - **Section 303 (Comprehensive Emergency Response Plans)**—Requires covered facilities to provide the LEPC with the name of a facility representative who will participate in the emergency planning process as a facility emergency coordinator. Facilities are also required to provide to the LEPC information necessary for developing and implementing an emergency plan.
   - **Section 305 (Emergency Notification)**—Requires covered facilities to notify appropriate authorities upon the release of certain hazardous chemicals.
   - **Section 311 (Material Safety Data Sheets)**—Requires each facility which must prepare or have available Material Safety Data Sheets (MSDSs) under OSHA regulations to submit either copies of MSDSs or a list of MSDS chemicals to the SERC, LEPC, and local fire department with jurisdiction over the facility.
   - **Section 312 (Emergency and Hazardous Chemical Inventory Forms)**—Requires each facility having MSDSs to submit an inventory form annually to the SERC, LEPC, and local fire department, including information aggregated by categories on estimated amounts and locations of chemicals in the facility. The facility may also be requested to provide this information on a chemical-by-chemical basis.
   - **Section 313 (Toxic Chemical Release Form)**—Requires facilities to complete an annual inventory of toxic chemical emissions for specific chemicals.

Clean Air Act (CAA)—The Clean Air Act was passed to limit the emission of pollutants into the atmosphere; it protects human health and the environment from the effects of airborne pollution. The EPA established National Ambient Air Quality Standards (NAAQS) for several substances. The NAAQS provide the public some protection from toxic air pollutants. Primary responsibility for meeting the requirements of the CAA rests with each state. States must submit plans for achieving NAAQS. Under Section 112 of the CAA, the EPA has the authority to designate hazardous air pollutants and set National Emission Standards for Hazardous Air Pollutants.

1. *Common Air Pollutants—*
   - Ozone is the chemical reaction of pollutants, especially volatile organic compounds.
   - Nitrogen dioxide results primarily from burning gasoline, natural gas, coal, and oil.
   - Carbon monoxide results from burning gasoline, wood, coal, natural gas, and oil.
   - Particulate matter comes from burning wood, diesel, and other fuels. It also comes from industrial plants, burning fields, unpaved roads, and agricultural operations.
   - Sulfur dioxide comes primarily from burning high-sulfur coal and from metal and paper industrial processes.
   - Lead gasoline is being phased out. Other sources include paint, metal refineries, and manufacture of lead batteries.

2. *CAA and RCRA Interaction—*
   - Air emissions from incinerators regulated under the RCRA must comply with applicable ambient air standards and/or emission limitations of the CAA.
   - The EPA is using authority contained in Section 3004(n) of the RCRA to develop more stringent air emission standards.
   - Extraction of pollutants from air emissions under CAA controls such as scrubbers can create hazardous wastes or sludge containing such wastes.
   - Disposal of incinerator materials must also comply with the RCRA.

Clean Water Act (CWA)—The Clean Water Act of 1977 strengthened and renamed the Federal Water Pollution Control Act of 1972. The CWA has several major provisions:

- *National Pollutant Discharge Elimination Systems (NPDES) Permit Program—* This primary element is the CWA permit to discharge into the nation's waterways.
- *Direct Discharge*—Direct discharges into surface water pursuant to an NPDES permit.
- *Indirect Discharge*—Indirect discharge means that the waste is first sent to a publicly owned treatment works and then discharged pursuant to an NPDES permit.
- *Wastewater Treatment*—Sludge resulting from wastewater treatment and pretreatment under the CWA must be handled as RCRA waste and disposed of at an RCRA facility if it is hazardous.
- *RCRA-Permitted Facilities*—Discharges for an RCRA-permitted facility must be pursuant to an NPDES permit. This means that either the facility itself has obtained an NPDES permit or the wastes meet CWA pre-treatment standards and have been transported to a publicly owned treatment works.
- *Water Quality Standards*—States adopt water quality standards for all streams within their borders. The standards address designated use provisions and prohibit water degradation actions.

- *Discharge of Oil and Hazardous Substances*—The act prohibits the discharge of oil and hazardous substances into navigable waters of the United States.
- *Stormwater Discharges*—The act requires that certain industrial and municipal stormwater discharges be regulated under the NPDES permit system.

State Environmental Regulation—Most states have created environmental management agencies to address environmental issues. States operate programs approved by the EPA under memoranda of agreement and for air quality state implementation plans. The EPA still retains authority and relies on eight major environmental statutes to address hazardous waste and environmental problems. The RCRA allows day-to-day management of solid and hazardous wastes. CERCLA provides guidance for cleanup in the event of waste releases. The CAA and CWA statutes limit the amount of materials released into the air and waterways.

Federal Insecticide, Fungicide, and Rodenticide Act (FIFRA)—FIFRA was passed in 1947 and administered by the U.S. Department of Agriculture. The EPA became responsible for the act in 1970; a 1972 amendment included provisions to protect public health and the environment.

- FIFRA controls risks of pesticides through a registration system. No new pesticide can be marketed until it is registered with the EPA.
- The EPA can refuse to register a pesticide or to limit use if evidence indicates a threat to humans and the environment.
- All pesticides used in healthcare facilities should be approved and registered by the EPA.

Toxic Substances Control Act (TSCA)—The TSCA was enacted in 1976 to help control the risk of substances not regulated as drugs, food additives, cosmetics, or pesticides.

- Under this law, the EPA can regulate the manufacture, use, and distribution of chemical substances. The TSCA mandates that the EPA be notified prior to the manufacture of any new chemical substance.
- The EPA ensures that all chemicals are tested to determine risks to humans. The TSCA also allows the EPA to regulate polychlorinated biphenyls (PCBs) under 40 CFR 761.

D. FOOD AND DRUG ADMINISTRATION

The Food and Drug Administration (FDA) was created by the Appropriation Act of 1931 and is an agency of the Department of Health and Human Services. The FDA approves prescription and over-the-counter drugs, including labeling requirements. The Bureau of Drugs develops policy on the safety, effectiveness, and labeling of all drugs developed for human consumption. The FDA monitors biologic and synthetic drugs. Biologics are substances found in the human body; all other substances are classified as synthetics. Two separate monitoring groups have been established; each works independently and maintains separate approval policies. The FDA has established drug development procedures which include animal testing and provisions for human testing if the drug seems to be promising. The Bureau of Radiological Health issues standards for radiation exposure and develops methods to control exposures. The FDA supervises an electronic product radiation program that strives to protect the public by developing standards to control the emission of radiation from electronic products.

FDA Laws—The FDA enforces other laws enacted by Congress or promulgated under congressional authority. These laws include:

- The Food, Drug, and Cosmetic Act (1938), which gave the FDA authority to regulate drug safety
- The Fair Packaging and Labeling Act (15 U.S.C. 1451-1461)
- Portions of the Public Health Service Act relating to biological products (42 U.S.C. 262-263)
- The Radiation Control for Health and Safety Act relates to electronic products such as lasers, x-rays, and microwaves (42 U.S.C. 263)

Medical Device Regulation—The FDA also regulates and classifies medical devices. Regulations are found in 21 CFR 860 and 862 through 890. The FDA requires healthcare facilities to take action to protect safety of patients, residents, staff, and visitors whenever a hazardous product is used in the facility. Each healthcare facility must have procedures in place to obtain, evaluate, and take corrective action in all situations concerning hazardous equipment, drugs, and food substances.

1. *Biological Products*—Biological products such as toxins, antitoxins, vaccines, blood products, and therapeutic serum products are controlled under standards and described in 21 CFR Parts 600 through 611. Drugs are regulated by dividing them into the following categories:
   - New drugs (application for approval is submitted under procedures found in 21 CFR 314).
   - Investigational drugs are regulated under 21 CFR 312.
   - Antibiotics must be certified under Section 507 of the law and are regulated much like all new drugs.
   - Insulin is subject to Section 506 of the law and must be certified by the FDA.

2. *Prescription Drugs*—The FDA requires prescription drugs to be dispensed or prescribed by a licensed practitioner. The label required reads "Caution—Federal Law Prohibits Dispensing without Prescription."
   - All establishments that manufacture or process drugs are required to register with the FDA.
   - Drug listing and registration requirements are in Section 510 of the Federal Food, Drug, and Cosmetic Act and are explained in detail in 21 CFR 207.

Safe Medical Device Act of 1990 (SMDA)—This law expanded FDA authority to regulate medical devices more effectively. The act permits the FDA to learn quickly about any medical product that has caused a serious illness, injury, or death and to take action to track and/or recall the product for further action.

1. *SMDA Reporting*—Healthcare facilities must make regular reports to the FDA on medical device failure.
   - A medical device is any instrument, apparatus, implant, *in vitro* agent, or other similar or related article or component which is used in the prevention, diagnosis, treatment, and care of disease.
   - User facilities such ambulatory care agencies, hospitals, and physicians' offices are required to report patient incidents involving medical devices and drugs to manufacturers and the FDA.
   - Healthcare personnel are expected to report any adverse experiences and events relating to the use of medical devices.
   - Hospital personnel are required to report any medical device incident even if they did not personally observe it.
   - Events involving patient injury or death must be reported if the probability exists that it was caused or contributed to by a medical device or drug interaction.
   - The device must be removed from use immediately.
   - The device must be delivered to a biomedical engineering department for evaluation.
   - Package information must be provided if available.
   - An event report must be completed, listing the patient's name, age, gender, and diagnosis.
   - The date of the event, the name of the person operating the device, and the type of injury incurred must be documented.
   - The specific location of the incident, the department affected, and any specific information required by the local facility must be reported.
   - The report must be forwarded to the appropriate action office within 48 hours.

2. *SMDA Definitions*—
   - **Device Failure**—When a device fails to perform as required or is not functioning within proper specifications of intended use.
   - **Device-Related Incident**—Results from the device being operated in an incorrect manner. Such an incident could be caused by a power failure or other extenuating circumstances.
   - **Patient Adverse Reaction**—If the event was caused by the patient's adverse reaction to proper use, the incident is not reported under the SMDA.

3. *The Medical Devices Amendments of 1992*—This legislation amended the SMDA to require organizations to report *user errors*.

E. OTHER GOVERNMENT AGENCIES

National Institute of Occupational Safety and Health (NIOSH)—NIOSH, an agency of the Department of Health and Human Services, conducts research on workplace hazards and safety concerns. The agency publishes studies and recommends new or improved standards to be adopted by OSHA. NIOSH also investigates specific workplace hazards. Currently, NIOSH is conducting a comprehensive study of lifting hazards found in long-term nursing facilities.

- The agency has the same entry rights as OSHA and conducts workplace safety and health evaluations.
- NIOSH can recommend hazard controls and abatement procedures but does not possess enforcement or citation authority.
- NIOSH conducts industry-wide research, evaluates hazard control measures, develops standards, and publishes occupational safety studies.
- The agency offers occupational safety, health, and hygiene classes throughout the country by working closely with colleges and universities.
- NIOSH is a leader in assessing and documenting the effectiveness of new hazard control technologies. It often assists other agencies in investigating fatal workplace accidents.
- NIOSH tests and approves respiratory devices including gas masks, respirators, and self-contained breathing apparatus. NIOSH also publishes policy documents on numerous hazardous substances.
- The 1988 publication "Protecting the Healthcare Worker" has received wide acclaim as one of the first comprehensive publications on hospital safety.

Nuclear Regulatory Commission (NRC)—The NRC adopts and enforces standards for the department of nuclear medicine in healthcare facilities. Some states have agreements with the government to assume these regulatory responsibilities.

1. *NRC Licenses*—The NRC issues five-year licenses to qualified healthcare organizations that follow prescribed safety precautions and standards.
   - Information that must be provided to the NRC includes identifying authorized users, designating a radiation safety officer, and the address/location of radioactive materials.
   - Any changes require the organization to file for a licensure amendment.

2. *Written Radiation Control Programs*—Each licensee must have a written radiation control program that covers the following:
   - Required participation in the program by all users, organizational administrators, and the radiation safety officer
   - Procedures to inform all workers about the types and amounts of materials used including dosing information, safety precautions, recurring training, and continuing education
   - Information and guidelines to be used to keep doses as low as reasonably achievable (ALARA levels)
   - All radiation safety officers must meet the training and education requirements of Section 35.900 of the NRC Rules

Department of Transportation (DOT)—The DOT regulates the transportation of hazardous materials under the Hazardous Materials Transportation Uniform Safety Act of 1990. The act requires those engaged in the transportation of hazardous chemicals to register with the secretary of transportation. Safety permits are required by carriers transporting class A or B explosives, liquefied natural gas, or any hazardous material that is extremely toxic by inhalation. The act requires training for all hazardous materials handlers, including those who unload, handle, store, or transport such materials. Recent changes require other workers involved in the hazardous materials transportation industry to receive training.

Centers for Disease Control and Prevention (CDC)—The CDC is a federal public health agency with primary facilities in Atlanta. It works to protect the health of the American people by tracking, monitoring, preventing, and researching disease. It is also responsible for surveillance and investigation of infectious disease in healthcare facilities. The CDC conducts research and publishes results in its *Morbidity and Mortality Weekly Report*. This weekly publication provides healthcare facilities with timely information on topics such as infection control, isolation procedures, bloodborne pathogens, tuberculosis management, infectious waste disposal recommendations, and how to protect workers.

- *Guidelines*—The CDC also publishes guidelines and recommendations in book and pamphlet form. Many of the guidelines are also published in the Federal Register. The bloodborne pathogens control procedures developed by the CDC were promulgated in the OSHA bloodborne pathogens standard (29 CFR 1910.1030). The CDC recently published updated guidelines on the control and management of tuberculosis in healthcare settings.
- *Isolation Procedures*—The CDC is well known for its recommendations for hospital isolation procedures and infection control. OSHA and other agencies such as the Joint Commission and the EPA rely on the research, statistics, and recommendations provided by the CDC.

Occupational Safety and Health Review Commission (OSHRC)—OSHRC is a quasi-official, three-member board appointed by the president and confirmed by the Senate. It is an independent agency of the executive branch. OSHRC adjudicates cases that have been brought by OSHA and contested by the employer or employees. The commission may conduct investigations and can uphold, change, or dismiss OSHA's findings. Some cases are determined by an administrative law judge, but the three-member board has final rule. Rulings by OSHRC can be reviewed further by the courts.

Health Resources Services Administration (HRSA)—This agency, formerly known as the Health Resources Administration, publishes Minimum Requirements for Construction and Equipment for Hospital and Medical Facilities, which provides information about hospital construction. Healthcare facilities that receive federal funds must comply with HRSA regulations.

Health Care Financing Administration (HCFA)—This agency, under the Department of Health and Human Services, sets rules and guidelines for healthcare facilities that receive federal funding through Medicare and Medicaid programs. The agency also publishes guidelines governing long-term nursing facilities. The guidelines emphasize residents' rights and quality of care. The Omnibus Budget Reconciliation Act (OBRA) of 1987 gave the HCFA the power to regulate facilities receiving federal funds.

Federal Emergency Management Agency (FEMA)—FEMA is a federal agency responsible for coordinating assistance to areas hit by catastrophic events or natural disasters. FEMA works with local governments, industries, and response agencies to coordinate emergency planning activities within a geographic area or region.

State, County, and Municipal Health Agencies—State health departments adopt and enforce regulations and licensing requirements in areas such as radiation, nuclear medicine, infectious disease control, hazardous waste disposal, and food handling. In some states, the health department and the Joint Commission accredit hospitals through a joint effort. The state departments of health and the Joint Commission focus primarily on patient safety rather than worker safety. County and city health and fire departments may also have jurisdiction in food handling, fire safety, and other hospital functions.

F. AMERICANS WITH DISABILITIES ACT

Americans with Disabilities Act of 1990 (ADA)—This legislation was passed to protect the civil rights of persons with disabilities. The act consists of five titles:

- *Title I (Employment)*—This title, detailed in 29 CFR Part 1630, protects qualified workers from discrimination in employment.
- *Title II (Public Services and Transportation)*—This title prohibits state and local governments and transportation systems from discriminating against persons with disabilities.
- *Title III (Public Accommodations and Commercial Facilities)*—This title prohibits privately owned establishments from discriminating against persons with disabilities.
- *Title IV (Telecommunications)*—This title requires telephone companies to provide special equipment and services to the hearing and speech impaired.
- *Title V (Miscellaneous)*—This title contains legal and implementation information.

ADA Title I (Employment)—Title I, Employment, is detailed in 42 USC and 29 CFR Part 1630. This act protects qualified persons with disabilities from employment discrimination. Other laws protect persons from discrimination on the basis of race, color, sex, age, and national origin. Employers must understand two very important terms: disability and qualified. Each decision is made on a case-by-case basis.

1. *Physical Impairment*—The ADA defines physical impairment as any physiological disorder, condition, cosmetic disfigurement, or anatomical loss affecting one or more of the following body systems:
   - Neurological
   - Musculoskeletal

- Special sense organs
- Respiratory (including speech organs)
- Cardiovascular
- Reproductive
- Digestive
- Genitourinary
- Hemic
- Lymphatic
- Skin
- Endocrine

2. *ADA Definitions—*
   - **Mental Impairment**—The ADA defines mental impairment as any mental or psychological disorder such as retardation, organic brain syndrome, emotional or mental illness, and specific learning disabilities. The definition of impairment does not include simple physical characteristics such as eye or hair color, left handedness, or weight within a normal range; pregnancy or a predisposition to a certain disease; personality traits such as a quick temper; a lack of education; or the existence of a prison record.
   - **Qualified Individual with an Impairment**—A person must not only be disabled but must also be qualified. An employer is not required to hire or retain an unqualified person.
     a. A qualified individual with a disability is a person who satisfies the requisite skill, experience, education, and other job-related requirements of the employment position and who can perform the essential functions of the position with or without accommodation.
     b. The ADA requires an employer to focus on the essential functions of a job to determine whether a person with a disability is qualified. An employer may establish physical or mental qualifications that are necessary to perform specific jobs or to protect health and safety. If a physical or mental qualification standard screens out a person with a disability, the employer must be prepared to show that the standard is job related and consistent with business necessity.
   - **Essential Job Functions**—Under guidelines of the ADA, employers must determine if an individual with a disability can perform the essential functions of the job with or without reasonable accommodation.
     a. The law protects people with disabilities who can perform essential job functions even if they cannot do things that are only marginal to the job.
     b. An applicant with a disability is considered qualified if he or she can perform the essential job functions.
     c. Reasonable accommodation must be provided in the job application process to enable a qualified applicant to have an equal opportunity to be considered for a job.

3. *Reasonable Accommodation*—Making facilities readily accessible to and usable by an individual with a disability includes restructuring a job or redistributing marginal job functions.

- Accommodation is altering when or how an essential job function is performed to include developing part-time or modified work schedules.
- Accommodation is also obtaining or modifying equipment, providing reserved parking for a person with a mobility impairment, and allowing an employee to provide equipment or devices that the employer is not required to provide.
- Accommodation must consider the specific abilities and functional limitations of a particular applicant or employee with a disability.
- Accommodation must also consider the specific functional requirements of a particular job.

4. **Essential Job Functions**—Sometimes it is necessary to identify the essential functions of a job in order to know whether an individual with a disability is qualified to do the job.
   - The regulations provide guidance on identifying the essential functions of a job. The first consideration is whether employees in the position are actually required to perform the function.
   - The ADA does not require an employer to develop or maintain job descriptions.

ADA Standards for Safety and Health (Direct Threat)—An employer may require as a qualification standard that an applicant not pose a "direct threat" to the health or safety of the individual or others, if this standard is applied to all applicants for a particular job. The employer must meet very specific and stringent requirements under the ADA to establish that such a direct threat exists. An employer must be prepared to show that:

TABLE 4.2   The Americans with Disabilities Act Records

Types of Records	*Retention/Filing Requirements*
Any personnel or employment record made or kept by the employer, including requests for reasonable accommodation, application forms submitted by applicants, and records concerning hiring, promotion, demotion, transfer, layoff or termination, rates of pay or other terms of compensation, and selection for training or apprenticeship	One year from date record made or personnel action taken, whichever is later
Personnel records relevant to charge of discrimination under Title VII of the ADA or action brought by the attorney general against the employer, including records relating to charging party and to all other employees holding similar positions and application forms or test papers completed by unsuccessful applicants and by all other candidates for the same position	Until final disposition of charge or action
Copy of EEO-1, Employer Information Report	Retention period not specified; must file EEO-1 Report with EEOC on or before September 30 of each year

- There is significant risk of substantial harm
- A specific risk can be identified
- The risk is not speculative or remote
- The assessment of risk is based on objective evidence
- The risk cannot be eliminated or reduced by reasonable accommodation

ADA and Laws/Regulations—The ADA does not override health and safety requirements established under other federal laws. If a standard is required by another federal law, an employer must comply with it and does not have to show that the standard is job related and consistent with business necessity. An employer has the obligation under the ADA to consider whether there is a reasonable accommodation, consistent with the standards of other federal laws, that will prevent exclusion of qualified individuals with disabilities who can perform jobs without violating the standards of those laws.

1. *State/Local Laws*—The ADA does not override state or local laws designed to protect public health and safety, except where such laws conflict with ADA requirements. If there is a state or local law that would exclude an individual with a disability from a particular job or profession because of a health or safety risk, the employer still must assess whether a particular individual would pose a direct threat to health or safety under the ADA standard.

2. *Inquiries About Disabilities*—The ADA prohibits pre-employment inquiries about a disability. This prohibition is necessary to assure that qualified candidates are not screened out because of their disability before their actual ability to do a job is evaluated.
   - Such protection is particularly important for people with hidden disabilities, who frequently are excluded, with no real opportunity to present their qualifications, because of information requested in application forms, medical history forms, job interviews, and pre-employment medical examinations.
   - The prohibition on pre-employment inquiries about disability does not prevent an employer from obtaining necessary information regarding an applicant's qualifications, including medical information necessary to assess qualifications and assure health and safety on the job.

3. *ADA and Medical Examinations*—After making a conditional job offer and before an individual starts work, an employer may conduct a medical examination or ask health-related questions, providing that all candidates who receive a conditional job offer in the same job category are required to take the same examination and/or respond to the same inquiries.
   - An employer may not make any pre-employment inquiry about a disability, including the nature or severity of a disability, on application forms, during job interviews, or in background or reference checks. An employer may ask questions to determine if an applicant can perform specific job functions.
   - The ADA prohibits medical inquiries or medical examinations before making a conditional job offer to an applicant.
   - This prohibition is necessary because the results of such inquiries and examinations frequently are used to exclude people with disabilities from jobs they are able to perform.

4. ***ADA and Substance Abuse***—After a conditional offer of employment, an employer may ask any questions concerning past or present drug or alcohol use. However, the employer may not use such information to exclude an individual with a disability, on the basis of a disability, unless the employer can show that the reason for exclusion is job related and consistent with business necessity and that legitimate job criteria cannot be met with reasonable accommodation.
   - An employer may conduct tests to detect illegal use of drugs. The ADA does not prohibit, require, or encourage drug tests. Drug tests are not considered medical examinations, and an applicant can be required to take a drug test before a conditional offer of employment has been made. An employee can be required to take a drug test whether or not such a test is job related and necessary for the business.
   - The ADA specifically permits employers to ensure that the workplace is free from the use of illegal drugs and alcohol and to comply with other federal laws and regulations regarding alcohol and drug use. At the same time, the ADA provides limited protection from discrimination for recovering drug addicts and alcoholics.

5. ***ADA and Injured Workers***—Whether an injured worker is protected by the ADA will depend on whether or not the person meets the ADA definitions of an "individual with a disability" and a "qualified individual with a disability." The person must have an impairment that "substantially limits a major life activity" and have a "record of" or be "regarded as" having such an impairment. The individual must also be able to perform the essential functions of a job currently held or desired with or without accommodation.
   - The ADA allows an employer to take reasonable steps to avoid increased workers' compensation liability while protecting persons with disabilities against exclusion from jobs they can safely perform. Filing a workers' compensation claim does not prevent an injured worker from filing a charge under the ADA.
   - Exclusivity clauses in state workers' compensation laws bar all other civil remedies related to an injury that has been compensated by a workers' compensation system. However, these clauses do not prohibit a qualified individual with a disability from filing a discrimination charge with the Equal Employment Opportunity Commission (EEOC) or filing a suit under the ADA, if issued a "right to sue" letter by the EEOC.

ADA Title I Enforcement—Job applicants or employees who believe they have been discriminated against on the basis of disability in employment by a private, state, or local government employer, labor union, employment agency, or joint labor management committee can file a charge with the EEOC. Individuals also may file a charge if they believe they have been discriminated against because of an association with a person with a known disability or suffered retaliation because of filing a charge or assisting in opposing a discriminatory practice. Remedies for violations of Title I of the ADA include hiring, reinstatement, promotion, back pay, front pay, restored benefits, reasonable accommodation, attorneys' fees, expert witness fees, and court costs. Compensatory and punitive damages also may be available in cases of intentional discrimination or where an employer fails to make a good faith effort to provide a reasonable accommodation. Employers may not retaliate against any applicant or employee who files a charge, participates in an EEOC investigation, or opposes an unlawful employment practice.

ADA Title III (Accessibility)—Accessibility laws were in effect before the ADA. Since the ADA became law, attention has turned to the issue of accessibility. Property owners and managers should focus on the impact the ADA has on life safety. The ADA contains life safety requirements which sometimes differ from provisions of current building codes. Title III covers accessibility in public and commercial facilities. Public facilities include hotels, restaurants, retail stores, private schools, and healthcare facilities. Commercial facilities are those intended for non-residential use by a private enterprise and whose operations affect commerce.

1. *Entrance Ways and Parking Lots—*
   - Ensure handicap parking spaces are 96 inches wide.
   - Parking spaces should be next to entry pathway.
   - Mark all spaces with highly visible stand-up marker or sign.
   - Provide at least one accessible route to the building.
   - Buildings should have at least one accessible entrance.
   - Ensure entrance width is a minimum of 32 inches.
   - Ramps must be at least 36 inches wide.
   - Side rails on entrances ways and ramps must be 36 inches apart.

2. *Common Areas—*
   - Lobbies, reception areas, and personnel offices should be accessible to all applicants and visitors.
   - Ramps, aisles, and halls should be slip resistant and kept free of obstructions.
   - Doors must be wide enough to accommodate wheelchairs.
   - Eating and lunch areas should have clear and clean pathways.
   - Serving lines should be accessible to those with mobility disabilities.
   - Restrooms should be convenient and accommodate those with disabilities.
   - Electrical switches, telephones, and fountains must be accessible.
   - Elevators should be equipped with braille markings and audible signals.

3. *Office and Work Areas—*
   - Chairs with arms should be provided to assist those with disabilities to lift themselves.
   - Desktops should be high enough from the floor to accommodate those in wheelchairs.
   - Office equipment should be placed and designed for workers with mobility disabilities.
   - A wheeled cart should be available to move books, reports, and other materials.
   - Floors should be free from extension cords and other hazards.
   - Space, exercise areas, and water should be provided for working dogs.

4. *Accessibility and Life Safety*—The ADA's accessibility guidelines affect life safety requirements.
   * Accessible means methods of egress are required to be provided in a number equal to that of the number of exits required by the local building/life safety regulations. The ADA's accessibility guidelines do not recognize stairs as an accessible means of egress.
   * Buildings that have inaccessible exits on levels above or below a level with accessible exits have to be equipped with automatic sprinklers, horizontal exits, or areas of rescue assistance.
   * Areas of rescue assistance are required to be provided with signs displaying the International Symbol of Accessibility and stating "AREA OF RESCUE ASSISTANCE." Signs are required to be illuminated when exit sign illumination is required. Signs are required to be provided at inaccessible exits to provide clear direction to the areas of rescue assistance.
   * Other areas addressed by the ADA accessibility guidelines that directly affect life safety are door hardware mounting height, stairway design, and ramp design.
   * Accessibility features should not be limited only to those that are part of everyday operations, but must be extended to those features that may only come into play during emergency situations. While efforts should be made to get people with disabilities into a building, the same efforts should be made to ensure that all can safely get out during emergency situations.

G. JOINT COMMISSION STANDARDS

Joint Commission on Accreditation of Healthcare Organizations (JCAHO)— This organization, commonly called the Joint Commission, attempts to standardize practices in healthcare facilities. The Joint Commission incorporates standards of other agencies such as OSHA, the EPA, and the National Fire Protection Association into its accreditation process. Any facility, including nursing homes, can apply for accreditation, which is valid for three years.

1. *JCAHO Certification*—The Joint Commission certifies that healthcare organizations and facilities meet a minimum standard for patient and/or resident care. The Joint Commission's *1995 Accreditation Manual for Hospitals* reflects three major changes:
   * Reformulation of Joint Commission standards, which emphasize actual organizational performance factors
   * Redesign of the entire survey process
   * Development of the Indicator Measurement System, which is a performance measurement system

2. *Standards Revision*—The revised standards emphasize that healthcare organizations exist to improve the health of patients and to utilize all resources efficiently.
   * The standards now focus on hospital performance aimed at continuously improving outcomes of patient care.

- The 1995 standards introduced six new chapters:
  a. Patient Rights and Organizational Ethics
  b. Care of Patients
  c. Continuum of Care
  d. Management of the Environment of Care (includes safety)
  e. Management of Human Resources
  f. Surveillance, Prevention, and Control of Infection

3. *Scoring*—Scoring emphasizes performance, and reduced scores will be earned when performance evidence is lacking. The Joint Commission publishes a number of manuals and books, including the *Accreditation Manual for Healthcare Organizations.* This manual has information on quality care as well as safety practices and standards.

4. *Accreditation*—
   - Joint Commission accreditation is voluntary, and the decision to be surveyed is made by the individual organization. The achievement of accreditation means that the organization has shown substantial compliance with standards and is making an effort to provide even better services.
   - The accreditation process benefits a healthcare organization by:
     a. Granting access to Joint Commission information resources
     b. Identifying areas in which performance needs improvement
     c. Receiving on-site education and consultation services
     d. Documenting the responsibility for delivering quality care
     e. Giving public recognition of the organization's commitment to quality
   - Major factors that affect accreditation decisions include evidence of overall compliance with standards, progress toward more complete compliance, and the absence of any serious deficiencies in such areas as assuring patient safety or maintaining acceptable quality of care.
   - Accreditation is not automatically renewed. Several months before the organization's accreditation is due to expire, the Joint Commission sends an Application for Survey. After receipt of an organization's application, a survey is scheduled.
   - Joint Commission surveyors are thoroughly trained to conduct the different types of surveys offered by the accreditation program to which they are assigned.

JCAHO Survey Process—The formal survey begins when an organization applies to the Joint Commission. The length of the survey and the composition of the survey team are determined on the basis of information provided in the organization's Application for Survey.

1. *General Information*—Survey fees are composed of a base charge and an added charge that depends on the volume of both inpatient and outpatient encounters. Organizations may pay the full survey fee when billed or pay the fee in two installments. An organization is notified of its survey date approximately four to six weeks before the team arrives.

2. *Initial Conference*—On arrival, the surveyors hold an initial conference to discuss the survey schedule. Accreditation surveys are designed to enable the surveyors to assess the extent of an organization's compliance with Joint Commission standards. This is accomplished through one or more of the following means:
   - Documentation of compliance provided by the surveyed facility
   - On-site observations by Joint Commission surveyors
   - Verbal information concerning the implementation of standards, or examples of their implementation, that will enable a judgment of compliance to be made

3. *Document Review*—Following an opening conference, the team privately reviews key indicator documents, including policies, procedures, rules, and regulations.
   - The interactive part of the process allows the team to interview managers, care providers, medical staff members, and patients.
   - On the last day of the survey, the team conducts a multi-disciplinary patient care conference that focuses on patient care areas, with an emphasis on improving care and enhancing compliance. Hospital and medical staff will be invited to attend.
   - Next, the team conducts an exit conference with organizational leaders.

4. *Organizational Performance*—The 1995 survey process emphasized organizational performance as opposed to evaluation of specific departments or services. The focus in every area is on quality patient care. The key changes are as follows:
   - The Joint Commission has established an Organization Liaison Unit to provide facilities with assistance during the survey process.
   - Surveyors will strive to gain a better understanding of the organization through more thorough review and on-site document review before the survey begins.
   - The survey will use an interactive team-based approach.
   - New survey protocols will help standardize the process and give more consistent results.
   - The survey process will be tailored to the facility's organizational characteristics.

5. *JCAHO Accreditation Decisions*—
   - **Accreditation with Commendation**—This is the highest accreditation, awarded to an organization that has demonstrated exemplary performance.
   - **Accreditation**—This decision indicates that an organization is in overall compliance with applicable standards.
   - **Conditional Accreditation**—This decision indicates that multiple substantial standards compliance deficiencies exist in an organization. Correction of deficiencies, which serves as the basis for further consideration of continuing accreditation, must be demonstrated through follow-up survey.
   - **Provisional Accreditation**—An accreditation decision that results when an organization has demonstrated substantial compliance in the first of two sur-

veys conducted. A second survey is conducted approximately six months after the first to allow the organization sufficient time to demonstrate a track record of performance.

- **Not Accredited**—An accreditation decision that results when an organization has been denied accreditation or withdraws from the process. This designation also describes organizations that have never applied for accreditation.

6. *Survey Validation*—Findings are analyzed and aggregated at the Joint Commission. Results are reviewed by professional staff members, and decision rules are then applied to reach the final accreditation decision. An accreditation decision is made on the basis of the survey findings and any other relevant information. Findings that raise specific issues are reviewed by the Joint Commission's Accreditation Committee, which makes the final decision.

7. *Recording Survey Results*—Once all pertinent information concerning an organization has been considered, surveyors record the results of their findings on survey report forms.
   - Scoring guidelines are used by surveyors to promote consistency in assessing and ranking an organization's degree of compliance.
   - Implementation monitoring enables the Joint Commission to provide additional assistance to healthcare organizations in interpreting and meeting standards requirements.
   - Once the survey report form has been completed, the surveyors return their findings to the Joint Commission.
   - Following detailed analysis of the survey findings, a determination is made concerning the accreditation status of the organization.

Indicator Measurement System—The Joint Commission is implementing this program for objective measurement of key patient outcomes; when fully implemented, the system will:

- Continually collect objective data in each organization
- Allow performance data to be analyzed
- Provide comparative performance information for internal use by accredited organizations
- Identify patterns and/or trends in performance factors that need immediate attention
- Provide an indicator, which is a valid and reliable quantitative outcome or process measure that directly relates to performance, to measure a specific process or the cumulative efforts of patient outcomes
- Eventually contribute to the Joint Commission's ability to base accreditation decisions on actual performance instead of ability to perform
- Also allow the Joint Commission to identify standard compliance deviations or problems

JCAHO Safety Standards—The safety management program is covered under the Management of the Environment of Care standard.

1. **Basic Criteria**—The organization must use the following criteria when meeting Environment of Care standards:
   - The organization designs a safe, accessible, effective, and efficient environment of care in accordance with its mission and services, as well as applicable laws and regulations.
   - The organization must adhere to Life Safety Code 1992 for newly constructed and existing environments of care.
   - The organization must follow *Guidelines for Construction and Equipment of Hospitals and Medical Facilities,* 1993 edition, as published by the American Institute of Architects, for newly constructed, renovated, or altered environments of care.

2. **Safety Program Components**—
   - Safety management
   - Security management
   - Hazardous materials and wastes
   - Emergency preparedness
   - Life safety
   - Medical equipment management
   - Utility systems management

3. **Information Collection and Evaluation System (ICES)**—The facility develops and operates an ICES for identifying and evaluating conditions in the environment of care.

4. **Safety Officer Appointment**—The organization appoints a safety officer to direct ICES efforts. The safety officer works with appropriate staff to implement the safety management program or safety committee recommendations.

5. **Safety Management Program**—A safety program must be developed to ensure a physical environment free of hazards and to manage staff activities to reduce the risk of human injury.
   - A life safety management program must be designed to protect patients, personnel, visitors, and property from fire and the products of combustion.
   - The program must also provide for the safe use of buildings and grounds.
   - A utilities management program must be designed to assure operational reliability, assess special risks, and respond to failures of utility systems that support the patient care environment.
   - A standard is developed when there is a need to assess or enhance the quality of a particular aspect of healthcare service.

6. **Standards**—The department of standards and the accreditation staff work together to identify needs and to test substantive changes in standards and the survey/accreditation processes.
   - If extensive changes are anticipated, a task force of experts in the area under consideration studies the issues and establishes principles to guide the development of new or revised standards.
   - Findings of the task force are reviewed by the appropriate Joint Commission

Professional and Technical Advisory Committees and the Standards and Survey Procedures Committee of the Board of Commissioners.
- If these committees determine that standards based on the guiding principles would improve the quality of care, the task force develops proposed standards.

7. **Standards Development**—In the process of standards development, consideration is given to whether:
   - Proposed standards effectively address the identified need
   - Proposed standards are likely to improve the quality of care
   - Implementation of the proposed standards is likely to affect the cost of providing care
   - The anticipated increase in cost is justified by any anticipated improvement in the quality of care
   - There are barriers to the implementation of the proposed standards
   - The Joint Commission has appropriate resources and services available to help organizations interpret and implement the proposed standards

H. VOLUNTARY COMPLIANCE AGENCIES

National Fire Protection Association (NFPA)—The NFPA is an independent nonprofit organization that was founded in 1896. The association publishes more than 280 codes and standards covering fire and electrical-related topics. NFPA codes and standards are a guide in the cor .:ruction and maintenance of buildings throughout the country. They are so widely recognized that most governmental bodies use the codes as the foundation for their fire prevention and construction codes.

1. **NFPA 99**—Standards for Healthcare Facilities, along with the life safety criteria for safeguarding patients/residents and employees from fire, electrical, and explosion hazards.

2. **NFPA 101**—Life Safety Code establishes codes for life safety, including a number of elements such as warning detection systems, fire alarms, fire partitions, exit availability, sprinkler systems, and storage procedures for occupied buildings. The code is comprehensive and has guidance on emergency preparedness, fire plans and drills, fire extinguishers, and waste handling systems.

3. **Other NFPA Codes and Standards**—The NFPA publishes almost 300 codes and standards, developed by committees of professionals, that affect fire and electrical safety. Most NFPA standards have been adopted by local, state, and federal agencies. This gives them the force of law in many instances. NFPA standards can be group into the following categories:
   - Sprinkler Standards
   - Flammable and Combustible Liquids
   - National Fuel Gas Code
   - National Electrical Code
   - Life Safety Code

- LP-Gases Standard
- Hazardous Material Response Standards

4. ***Healthcare-Related Standards/Codes***—Some NFPA publications relevant to healthcare facilities include:
   - NFPA 10   Portable Fire Extinguishers
   - NFPA 13   Installation of Sprinkler Systems
   - NFPA 30   Flammables and Combustibles
   - NFPA 45   Laboratories Using Chemicals
   - NFPA 50   Bulk Oxygen Systems
   - NFPA 70   National Electric Code
   - NFPA 74   Installation, Maintenance, and Use of Household Fire Warning Equipment
   - NFPA 90A   Installation of Air Conditioning and Ventilating Systems
   - NFPA 99   Healthcare Facilities
   - NFPA 101   Life Safety Code
   - NFPA 220   Standard Types of Building Construction
   - NFPA 704   Fire Hazards of Materials

   Note—NFPA 99 and 101 are covered in Chapter 5 of this book.

5. ***NFPA Periodicals***—The NFPA publishes the *NFPA Fire Journal, Life Safety Code Handbook,* and *Fire Technology.*
   - NFPA standards are published as the National Fire Codes, with 11 volumes and more than 8000 pages. The NFPA also publishes the *Fire Protection Handbook,* which is an authoritative manual on fires and their control.

American National Standards Institute (ANSI)—The organization was founded in 1918 to consolidate voluntary standards. ANSI is a federation of more than 1500 professional, trade, governmental, industrial, labor, and consumer organizations. It publishes national consensus standards that have been developed by various technical, professional, trade, and consumer organizations. ANSI also serves as the coordinating agency for safety standards that have been adopted for international implementation.

- ANSI represents the United States as a member of the International Organization for Standardization and the International Electro-technical Commission.
- Many ANSI standards have been adopted by OSHA, and others are widely followed throughout complete industries. Any of the safety-related ANSI standards are developed by ANSI-accredited standards committees.
- ANSI provides members access to more than 9000 standards from around the world and publishes specifications for the following:
  1. Protective eyewear including safety glasses and goggles
  2. Hard hats, safety shoes, and fall-protection equipment
  3. Eyewash stations and emergency shower equipment

Underwriters Laboratories (UL)—This non-profit organization, founded in 1894, maintains laboratories for the examination and testing of systems, devices, and material to ensure compliance with safety and health standards. UL inspects or tests more than

70,000 products each year, including fire-fighting equipment, lockout/tagout supplies, lighting fixtures, and flammable liquid storage containers. UL certification only pertains to the area of safety and does not involve performance testing. UL has issued more than 500 standards, with many adopted by ANSI. UL publishes directories of companies whose products meet or exceed criteria outlined in appropriate standards. Some of these directories include:

- Building Materials
- Fire Protection Equipment
- Hazardous Location Equipment
- Electrical Appliance Equipment

American Society for Testing and Materials (ASTM)—This is the world's largest source of voluntary consensus standards, with research done by its more than 30,000 members. ASTM publishes more than 8000 standards yearly in a 68-volume set, with categories such as medical devices, occupational safety and health, environmental effects, energy, and security systems. The six major categories of standards are

- *Classification*—Information on materials grouped together by characteristics
- *Practices*—Procedures detailing how to accomplish a process or function
- *Test Methods*—Procedures developed to test certain products or materials
- *Guides*—A set of directional procedures
- *Specifications*—Precise information about material specifications
- *Terminology*—A compilation of definitions and terms

Factory Mutual Research Corporation (FM)—A nationally recognized testing laboratory and approval service organization recognized by OSHA, FM was established to focus on industrial loss control. It employs engineers and scientists to investigate fire losses and determine ways to prevent losses. It also does third-party testing on fire extinguishing equipment, sprinklers, building materials, and smoke detectors. FM lists approved equipment, materials, and other services in its annual 500-page guide.

Manufacturers can display a special symbol on approved items to inform users and buyers that the product or piece of equipment has been tested and approved by an independent laboratory. The FM lab is recognized by OSHA.

American Society of Healthcare Engineering (ASHE)—This American Healthcare Association affiliated organization has almost 6000 members worldwide. The ASHE promotes healthcare safety, emergency preparedness, engineering, and security issues and takes a leading role in providing members information about regulatory codes and standards.

1. *Society Information*—The society provides advice on a number of operational concerns including:
   - Facilities management
   - Plant design and engineering
   - Building maintenance and support services
   - Environmental and waste management

- Safety and security
- Clinical engineering

2. **Seminars and Publications**—The ASHE promotes healthcare education through professional development seminars and conferences. Monthly publication of technical documents keeps members informed on the latest changes and developments related to healthcare engineering and facility management. The society's more than 100 publications and innovative software programs help members meet new challenges.

Compressed Gas Association (CGA)—The CGA develops technical and safety standards for the compressed gas industry. Members work together through a committee system to develop technical specifications, safety standards, and educational materials and to promote compliance with regulations and standards in the workplace. Member companies represent manufacturers, distributors, suppliers, and transporters of gases, cryogenic liquids, and related products. The CGA publishes more than 100 technical standards, many of which are formally recognized by U.S. government agencies. The association publishes the *Handbook of Compressed Gases,* which is widely used and sets forth the recognized safe methods for handling, storing, and transporting industrial gases.

National Restaurant Association—This organization publishes policies to reduce accidents and hazards that affect the safety of food service employees and patrons. A major association activity is the preparation and distribution of educational materials including self-inspection guidelines on general safety concerns, OSHA requirements, and fire protection. Materials are also available for purchase through the association's Information Services Department. Safety-related information appears in the association's *Washington Weekly* report and monthly magazine, *Restaurants USA.*

National Sanitation Foundation (NSF)—The NSF provides clients and the public with objective, quality, and timely services, including development of consensus standards, voluntary product testing, and certification. Topics include:

- Food service equipment
- Drinking water treatment systems
- Wastewater treatment devices
- Biohazard cabinets
- Special categories of equipment and products
- Registries
- Bottled water
- Packaged ice
- Drinking water laboratory accreditation
- Sanitizers/disinfectants
- Standards and criteria

American Conference of Governmental Industrial Hygienists (ACGIH)—This conference actively promotes the administrative and technical aspects of worker safety.

The organization provides leadership and educational and professional development opportunities to personnel representing governmental and educational organizations. The ACGIH is well known for its publication *Threshold Limit Values/Biological Exposure Indices*. The conference also publishes a monthly journal, *Applied Occupational and Environmental Hygiene*. This publication provides peer-reviewed papers/articles and technical articles for the practicing professional. The ACGIH also has a publications catalog that lists hundreds of titles in the areas of occupational safety and industrial hygiene.

I. PROFESSIONAL ASSOCIATIONS

Board of Hazard Control Management (BHCM)—The BHCM was founded in 1976 to evaluate and certify the capabilities of practitioners engaged in the administration of safety and health programs. Certification is at the Master and Senior levels. Individuals achieving the Master Level attain the status of possessing the skill and knowledge to effectively manage comprehensive safety and health programs. The BHCM offers advice and assistance to those wishing to improve their status in the profession. It also strives to increase competence and stimulate professional development by providing recognition and status for those who, by education, experience, and achievement, are deemed qualified. The BHCM promotes the exchange of ideas and technology that will improve performance of practicing hazard control professionals.

Board of Certified Healthcare Safety—The board has a program to certify healthcare safety professionals through an evaluation and examination process. Healthcare organizations are constantly being challenged to deliver services safely, efficiently, and economically. This need is reflected in the more stringent requirements being reflected by the Joint Commission. There is a need for professionals who can understand and assist in controlling the many serious hazards found in healthcare facilities and activities. Healthcare administrators are seeking capable individuals who possess hazard control skills. These individuals are more readily identified by their designation as a Certified Healthcare Safety Professional (CHSP). The board's CHSP program has several objectives:

- Evaluate the qualifications of persons engaged in hazard control activities in healthcare facilities
- Certify as proficient individuals who meet the level of competency for this recognition
- Increase the competence of and stimulate professional development of practitioners
- Provide recognition and status to those individuals who by education, experience, and achievement are considered qualified
- Facilitate the exchange of ideas and technology that will improve performance

The CHSP at the Master Level should be knowledgeable about:

- JCAHO requirements
- Life Safety Code

- Disaster planning
- Biohazard controls
- Chemical safety
- Controlling physical hazards
- Maintenance, engineering, and ventilation
- Personal protective equipment
- Waste management
- Housekeeping safety
- Governmental legislation and regulation

American Society of Safety Engineers (ASSE)—The society was founded in October 1911; its original name was the United Society of Casualty Inspectors. This non-profit organization is the only organization of individual safety professionals. It works to promote the safety profession and foster the professional development of its members. The ASSE plays an important role in the development of many national programs and standards. The society continues to expand its focus in the United States and has chapters in the Middle East and Great Britain. It currently has 31,000 members. The ASSE is guided by a 25-member board of directors which includes 13 regional vice-presidents. The objectives of the society's 138 chapters are to promote, establish, and maintain standards for the safety profession; to develop educational programs; to conduct research in areas that further the purpose of the society; to provide forums for the exchange of information among society members; and to provide a liaison to related disciplines. The ASSE conducts an annual professional development conference and works closely with the Board of Certified Safety Professionals of America. The society also publishes a monthly journal, *Safety Professional,* and sponsors a number of educational conferences, seminars, and educational programs. The society has a healthcare division which publishes a periodic newsletter.

National Safety Council (NSC)—The NSC is the world's largest organization that devotes its entire efforts to safety promotion and accident prevention. The council is the largest non-governmental and not-for-profit organization promoting safety in the United States. It oversees the activities of 100 local safety councils, which work under the leadership of local citizens, industrial interests, responsible official agencies, and other important groups. Each council is self-supporting and its goal is to reduce accidents and injuries through prevention training.

SUMMARY

Government regulations that affect healthcare organizations are so numerous that hazard control professionals have an enormous task just keeping up with the latest developments. Some regulations are duplicated by several agencies; many conflict, and most are difficult to understand. Many people believe that healthcare facilities, especially those that provide long-term care, are the most regulated organizations outside the nuclear arena. Hazard control practitioners must take the time to become familiar with these agencies and their corresponding regulations or standards. Healthcare administrators, risk managers, care providers, and hazard control professionals should be aware of the numerous voluntary agencies involved in healthcare regulation. This chapter also looked at some of the voluntary compliance organizations involved in healthcare safety, including the Joint Commission and the NFPA.

REVIEW QUESTIONS

1   The federal publication that is a compilation of general and permanent rules of agencies of the government is the:
    a.   United States Code
    b.   Federal Register
    c.   Code of Federal Regulations
    d.   Public Law Update Letter

2   The Occupational Safety and Health Administration is an agency within the:
    a.   National Institute of Occupational Safety and Health
    b.   Department of Health and Human Services
    c.   Department of Transportation
    d.   Department of Labor

3   OSHA issues citations to employers for safety and health violations not covered by a specific standard by using its:
    a.   General duty clause
    b.   Notice of Proposed Rulemaking procedures
    c.   Intervention powers
    d.   OSHA does not write citations without referencing a specific standard

4   The OSHA injury/illness log must be maintained by employers for:
    a.   24 months
    b.   3 years
    c.   5 years
    d.   30 years

5   The *primary* statutory weapon used by the EPA to regulate and protect the environment from solid waste materials is the:
    a.   Resource Conservation and Recovery Act
    b.   Superfund Amendments and Re-Authorization Act
    c.   Comprehensive Environmental Response, Compensation and Liability Act
    d.   Emergency Planning and Community Right-to-Know Act (SARA Title III)

6   Hazardous waste incinerators regulated under the RCRA must also comply with ambient air standards and/or emission limitations of the:
    a.   Clean Air Act
    b.   Toxic Substances Control Act

  c. Clean Water Act
  d. Ambient Air Standards Act

7 The Food and Drug Administration does all of the following *except*:
  a. develop policy on the safety and labeling of all human drugs
  b. issue standards for radiation exposure levels
  c. develop methods to control radiation exposures
  d. regulate the disposal of polychlorinated biphenyls
  e. classify medical devices and equipment

8 The Safe Medical Devices Act of 1990:
  a. reduced FDA authority to regulate medical devices more effectively
  b. eliminated the requirement to report medical device failures
  c. required facilities to report patient deaths but not patient injuries
  d. was amended by the Medical Device Amendments of 1992 to require facilities
      to report user errors

9 Which of the following agencies adopts and enforces standards for the department
   of nuclear medicine in healthcare facilities?
  a. NIOSH
  b. NRC
  c. FDA
  d. OSHA

10 Which title of the Americans with Disabilities Act of 1990 deals with employment
    issues?
  a. Title I
  b. Title II
  c. Title III
  d. Title IV

11 The 1995 Joint Commission standards incorporated most healthcare safety re-
    quirements into the _____ chapter.
  a. Care of Patients
  b. Continuum of Care
  c. Management of the Environment of Care
  d. Management of Human Resources

12 Which of the following NFPA standards applies specifically to healthcare fa-
    cilities?
  a. NFPA 10
  b. NFPA 30
  c. NFPA 99
  d. NFPA 704

13 Which of the following is the largest voluntary standards organization in the United States?
   a. American Society for Testing and Materials
   b. Factory Mutual Corporation
   c. Underwriters Laboratories
   d. American National Standards Institute

Answers: 1–c   2–d   3–a   4–c   5–a   6–a   7–d   8–d   9–b   10–a   11–c
              12–c   13–d

CHAPTER *5*

EMERGENCY PLANNING AND FIRE SAFETY

A. EMERGENCY PLANNING

Emergency planning involves developing a comprehensive facility-wide program that can be implemented to address both internal and external disasters. The plan should strive to accomplish the following:

- Reduce the confusion and panic that result in an emergency situation
- Allow the emergency management team to assess the situation and to make informed decisions
- Provide an environment of predictable behavior for the staff following a disaster or emergency event

Emergency Plan Development—Development of an emergency plan begins with the disaster committee or safety committee. The plan should address alternative sources of essential utilities and emergency communication.

- The plan should also have a current personnel call listing to request additional staff help if required.
- Effective communication and departmental coordination are the greatest challenges in developing an effective plan. A healthcare facility must have a disaster plan that details procedures to be followed in an emergency.
- The plan should be developed and maintained with the help of qualified fire, safety, and other professional personnel. The plan must include procedures for prompt transfer/transport of casualties and records.
- The plan must have instructions regarding the locations and use of alarm systems, signals, fire-fighting equipment, methods of fire containment, procedures for notification of appropriate persons, and evacuation routes/procedures.
- The plan must address procurement of food, water, and medical supplies.

- The plan must detail personnel responsibilities in moving patients in and out of the facility during a variety of situations.

Scope of the Plan—Whether the emergency preparedness program developed for multi-hazard situations is comprehensive or austere, there is a certain minimum legal obligation to life safety, as well as moral obligations.

- It is important that planning reflect the unique capabilities and limitations of a facility. When defining the scope of an emergency preparedness program, a critical self-assessment of capabilities and limitations should be performed before defining specific procedures.
- Factors to consider include the size of a facility in relation to the size of the surrounding community and accessibility to other facilities and utilities.
- Climate, weather, and geographical location should be considered when deciding what to include in the plan.
- Are industries with potential hazardous or dangerous operations or substances located in the area?
- Is the facility near major transportation sources such as airports, railroads, and major highways?

Emergency Assistance Agencies—During a community-wide emergency, a number of governmental and private agencies are available to assist:

- Federal Emergency Management Agency (FEMA)
- U.S. Army Corps of Engineers
- Salvation Army
- American Red Cross
- U.S. Public Health Service
- National Weather Service
- National Oceanic and Atmospheric Administration (NOAA)
- National Guard (if activated by proper authorities)

Coordinating Response—Coordinating emergency plans is vital to coping with disaster in an effective manner. The efforts of all affected and participating emergency agencies must be coordinated and directed toward creating a workable plan. Safety and emergency planning directors should meet periodically to discuss mutual concerns, responsibilities, and disaster response procedures. Emergency planning directors should become familiar with the authority, organization, and emergency procedures that become effective in time of a civil emergency. Area healthcare facilities and hospitals should take on responsibilities in accordance with their capabilities for providing emergency services. Most large communities have an emergency management agency that serves as the coordinator for all community emergency action plans.

Emergency Coordinator Responsibilities—The facility emergency coordinator and alternates should be responsible to top management for:

- Emergency command center effectiveness
- Communication and security

- Public information and media releases
- Coordination of facilities and engineering functions
- Sheltering, feeding, and counseling
- Morgue establishment and notification of next of kin
- Emergency medical services

Joint Commission Emergency Planning Requirements—The Joint Commission on Accreditation of Healthcare Organizations requires facilities to develop comprehensive emergency preparedness plans that outline pre-established procedures to follow during disasters or emergency situations. Emergency planning must strive to develop plans that will ensure continued patient care in a variety of situations. The comprehensive plan or program must allow for implementation in the event of a community emergency situation or area disaster. Healthcare facilities must plan for all anticipated and/or possible situations.

Planning Categories—Planning should define and address disasters in two major categories:

- *Internal Emergencies*—An internal situation is an event that affects patient care and facility operation. This could include electrical failure, storm damage, fire, or chemical release.
- *External Disasters*—This is an event that happens in the community, and healthcare facilities have responsibilities to assist in dealing with the situation. The level of responsibility depends on the type and size of the facility. External disasters such as major storms, industrial or transportation accidents, and earthquakes could easily result in patient overload, which could affect quality of care.

Definition of Disaster—The Joint Commission considers a disaster as any incident that interferes with a facility's ability to operate in a normal manner. Planning for disasters must consider all potential situations that might affect the operational readiness of a facility. Some situations to be considered when developing a program include:

- Natural phenomena or storms (tornadoes, floods, etc.)
- Fire or explosion
- Hazardous material spills or radiological contamination
- Transportation accidents
- Civil disturbances

Elements in the Emergency Planning Process—Coordination with community emergency response agencies and other healthcare facilities is vital to establishing a plan to address both internal and external disasters.

- Internal coordination is required to ensure that all staff members understand their responsibilities under the plan.
- External coordination is necessary to ensure that community agencies understand the role of the healthcare facility.
- External agencies include law enforcement departments, fire departments, emergency medical response councils, volunteer agencies, the Red Cross, and emergency management organizations.

- Planning documents should include responsibilities for all organizations.
- Healthcare facilities in a community should coordinate their respective plans to prevent one organization from becoming overburdened.
- Community planning can be a difficult challenge, but it is an area that cannot be overemphasized.

Facility Management—Joint Commission guidelines suggest that organizations focus on facility preparation, employee readiness, and patient control measures. Planning in these areas is critical to the success of any effective plan. Refer to the *Joint Commission Plant Technology Safety Management Handbook* and NFPA 99 for additional guidance.

- *Facility Readiness*—Several elements must be addressed if staff members are to respond to an emergency situation. Important considerations are available space, supplies and materials, adequate utilities, effective communication systems, and security.
- *Employee Preparedness*—The preparedness of the facility staff considers manpower requirements, assigning responsibilities, designating staff functions, determining staff roles, and providing the necessary education and training.
- *Patient Control*—The final element of facility management is concerned with maintaining patient care at proper levels, which could include modifying treatment and surgical schedules. Patient moving and evacuation procedures must be developed, with an emphasis on discharge planning and medical record documentation.

B. EXTERNAL DISASTERS

External Disaster Planning—Planning for a disaster requires organizations to evaluate facility space to include considering the capacity of the emergency room and surgical areas, bed capacity, and areas to meet specialized needs of a disaster. Provisions should be made to acquire or have access to critical supplies such as food, medicine, and surgical supplies. Many facilities store emergency materials and have implemented a system to rotate, replace, and revitalize medical supplies in the event of an emergency. Arrangements should be made with suppliers to ensure the availability of critical utilities to meet high-priority emergency needs. Disasters and emergency situations would require additional security personnel to control human and vehicular traffic. The plan must address policies and procedures necessary to ensure that communications remain in operation during an emergency. A public information center and a medical command center should also be established. The community effects of an external disaster are presented in Table 5.1. The following is an outline of the basic elements that should be covered in the external disaster plan:

1. *General Information*—This section provides an overview of general procedures such as personnel utilization, material placement, and space requirements.

2. *Initial Response Procedures*—This section outlines the actions to take to initiate the disaster response plan. The response procedures should address notification

TABLE 5.1 External Disaster Effects

Disaster/Example	Maximum Magnitude/Impact	Immediate Response
Local, single family Fires Power outages Small HAZMAT spills Bomb threats and/or explosions	Affects one plant only; threats to personnel, equipment, product, raw materials, or production; little impact outside facility; suppliers and repair and recovery resources unaffected	In-house emergency response team; local fire department; building and facility evacuation of threatened personnel; on-site medical; ambulance and local hospital; does not overload local emergency response system
Local, multi-facility (including residences) Fires Power outages Large-scale HAZMAT spills Bomb threats and/or explosions	Affects several plants, commercial facilities, homes, or apartments; threats the same as above; little impact outside facilities involved; suppliers and repair and recovery resources unaffected	In-house emergency response team; local fire department; building and facility evacuation of threatened personnel; on-site medical; ambulances and hospitals may overload local emergency response; may need mutual aid from adjacent communities
Regional, community-wide Earthquake Hurricane Tornado Flood Dam failure Area-wide major storm Nuclear terrorism	Affects one or more communities; threats include all the above and may also include considerable disruption of entire community and its operation (e.g., sanitation and health, law enforcement, the ability to cope with another emergency, food supplies, potable water); some local suppliers and repair and recovery resources affected.	All local emergency response agencies overtaxed; mutual aid scaled up to involve several adjacent communities; relocation of dispossessed; ambulances and helicopters; national guard; multi-community hospitals; hospital evacuations; people isolated; local resource shortages
National Nuclear attack	Everybody affected; most resources affected; all the problems listed above multiplied; radiation fields will curtail most normal activities for a while; priorities drastically changed for everyone	All response systems overtaxed by fire, HAZMAT spills, need for rescue work, tending injured, disrupted transportation routes; mutual aid limited to shouting distance; small groups essentially on their own for a while

Adapted from Federal Emergency Management Agency Publication FM-41.

of all department heads, emergency staff, and other essential personnel, depending on the type of disaster.

3. ***Departmental Responsibilities***—The plan should require each department or response agency to have written procedures outlining their actions, duties, and responsibilities.

4. ***Departments/Systems Activated***—This section of the plan outlines general duties and responsibilities of the following departments or systems:

- Emergency department
- Medical staff
- Clerical support
- Environmental services
- Maintenance department
- Anesthesia services
- Security of buildings and grounds
- Command and information center
- Utilities and materials management
- Patient management

5. *Triage Areas*—This section of the plan describes the procedures to be followed in both primary and secondary locations. It also outline the duties and responsibilities of all triage officers.

C. INTERNAL DISASTERS

Internal Disaster Planning—Healthcare facilities must be able to respond to any internal emergency and also maintain quality patient care. A well-written plan that is exercised frequently minimizes the impact a disaster will have on the facility's operation. The Joint Commission requires that facilities conduct semi-annual disaster drills. The plan should stress maintaining and improving the reliability of critical systems.

1. *Definition of Internal Disaster*—This should be determined by the facility and should address the need to evacuate patients and loss of life support systems. This section should give examples of emergency situations and provide guidance on implementing the plan.

2. *Planning Considerations*—The following situations should be considered when developing an internal disaster plan:
   - Bomb threats or terrorist activity
   - Chemical or hazardous material release or spill
   - Computer, communications, or power failure
   - Oxygen or other medical air failure
   - Fire or explosion
   - Steam or heat loss
   - Water or other utility loss
   - Biohazard spill or exposure
   - Medical equipment failure

3. *Developing an Internal Plan*—Facilities must develop a plan that will be realistic when followed. The key to developing a comprehensive plan is to identify all possible contingencies and then write response actions for each situation that might occur. This cannot be done effectively without coordination among departments. All major operational departments must be involved.

Planning Committees—Some facilities have a disaster planning committee that meets regularly to assess the effectiveness of the plan. Response actions, responsibilities, and lines of authority must be outlined thoroughly. The final plan must be in a format that is easy to read and understand. A plan is most effective when it can serve as a guide.

1. *Implementing Plans*—A completed or revised plan should be implemented only after all personnel are familiar with the procedures and understand their specific responsibilities. Training can be difficult because responsibilities will vary greatly. A survey of all personnel can help management determine if the plan is ready to be implemented.

2. *Plan Evaluations*—The plan should be evaluated within 24 to 48 hours after each actual emergency or planned drill. Key personnel and/or departments should provide input on an evaluation form. Forms should compare actual responses to required responses. Deviations and variances should be noted, and a final report should be presented to the disaster planning committee and top management.

3. *Plan Changes*—The plan should be modified or changed as necessary based on results and/or recommendations. Actual disasters provide the most accurate picture of the effectiveness of the plan.

Emergency Response Procedures and Announcement Codes—Healthcare facilities should develop emergency codes to notify personnel of the current situation.

1. *Sample Emergency Codes*—

Code	Meaning
Dr. Redburn	Fire disaster or fire alarm
Dr. Coolsport	Fire alarm clear
Dr. Blueplan	Disaster or disaster drill
Dr. Watchout	Tornado or severe storm watch
Dr. Badstorm	Tornado or severe storm warning
Dr. Ethyl	Ethylene oxide spill

Note—Other codes can be added to meet local requirements.

2. *Emergency Response Procedures*—The emergency code response system is activated by contacting the switchboard, security, or other designated location. The following information should be provided:
   • Name and department of caller
   • Type of emergency
   • Exact location of the emergency
   • A callback number if appropriate

3. *Immediate Actions (RACE)*—Some facilities use the following acronym for emergency response:
   • Remove all patients and others to safety if possible.
   • Activate emergency response system by following facility procedures.
   • Confine the disaster or fire if possible, but only if trained to do so.
   • Extinguish the fire or contain the chemical spill by following facility procedures if trained to do so.

4. ***Plan Notification Procedures***—The plan should give detailed instructions for notifying personnel of the situation. It also should cover procedures to be used to recall essential personnel. This section of the plan should be reviewed and evaluated on a regular basis.

5. ***Duties and Responsibilities***—This part of the plan outlines overall reporting requirements and assigns responsibilities to key staff members. It should also specify who will assume control of the situation to serve as the triage officer. Departmental guidelines should be published to guide other personnel in accomplishment of duties to support the plan.

6. ***Medical Staff Duties***—Specific duties and responsibilities of staff and resident physicians should be outlined in this section.

7. ***Evacuation Procedures***—The triage officer should make the decision to evacuate patients to another facility. The plan should have a priority listing of institutions to which patients will be evacuated. The disaster command center should assist in coordinating the evacuation and transport of patients to other facilities.

8. ***Patient Holding Areas***—Evacuation requirements will dictate that patients be brought to a pre-determined holding area. This section of the plan should specify the exact locations where patients in various parts of the facility will be held pending further decisions. This part of the plan should address procedures covering the movement of patients from "special care" units.

Fire Extinguishers—Personnel expected to use portable fire extinguishers must be trained annually in their operation and safe use. PASS guidelines should be followed:

- Pull the pin on the extinguisher.
- Aim the nozzle at the base of the fire.
- Squeeze the handle firmly.
- Spray in a sweeping motion.

Practice Drills—The emergency or disaster preparedness plan should be implemented every six months. Experience shows that training is lost if a plan is not realistically exercised. The high turnover of personnel and changes in the facility necessitate that the plan be practiced frequently.

- The implementation may be a planned drill or an actual emergency that effectively activates the plan.
- If the facility is designated to receive patients in an emergency or is likely to receive patients because of location or situation, at least one mass casualty exercise should be run annually, with enough victims to sufficiently test the program.
- Paper and conference room drills that require no disruption of normal activity are useful but cannot be considered an active drill.
- A drill that moves surrogate patients through the facility can be considered a real exercise of the plan.
- Staff preparedness requirements must address how workers are notified and by whom. Facilities must keep an up-to-date recall roster and develop backup noti-

fication methods. The plan must ascertain whether sufficient staff can be on the scene in a reasonable time. The plan should also address emergency transportation for emergency workers if necessary.

- Education and training are crucial for the director and staff. Training for each type of disaster is essential in developing a control plan. Workers must be taught that an emergency plan is vital and real and is of no value if it remains simply an idea. Training and rehearsal are time consuming but keep the program in good working order.

- Implementation of the plan should be observed carefully and staff should be debriefed. The most important issue is identification of problems. After a drill or a real disaster, there is a natural tendency to gloss over problems and emphasize the good points. Follow-up of the drill is also important. Each problem identified should be addressed by training or by changes in the system, equipment, or plan to assure that things can run more smoothly next time.

Fire Response Planning—Planning to respond to fires is an important aspect of the emergency planning process. Fire safety and procedures, including life safety, are covered more thoroughly later in this chapter.

1. *Basic Fire Planning*—
   - Provide information about alarm systems or signals.
   - Provide information on the location and use of fire-fighting equipment.
   - Explain basic fire containment procedures.
   - Detail evacuation routes necessary to meet the needs of the facility.
   - Establish procedures to ensure that fire drills meet regulatory standards.
   - Nursing homes must exercise the plan quarterly on all shifts.
   - Drills should be held at various times and should be as realistic as possible.
   - Each drill should be documented to identify problems and corrective procedures.
   - Fire planning is addressed by Joint Commission standards and OSHA Standard 29 CFR 1910.38.

2. *Master Fire Plan*—The primary or master plan should cover the following items:
   - Fire department notification/follow-up procedures
   - Procedures for announcing fire location using an interco n or public address system
   - An emergency command center where key personnel assemble and manage decisions associated with the fire
   - Procedures for determining who will keep facility departments apprised of the situation
   - Designated personnel who will meet and direct fire department personnel
   - Procedure for holding non-emergency calls and giving an "all-clear" signal when the emergency is over
   - Fire response as a unique challenge to those working in a healthcare facility
   - Because patients and residents will not be able to move quickly, healthcare workers must trained in fire safety prevention and response actions

D. NATURAL DISASTERS

Weather-Related Disasters—An emergency plan must assume that most disasters will arrive with very little warning, develop rapidly, and carry the potential for substantial destruction. The likelihood that the kinds of disasters cited in this section would ever strike may be very small, but a facility must plan to react to, cope with, and recover from any emergency situation that occurs.

1. *Storm Preparedness Measures*—
   - Contact local government or the National Weather Service and learn about storm warnings that pertain to the area.
   - Inform employees of storm safety rules.
   - Establish a system for early release from work and "employee-stay-home" announcements.

2. *Hurricane Preparation*—The National Weather Service is responsible for issuing warnings when a hurricane appears to be a threat to the U.S. mainland, Puerto Rico, the Virgin Islands, Hawaii, or the Pacific Territories. As soon as conditions intensify to the tropical storm level, even though the storm may be thousands of miles away, the storm is named, and the Weather Service begins issuing advisories. Advisories are issued every three hours or less as the storm nears. Location, wind intensity, speed, and direction are given. As a hurricane moves toward the mainland, hurricane watch notices are issued.
   - **Hurricane Watch**—Indicates a threat to coastal areas. Persons in the watch area should listen for further advisories and be prepared to take precautionary actions, including evacuation if directed. When forecasters determine that a particular section of the coast will receive the full effect of the hurricane within 24 hours, a hurricane warning is issued.
   - **Hurricane Warning**—Specifies coastal areas where winds of 74 mph or higher or a combination of dangerously high water and very rough seas is expected.
   - **Tropical Storm Warning**—Issued for those areas that are expected to receive gale-force winds (greater than 40 mph).

3. *Thunderstorms*—The National Weather Service issues severe weather warnings using the following terms:
   - **Severe Thunderstorm**—Indicates the possibility of frequent lightning and/or damaging winds greater than 50 mph, hail $3/4$ inch or more in diameter (about the size of a dime), and heavy rain.
   - **Severe Thunderstorm Watch**—Indicates the possibility of tornadoes, thunderstorms, frequent lightning, hail, and winds greater than 75 mph.
   - **Tornado Watch**—Tornadoes could develop in the designated area.
   - **Tornado Warning**—A tornado has been sighted in the area or is indicated by radar.

4. *Tornadoes*—These violent local storms have whirlwinds of tremendous speed that can reach 200 to 400 mph. An individual tornado appears as a rotating, funnel-shaped cloud which extends toward the ground from the base of a thundercloud varying from gray to black in color. A tornado spins like a top and may

sound like the roaring of an airplane or locomotive. These short-lived storms are the most violent of all atmospheric phenomena. The width of the path of a tornado generally ranges from 200 yards to 1 mile. Tornadoes can travel 5 to 50 miles along the ground at speeds of 30 to 75 mph. Tornadoes sometimes reverse direction or move in circles. Others remain motionless for a while before moving on. Tornadoes occur primarily in the central plains and southeastern states. They are highly localized, and facilities in tornado-prone regions should participate in a Tornado Watch System (Sky Warn), in coordination with the local emergency management agency.

- Since tornadoes occur with little or no warning, there is very little time for preparation. The best protection is an underground area. In buildings without basements, interior hallways on the lowest floor should be designated tornado shelter areas. News broadcasts should be continuously monitored following a tornado watch announcement.
- When a tornado warning is issued, employees should be directed to take shelter immediately, crouch down, and cover their heads with their arms. All doors to outside rooms should be closed. Periodic drills should be scheduled to ensure that employees know where and how to best protect themselves.
- Employees who work outside should be advised to lie flat in the nearest ditch, ravine, or culvert with their hands shielding their heads if there is insufficient time to reach an indoor shelter. In the aftermath of the tornado, all damaged facilities should be checked for survivors. Avoid downed power lines, check for gas leaks, and contain small fires.

5. *Winter Storms*—Winter storms vary in size and intensity, from minor ice storms to full-blown blizzards. Freezing rain or sleet, ice, heavy snow, or blizzards can be serious hazards.
   - **Winter Storm Watch**—Indicates severe winter weather conditions may affect the area (freezing rain, sleet, or heavy snow may occur either separately or in combination).
   - **Winter Storm Warning**—Indicates that severe winter weather conditions are imminent.
   - **High Wind Watch**—Indicates sustained winds of at least 40 mph, or gusts of 50 mph or greater, are expected to last for at least one hour. (In some areas, this means strong gusty winds occurring in shorter time periods.)
   - **Heavy Snow Warning**—Indicates snowfall of at least 4 inches in 12 hours or 6 inches in 24 hours is expected. (Heavy snow can mean lesser amounts where winter storms are infrequent.)
   - **Blizzard Warning**—Issued when sustained wind speeds of at least 35 mph are accompanied by considerable falling and/or blowing snow. Visibility is dangerously restricted.
   - **Travelers' Advisory**—Issued to indicate that falling, blowing, or drifting snow; freezing rain or drizzle; sleet; or strong winds may make driving difficult.

6. *Floods*—Except in the case of flash flooding from thunderstorms, coastal storms, or dam failure, the onset of most floods is a relatively slow process, which allows adequate warning. The buildup usually takes several days.

- **Flood Report**—Progressive situation reports are available from NOAA through its Weather Service River Forecast Centers and River District Offices.
- **Flash Flood Warning**—This is the most urgent type of flood warning. It is transmitted to the public over radio and television and should also be transmitted through local warning systems by sirens, horns, or whistles; through telephone alerts; or by police cars using loudspeakers.

Earthquakes—Although science is searching for the means to predict impending earthquakes, accurate predictions are not yet possible. However, it can be assumed that earthquakes will continue to occur in areas where they have been relatively common in the past. They can range in intensity from slight tremors to strong shocks and can last from a few seconds to as long as five minutes and in a series over a period of several days. The actual movement of the ground in an earthquake is seldom the direct cause of injury or death. Most casualties result from falling materials. Severe quakes usually destroy utility lines and gas, sewer, or water mains; they also trigger landslides, rupture dams, and generate seismic waves (tsunamis).

1. During the shaking, employees should stay indoors if already there.
   - Take cover under sturdy furniture such as a work table, brace oneself in a doorway, or move into a corner and protect the head and neck in any way possible.
   - Stay near the center of the building, away from glass windows, skylights, and doors.
   - Never run through or near buildings where there is danger of falling debris.
   - If outside, stay in the open, away from buildings and utility wires.

2. After the shaking, employees should exit the building through the stairways and never use elevators.
   - Stay out of damaged buildings; the aftershock can shake them down.
   - Facility officials should check utilities for damaged water pipes and electrical malfunctions.
   - If gas leakage is detected, shut off the main valve, open windows, and keep the building cleared until utility officials determine it is safe.

E. OTHER EMERGENCY SITUATIONS

Transportation Accidents—The U.S. Department of Transportation regulates the movement of hazardous chemicals. When chemicals that pose a significant hazard to the public if released from their packing are transported interstate, they must be labeled with appropriate words of identification and caution. Shipping papers identifying the hazardous material being transported are required to be in the vehicle or vessel.

- Major transportation accidents often cause chemical spills, fires, explosions, and other problems, which call for special operations such as rescue and evacuation. Usually, transportation accidents affect only relatively small areas and involve only a small number of people.
- Regardless of the type of transportation accident, the first consideration should be

to save lives. This can be accomplished through quick response and coordination with emergency services and local police, fire, and medical services.

Public Demonstrations/Civil Disturbances—In recent years, there have been a variety of demonstrations for different purposes in many locations throughout the country. Some demonstrations develop slowly, allowing authorities to assess the problem, conduct negotiations with the organizers, and arrange for control measures. On other occasions, violence may flare up with little advance notice, but even these incidents are usually preceded by earlier indications of a buildup of tensions and pressures.

- In a situation where there is a sudden eruption of violence, perhaps perhaps by attempted arson and assaults, company security personnel usually serve as the source for information regarding the characteristics and extent of the disturbance.
- Company security personnel should cooperate completely with local law enforcement agencies, which will provide the information needed to make appropriate decisions. An effective employee notification/recall system is a must.

Bomb Threats and Terrorism—Compared with other facility emergencies, the covert and criminal nature of terrorism, including bombing incidents, bomb threats, and the taking of hostages, is a highly complex problem for management and emergency service personnel. Consequently, planning to meet such threats must include prompt contact with local or neighboring city law enforcement agencies, particularly if they have a bomb disposal unit, and the local office of the FBI. Arrangements should be made to obtain the assistance of experienced personnel. Experience shows that over 95% of all written or telephoned bomb threats are hoaxes. However, there is always a chance that a threat may be authentic. Appropriate action should be taken in each case to provide for the safety of employees, the public, and property and to locate the suspected explosive or incendiary device so it can be neutralized.

1. The first line of response is threat analysis. In order to do this effectively, it is important to obtain as much information as possible about the person or group making the threat and the size and location of the bomb. Industries concerned about such threats will want receptionists who can remain calm and who have had training in the type of questions to ask. Placing a questionnaire with relevant questions near the phone can be very useful.
2. The information gathered on this questionnaire may be sufficient to discount the threat or may direct that actions other than evacuation be taken. It is not as unlikely that the caller will give his or her address as you may think. According to the FBI, many of these callers are seeking someone to talk to. A simple ruse such as "We'll have the manager get back to you. Could I have your name and address so he or she can do so?" may be all that is necessary to get such information. Therefore, this questionnaire may be the single most important resource in dealing with a bomb threat.
3. If a suspicious object is located and thought to be a bomb, and the local law enforcement personnel cannot dispose of it, the local emergency management agency should be contacted to secure the services of the nearest explosive/bomb disposal team for assistance.

Sabotage—No facility is immune to sabotage. However, the types of targets for sabotage can usually be predicted with reasonable accuracy. The saboteur will generally look for a target that is critical, vulnerable, accessible, and at least partially conducive to self-destruction. Saboteurs in general are enemy agents, disgruntled employees who commit sabotage for revenge, individuals who have been duped by enemy propaganda, or individuals who are mentally ill. Sabotage may be linked to or be another form of terrorism.

1. The types of sabotage can be classified as follows:
   - *Chemical*—The addition or insertion of destructive or polluting chemicals
   - *Electric or Electronic*—Interrupting or interfering with electrical or electronic processes or power, and jamming communications
   - *Explosive*—Detonating explosive materials or damaging or destroying by explosives
   - *Incendiary*—Fires ignited by chemical, electrical, electronic, or mechanical means or any ordinary means of arson
   - *Mechanical*—Breaking or omitting parts, using improper or inferior parts, or failing to lubricate or properly maintain parts
   - *Psychological*—Inciting strikes, boycotts, unrest, or personal animosities or causing slowdowns or work stoppage by excessive spoilage or inferior work

2. Sabotage may be prevented by reducing target accessibility and vulnerability.
   - Permit only authorized access to the potential target area.
   - Screen and place employees in accordance with security requirements.
   - Design, construct, and modify equipment with built-in protection against sabotage.
   - Conduct continuing education for employees on prevention of sabotage.
   - Develop a plan with organizational procedures for handling potential or actual sabotage.

Other Disasters—A wide range of disasters may strike communities at any time, with or without warning. The probability of secondary disasters should be anticipated. An earthquake could cause structural fire which may, in turn, burn out circuits, resulting in a power failure. Severe winds may damage a chemical plant and spilled chemicals could start a fire, which would release toxic smoke.

Intense rains can cause flooding, which may set off landslides. The multi-hazard situation creates an environment in which resource availability and priorities may be radically different from the norm. In large-scale events, services and options would likely be severely curtailed and/or adversely affected. A company would still be responsible for doing everything possible to protect life and property.

F. HAZARDOUS MATERIALS

Hazardous Substance Release—The Superfund Amendments and Re-Authorization Act of 1986 (SARA) included Title III, the Emergency Planning and Community Right-to-Know Act. Title III requires community response to emergency incidents involving hazardous chemicals.

- OSHA promulgated a final rule on Hazardous Waste and Emergency Response in March 1989. This rule was issued under SARA Titles I and III. The OSHA rule established safety and health requirements for employers involved in hazardous waste operations and required employers to develop and implement emergency response plans.
- The Clean Air Act (CAA) became law in 1990, and both OSHA and the EPA received additional responsibilities for preventing chemical emergencies.
- The standard covers employers with workers engaged in emergency response to substance releases. Those emergency response plans designed to meet the requirements of Section 303 of SARA will be considered to have met the requirements of this paragraph.
- In October 1993, the EPA proposed Risk Management Programs for Chemical Accidental Release Prevention, 40 CFR Part 68.
- The new standard mandated by the Clean Air Act Amendments of 1990 intends to prevent the accidental release of regulated substances and extremely hazardous materials into the air.
- Facilities that have listed chemicals above the threshold quantities must register with the EPA and develop and implement a risk management program.
- The risk management program must include a hazard assessment, prevention plan, and emergency response procedures.

Emergency Response Planning—Facilities are also required to publish a plan that summarizes the results of the assessment and analysis of potential hazardous releases or spills.

1. *Contingency Planning*—
   - Prepare written emergency response plans.
   - Provide for realistic drill and exercises.
   - Coordinate all plans with public emergency response agencies.
   - Develop information on emergency medical aid.
   - Training should cover all emergency response actions.
   - This requirement will expand the emergency release planning requirements to more facilities than currently covered under X Title III and the OSHA HAZWOPER Standard, 29 CFR 1910.120.
   - The chemical list has not yet been published by the EPA, but it should be similar to the list in the CAA, Section 112(r)(3).
   - Hazard analysis would have to be based on worst-possible release instead of worst-credible release.

2. *OSHA Requirements*—Current OSHA standards require all facilities engaged in storing, treating, and disposing of hazardous wastes to develop an emergency response plan that addresses:
   - Pre-emergency planning with outside parties
   - Personnel roles, lines of authority, training, and communication

- Emergency recognition and prevention
- Safe distances and places of refuge
- Site security and control procedures
- Evacuation routes and procedures
- Decontamination and emergency medical treatment procedures
- Emergency alerting and response procedures
- Personal protective equipment and emergency equipment
- Local emergency organizations may use local/state plans

Chemical Emergency Response Training (OSHA 29 CFR 1910.120)—The training requirements of 29 CFR 1910.120 require that training be based on the duties and functions to be performed. All newly hired personnel who will be required to perform emergency responses actions must be trained before participating in actual emergencies.

1. *Training Levels—*
   - **First Responder Awareness Level**—These workers are likely to witness or discover a hazardous substance release and have been trained to notify emergency personnel. They must have an understanding of the hazardous substances present and the hazards posed by their release.
   - **First Responder Operations Level**—These workers must receive eight hours of training which allows them to contain small spills and protect others and the environment.
   - **Hazardous Materials Technicians**—These persons respond to releases for the purpose of stopping them. They receive 24 hours of training and must know how to implement the emergency response plan and understand actions to take to control, confine, and contain the spill.
   - **Hazardous Materials Specialists**—These persons usually receive 40 hours of training and are qualified to support the technicians on the scene. They possess more knowledge about specific substances and containment methods necessary to resolve the situation.

2. *Non-Mandatory Guidelines*—OSHA recently issued Appendix E, Non-Mandatory Training Guidelines, which can be helpful to facilities in complying with the training requirements of 29 CFR 1910.120. This appendix follows guidelines recently issued by the EPA concerning contingency planning and emergency response actions. Healthcare facilities should evaluate their hazards and ensure that training at the appropriate level is accomplished.

G. FIRE SAFETY

The National Fire Protection Association (NFPA) reports that one-third of hospital fires originate in patient rooms or staff quarters. Matches and smoking are the most frequent causal factors. Fires also originate from malfunctioning or misused electrical equipment such as hot plates, coffeepots, and toaster ovens. Deaths during hospital fires are overwhelmingly due to inhaling toxic smoke and fumes and not direct exposure to the fire. Fire safety programs should include a facility fire plan, evacuation plan, chemical response plan, and training requirements. The safety and/or disaster committee should approve all plans used to train employees in emergency fire procedures. The facility should have an active fire prevention plan that requires regularly scheduled fire drills for all shifts. Written reports should be maintained on the effectiveness of each drill performed. A quarterly fire inspection must be performed for each fire zone.

Emergency Fire Planning—Fire response is a unique challenge to those working in a healthcare facility. Patients and residents will not be able to move quickly. Employees must be trained in fire safety prevention and actions to take in the event of a fire. The engineering precautions normally built into healthcare facility plants include side corridors, wide exits, and panic hardware.

1. *Fire Safety Planning*—Development of sound fire prevention plans and response procedures begins with the disaster committee or safety committee.
   - Fire planning should be a component of the internal disaster plan.
   - The plan must have instructions regarding the locations and use of alarm systems, signals, and fire-fighting equipment.
   - The plan should detail methods of containing fire, procedures for notification of appropriate persons, and evacuation routes/procedures.

2. *Worker Obligations*—Each employee should recognize and report all common hazards and fire safety hazards. The department head should rectify the situation immediately by making the proper adjustments and notify the administrator.

3. *Hazardous Conditions*—When planning for fire emergency, the following hazardous situation and conditions should be considered:
   - Smoking hazards
   - Electrical equipment and appliances
   - Spontaneous combustion hazards
   - Use and storage of flammable materials
   - Oxygen and other compressed gases

4. *Basic Fire Planning Considerations*—Fire prevention and protection measures must be documented in written policies and procedures.
   - Buildings must have a fire alarm or fire detection system. The system should automatically activate an alarm in the event of fire.
   - Air-conditioning and heating ducts and related equipment must be installed in accordance with NFPA 90A.
   - The sound of the alarm must be distinct and loud enough to be heard over normal operational noise levels.
   - Manual fire alarm stations must be located near each exit.
   - Electrical monitoring devices for automatic sprinkler systems must be connected to the fire alarm system. These must be tested at least annually.
   - Fire extinguishers must be located so that distance to a unit never exceeds 75 feet. The fire extinguisher provided must be appropriate for the type of fire likely in that area and must be clearly identified.

- Fire extinguishers must be inspected at least monthly and maintained regularly, and records of inspections and maintenance must be kept.
- All fire alarm/detection systems should be tested quarterly.
- New materials such as bedding, draperies, furnishings, and decorations should be flame resistant.
- Fire drills for all personnel on all shifts should be conducted as required by standard, regulation, or code.
- A facility-wide smoking policy should be enforced.
- Electrical safety policies must be written and enacted.
- Personnel must be regularly trained in fire safety, including response plans. Everyone should know the location of fire-fighting equipment in their area and be observant as to location of all equipment in the facility. Locations of fire extinguishers should be shown on diagrams placed at each nursing station, and everyone should be familiarize with them.

5. *Inspections*—Quarterly fire inspection must be performed for each fire zone. Inspections should emphasize:
   - Assessment of all equipment
   - Testing of alarms, detectors, and pull stations
   - Evaluation of housekeeping practices and security alarms
   - Inspection of sprinkler pressure
   - Water availability and hydrant operation
   - Annual suppression, detection, and activation systems checks
   - Coordination with fire marshal and engineering/maintenance department

6. *Fire Drills*—Fire drills should be conducted regularly. Good housekeeping measures are crucial in preventing fires. Employees must be alert for fire hazards. Each department should have a copy of the facility fire plan. General procedures are as follows:
   - Be prepared to give information on location of fire, type of fire, and equipment failure.
   - Facilities should have an effective and convenient fire alarm system.
   - A signal should sound to alert all personnel.
   - Shut off oxygen and gas valves if possible.
   - Do not use telephone except for emergency calls.
   - Reassure patients and visitors that emergency plans are in effect.
   - Disconnect unnecessary electrical equipment.
   - Evacuation of patients will be last resort according to procedures.

Fire Response Actions—In the event of a fire, personnel should

- Pull the fire alarm immediately when a fire is discovered. This may be delayed only to remove another person from immediate danger. The fire alarm sounds immediately and automatically at the monitoring service, which notifies the fire department.
- Be alert and immediately report to their department.
- Carry out specific duties under fire in their area or fire in other departments.
- Escort public and patients to safety.

- Take all measures to protect and preserve property and equipment.
- If not needed to fight fire, remain for further instructions.

Emergency Command Center Functions—The administrator should report to the command center location and assume full responsibility.

- A central area should be designated as the command center.
- All communications to and from the fire area are reported here.
- Decisions as to the need for extra personnel are reported here.
- Extra personnel are sent out from here.
- Progress of the fire is reported here.
- Decision for evacuation is made here.
- Reports to other units in danger are issued from here.

H. EVACUATION PROCEDURES

General Evacuation Responsibilities—Evacuation is the removal of patients and equipment from an unsafe area to a safe area during a fire or other emergency situation. Patients are moved primarily to protect them from smoke. The liberated heat may render some corridors impassable; therefore, it is necessary to move to safety first those patients located beyond the point of origin of the fire, while the fire is contained within the room of origin. Any patient who is in immediate danger should be moved, without waiting for special instructions. Patients should be taken to the nearest safe place as calmly as possible. If other patients on the floor do not appear to be in danger of fire, smoke, or panic, personnel should wait for evacuation orders and try to reassure patients and visitors. Evacuation may be ordered as a necessity or as a general precaution and may be partial or complete.

1. *Partial Evacuation*—Moving patients to a neighboring safe area (i.e., on the same floor but on the other side of the building).

2. *Total Evacuation*—Removing patients and equipment from one or more wings to an adjoining area or outside the facility.

3. *Handling Patients During Evacuation*—
   - Ascertain which plan is to be used.
   - Know which exits are to be used to find the most secure and quickest way to safety.
   - Evacuate patients according to their physical condition.
     a. **Stretcher Patients**—These patients should be evacuated first. Patients will yield their stretchers and litters after reaching safety if there is a need for them to do so.
     b. **Wheelchair Patients**—Wheelchairs should be used to remove these patients to a safe area on the same floor. The wheelchairs should be taken back for the removal of additional patients. Those patients who are confined to chairs in their rooms may be pushed out in these chairs instead of taking time to transfer them to wheelchairs.

c. **Ambulatory Patients**—These patients should be led as a group to a safe area. Someone must be put in charge of them; ambulatory patients should not be left without guidance, so as to avoid panic.

I. IGNITABLE MATERIALS

Flammable/Combustible Materials—Flammable and combustible liquids, vapors, and gases present a major fire hazard in all hospitals. Although workers usually recognize this potential hazard, they should also be aware of important facts about flammable liquids that can help to prevent fires. Many liquids have vapors that are flammable or combustible and can be ignited by a spark from a motor, friction, or static electricity. A liquid may be classified as either combustible or flammable, depending on its flash point, which is the temperature at which it gives off enough vapor to form an ignitable mixture with air. When a liquid reaches its flash point, contact with any source of ignition will cause the vapor to burst into flame.

Flash Points—OSHA and the NFPA have defined the limits for combustibility and flammability as follows:

- A combustible liquid has a flash point at or above 100°F (37.8°C).
- A flammable liquid has a flash point below 100°F (37.8°C).
- A flammable liquid can reach its flash point at room temperature, and any unrecognized leak can pose a potential hazard.

NFPA Definitions of Flammable/Combustible Liquids—

1. *NFPA 30 Definitions*—
   - **Flammable Liquids**—Those liquids with a closed-cup flash point below 100°F (37.8°C) and a vapor pressure not exceeding 40 psia (1276 kPa) at 100°F. Refer to NFPA 30 and NFPA 321 for additional information.
   - **Combustible Liquids**—Those liquids with a closed-cup flash point at or above 100°F (37.8°C) but below 200°F (93.3°C). They are divided into three classes (see below). Although they do not ignite as easily as flammable liquids, combustible liquids can be ignited under certain circumstances and thus must be handled with caution.

2. *Classes of Liquids*—
   - **Class I**—Flash point below 100°F (37.8°C) and vapor pressure less than or equal to 40 psia (1276 kPa) at 100°F (37.8°C); subdivided as follows:
     Class IA—Flash point below 73°F (22.8°C) and boiling point below 100°F (37.8°C)
     Class IB—Flash point below 73°F (22.8°C) and boiling point at or above 100°F (37.8°C)
     Class IC—Flash point at or above 73°F (22.8°C) and boiling point below 100°F (37.8°C)
   - **Class II**—Flash point at or above 100°F (37.8°C) and boiling point below 140°F (60°C)

- **Class III**—Flash point at or above 140°F (60°C); subdivided as follows:
  Class IIIA—Flash point at or above 140°F (60°C) and boiling point below 200°F (93.4°C)
  Class IIIB—Flash point at or above 200°F (93.4°C)

3. *DOT Definitions*—The U.S. Department of Transportation, in 49 CFR, Hazardous Materials Regulations, Parts 170–179, defines a flammable liquid as any liquid that gives off flammable vapors at or below a temperature of 80°F (26.7°C). This is important because the U.S. DOT Flammable Liquid Label is one means by which containers of flammable liquids can be identified for shipping, receiving, and transportation.

   - **Flash Point**—Flash point is the minimum temperature at which a liquid gives off vapor in a concentration high enough to form an ignitable mixture with air near the surface of the liquid within a vessel specified by appropriate test procedure and apparatus. Although other properties influence the relative hazards of flammable liquids, the flash point is the most significant factor. The relative hazard increases as the flash point decreases. The significance of this property becomes more apparent when liquids of different flash points are compared. When heated to a temperature at or above its flash point, any combustible liquid will produce ignitable vapors.
   - **Auto-Ignition Temperature**—Auto-ignition temperature is the minimum temperature at which a flammable gas–air or vapor–air mixture will ignite from its own heat source or a contacted heated surface without an open spark or flame. Vapors and gases will spontaneously ignite at a lower temperature in oxygen than in air. The auto-ignition temperature can be lowered by the presence of catalytic substances.
   - **Flammable Limits**—The minimum concentration of vapor or gas in air below which propagation of flame does not occur on contact with a source of ignition is known as the lower flammable limit (LFL). The maximum proportion of vapor or gas in air above which propagation of flame does not occur is known as the upper flammable limit (UFL).
   - **Flammable Range**—The flammable range lies between the lower and upper flammable limits, expressed in terms of percentage of vapor or gas in air by volume. It is sometimes referred to as the explosive range.

4. *General Safety Precautions*—
   - The transfer of flammable or combustible liquids from bulk stock containers to smaller containers must be done in a storage room as described in NFPA 30 or within a fume hood that has a face velocity of at least 100 ft/min (30.5 m/min).
   - Spills of flammable and combustible liquids must be cleaned up promptly. Cleanup personnel should use appropriate personal protective equipment. If a major spill occurs, all ignition sources should be removed and the area ventilated. Such liquids should never be allowed to enter a confined space such as a sewer because explosion is possible.
   - Flammable or combustible liquid storage and use containers must meet approval standards in NFPA 30. Flammable liquids must be kept in closed containers as required in 29 CFR 1910.106.

- Combustible waste material such as oily rags and paint rags must be stored in covered metal containers and disposed of daily per 29 CFR 1910.106.
- Storage areas must be posted as no smoking areas as required in 29 CFR 1910.106.
- Piping systems, including tubing, flanges, bolting, gaskets, valves, fittings, and all pressurized parts containing flammable and combustible liquids, must meet the requirements of NFPA 30.

5. *Caution Signs—*
   - Hydrogen storage locations must be permanently marked as follows: HYDROGEN—FLAMMABLE GAS—NO SMOKING—NO OPEN FLAMES.
   - Liquefied containers must be marked as follows: LIQUEFIED HYDROGEN—FLAMMABLE GAS.
   - Bulk oxygen storage locations must be permanently marked as follows: OXYGEN—NO SMOKING—NO OPEN FLAMES.
   - Storage cabinets should be labeled as follows: FLAMMABLE—KEEP FLAME AWAY.

6. *NFPA Metal Container Requirements—*
   - Metal cabinets must be constructed of sheet steel that is at least No. 18 gauge. They must be double-walled with a 1.5-inch (38.1-mm) air space.
   - Cabinets must have joints that have been riveted, welded, or otherwise made tight.
   - Doors must have a three-point latch arrangement, and the sill must be at least 2 inches (50.8 mm) above the bottom of the cabinet.

7. *Inside Storage Areas—*Each inside storage area should be prominently marked as a no smoking area. The NFPA National Fire Codes detail requirements for inside storage areas for flammable and combustible liquids, including the following:
   - Openings to rooms or buildings must be provided with non-combustible, liquid-tight, raised sills or ramps that are at least four inches high or are otherwise designed to prevent the flow of liquids to adjoining areas.
   - A permissible alternative to a sill or ramp is an open-grated trench that spans the width of the opening inside the room and drains to a safe location.
   - General exhaust ventilation, either mechanical or gravity, is required.
   - Electrical wiring and equipment used inside the rooms must conform to the requirements in NFPA 70, National Electrical Code. A fire extinguisher must be available.

8. *Outside Storage Requirements—*If flammable/combustible liquids are stored outside the storage area, the area must be graded to divert spills from buildings or surrounded by a curb at least 6 inches (152.4 mm) high.
   - Storage areas should be posted as no smoking and kept free of weeds, debris, and other combustible material.
   - A fire extinguisher should be available at the storage area.
   - Storage areas for liquid propane gas tanks should be posted as no smoking areas.

J. COMPRESSED GASES

Compressed Gas Safety—

1. *General Safety Guidelines*—
   - Storage areas for compressed gas cylinders should be well ventilated, fire-proof, and dry.
   - Compressed gas cylinders should never be subjected to temperatures higher than 125°F.
   - Cylinders should not be stored near steam pipes, hot water pipes, boilers, highly flammable solvents, combustible wastes, unprotected electrical connections, open flames, or other potential sources of heat or ignition.
   - Cylinders should be properly labeled.
   - The valve protection cap should not be removed until the cylinder is secured and ready for use.
   - Workers responsible for transferring, handling, storing, or using compressed gases should review the requirements in 29 CFR 1910.101–1910.105 and 49 CFR 171–179.
   - Workers should also consult the National Fire Codes and any applicable state or local regulations.
   - If a cylinder leaks and the leak cannot be fixed simply by tightening a valve gland or packing nut, close the valve, tag the cylinder unserviceable, and move it to a well-ventilated location.
   - Consult ANSI-Z48.1 and Compressed Gas Association (CGA) Pamphlet C-7 for information on marking cylinders.
   - Before returning empty cylinders, close the valve and ensure that cylinder valve protective caps and outlets plugs are replaced.
   - Never use cylinders as rollers, supports, or for any purpose other than to contain the content as received.
   - Never place cylinders in an area where they might become part of an electrical circuit.

2. *Moving Cylinders*—
   - Do not lift cylinders by the cap.
   - Never drop cylinders or allow them to violently strike against each other or other surfaces.
   - Avoid dragging or sliding cylinders.
   - Use a suitable hand truck, forklift, or roll platform for transporting and unloading cylinders.

3. *Storing Cylinders*—
   - Post cylinder storage areas with the names of the gases to be stored.
   - Flammable gases should not be stored near oxidizing gases.
   - Charged and empty cylinders should be stored separately.
   - Storage rooms should be cool, dry, and well ventilated.
   - Cylinders should not be stored at temperatures above 125°F or near heat sources.

- Cylinders should not be exposed to dampness or other corrosive chemicals or fumes.
- Cylinders should be protected from any object that will produce a cut or other abrasion in the surface of the metal.
- Cylinders may be stored in the sun except in localities where extreme temperatures prevail.
- If ice or snow accumulates on a cylinder, thaw at room temperature or use water not exceeding 125°F.
- Cylinders should be protected against tampering by unauthorized individuals.

4. *Withdrawing Contents from Cylinders—*
   - Compressed gases should be handled only by trained persons.
   - Ensure that all cylinders are properly labeled when received from supplier.
   - Removable-type valve protective caps should remain in place until ready to withdraw contents or to connect to a manifold.
   - Before using a cylinder, ensure it is properly supported to prevent it from falling.
   - Use suitable pressure-regulating devices in situations where gas is admitted to systems that have pressure rating limitations lower than the cylinder pressure.
   - Never force connections that do not fit. Ensure that threads on regulator connections are the same as those on the cylinder valve outlet.
   - Regulators, gauges, hoses, and other appliances provided for use with a particular gas must not be used on cylinders that contain gases with different chemical properties unless approved by the supplier.
   - Open cylinder valves slowly, and point the valve opening away from yourself and other persons.
   - Never use wrenches or tools other than those provided or approved by the gas manufacturer.
   - For valves that are hard to open because of corrosion, contact the supplier for instructions.
   - Never use compressed gas to dust off clothing; this may result in serious injury to the eyes or body and can create a fire hazard.
   - Close the cylinder valve and release all pressure before removing a regulator.

Flammable Gas Cylinder Safety—Cylinders should not be stored near highly flammable solvents, combustible waste material or similar substances, or near unprotected electrical connections, gas flames, or other sources of ignition.

- Never use a flame to detect flammable gas leaks; always use soapy water.
- Do not store reserve stocks of cylinders that contain flammable gases with cylinders that contain oxygen.
- Inside buildings, stored oxygen and gas cylinders should be separated by a minimum of 20 feet or a fire-resistant partition should be placed between the oxygen and fuel gas cylinders.

Dangers of Smoking and Electrical Fires—An effective and ongoing program to educate the staff about the hazards of smoking and electrical fires can help reduce these risks.

1. *Risk Reduction Efforts—*
   - Patients should be informed about the dangers of smoking when admitted and should be reminded frequently.
   - Some states prohibit ambulatory patients from smoking in bed and require that bedridden patients be supervised by either staff of family members while smoking.
   - The use of oxygen in patient areas is another obvious fire hazard. Fires can occur in an oxygen-enriched atmosphere because of patient smoking, electrical malfunctions, and the use of flammable liquids. Procedures should be developed and strictly enforced to prevent fire hazards in patient areas where oxygen is used.
   - The basic code for fire safety is the NFPA Life Safety Code (NFPA 101).
   - Many municipal, state, and federal agencies and non-government organizations have also issued regulations, codes, and recommendations for fire safety.
   - Fire drills should be held regularly and should include training to operate fire extinguishers, locate alarms and identify their codes, assign responsibilities for patient safety, and locate exits.

2. *Electrical Equipment Installations—*Electrical equipment should be maintained and installed to meet NFPA 70, National Electrical Code (NEC).
   - Use only Underwriters Laboratories (UL)-listed or Factory Mutual (FM)-approved equipment where flammable gases/vapors may be present.
   - Temporary or makeshift wiring, particularly if defective or overloaded, is a common cause of electrical fires and should not be used.
   - Overloaded or partially grounded wiring may heat up enough to ignite combustibles without blowing fuses or tripping circuit breakers.
   - Where flammable liquids are used, bonding and grounding with adequate and true grounds in accordance with the NEC should be provided.

K. UNDERSTANDING FIRES

Classes of Fire—

- *Class A*—For fires that involve ordinary combustible materials, such as wood, paper, or clothing, where the quenching and cooling effects of water are most effective, use a pressurized water extinguisher or A:B:C type dry powder extinguisher.
- *Class B*—For fires that involve flammable liquids, greases, and similar materials, use type B:C or A:B:C dry powder extinguishers. Carbon dioxide (CO_2) extinguishers may also be used.
- *Class C*—For fires in or near energized electrical equipment where the use of a non-conductive extinguishing agent is of first importance, use CO_2, or dry powder (B:C or A:B:C). **Water must never be used.** *Note*—Halon extinguishers have been declared an environmental danger and are no longer being produced.
- *Class D*—For fires that occur in combustible metals such as magnesium, titanium, lithium, sodium, or potassium, use special extinguishers filled with graph-

ite or some other special extinguishing agent. These fire extinguishers may be red or yellow. Sand, sodium chloride, and dry soda ash are also effective for extinguishing molten sodium and potassium.

- *Other Fires*—Fires that involve combustible metals or certain hazardous chemicals may require special extinguishing agents and/or techniques. Consult NFPA 49, Hazardous Chemicals Data or NFPA 325M, Fire Hazard Properties of Flammable Liquids, Gases, and Volatile Solids for additional information.

Fire Development Stages—Fire is a chemical combustion process created by the rapid combination of fuel, oxygen, and heat. Most fires develop in four distinct stages, as described below:

- *Incipient Stage*—No visible smoke, flame, or significant heat develops, but a large amount of combustion particles is generated over time. These particles, created by chemical decomposition, have weight and mass but are too small to be visible to the human eye. They behave according to gas laws and quickly rise to the ceiling. Ionization detectors respond to these particles.
- *Smoldering Stage*—As the incipient stage continues, the combustion particles increase until they become visible, a condition called smoke. No flame or significant heat has developed. Photoelectric detectors "see" visible smoke.
- *Flame Stage*—As the fire condition develops further, ignition occurs and flames start. The level of visible smoke decreases and the heat level increases. Infrared energy is given off, which can be picked up by infrared detectors.
- *Heat Stage*—At this point, large amounts of heat, flame, smoke, and toxic gases are produced. This stage develops very quickly, usually in seconds. Thermal detectors respond to heat energy.

Fire Confinement—Regardless of the type of building construction, stair enclosures are necessary to provide a safe exit path for occupants. They also retard the upward spread of fire. Confinement measures include the following:

- Divide a building so as to break up the total area into small cells. Fires will remain localized and can be more easily suppressed.
- To prevent fire from spreading from one unit to another, various building codes require that units be made structurally sound enough to withstand full fire exposure without major damage and that the boundaries of units be capable of acting as non-conducting heat barriers.
- Fire doors are a widely accepted means of protection. They are rated by testing laboratories and usually have a rating of 45 minutes to 3 hours. They may be constructed of metal or metal-clad treated wood materials and may be hinged, rolling (sliding), or curtain doors. Single or double doors may be specified.
- To assure proper protection of openings, fire doors should be installed in accordance with NFPA 80, Fire Doors and Windows.

Controlling Smoke—Smoke and gases generated by an uncontrolled fire can seriously impair fire-fighting operations and spread the fire under the roof for considerable distances from the point of origin. The movement of smoke within a structure is affected by many factors, including building height, ceiling heights, suspended ceilings, venting, external wind force, and direction of the wind.

- One way to control smoke is to use a physical barrier, such as a door, wall, or damper, to block the movement of the smoke. An alternative method is to use a pressure differential between the smoke-filled area and the protected area.
- Venting is another way to remove smoke, heat, and gases from a building. Vents are also useful in windowless and underground buildings. They are not, however, a substitute for automatic sprinkler protection. Many variables affect the burning of combustible material, and a formula can be used to compute the amount of venting required.

Fire Exits—Management and others entrusted with the safety of patients and employees must consider many problems when planning the emergency evacuation of buildings. Population of the building and degree of hazard are the major factors when designing exits. Designing exits involves more than studying flow rate and population density.

- Safe exits also require a safe path of escape from the fire. NFPA 101, Life Safety Code provides a reasonable and comprehensive guide for exit requirements.
- Exits and other safeguards should not be designed so as to depend solely on any single safeguard.
- Exit doors must withstand fire and smoke during the length of time for which they are designed to be in use.
- Alternative exits and pathways should be provided in case one exit is blocked by fire.
- Exits should have adequate lighting.
- Exits and access to exits should be marked with readily visible signs.
- To protect exiting personnel, equipment and areas of any unusual hazard that might spread fire and smoke should be safeguarded.

L. FIRE EXTINGUISHERS

Two basic types of equipment are used to fight and control fires. The first is fixed systems, which includes automatic sprinklers, standpipe hoses, and various pipe systems. Fixed systems must be supplemented by portable-type extinguishers. The second type is portable extinguishers (Table 5.2).

1. *Fire Extinguisher Requirements—*
   - Approved by an approved testing lab such as FM or UL
   - The appropriate type for class of fire that could occur in the area
   - Of sufficient size and/or quantity to protect the area
   - Accessible for immediate use and easy to reach
   - Inspected and maintained in top operating condition
   - Only used by trained personnel

TABLE 5.2   Portable Fire Extinguisher Types

Class A	Fires that involve ordinary combustible materials, such as wood, paper, cloth, rubber, and many plastics, where the quenching and cooling effects of water or solutions that contain large percentages of water are of prime importance.
Class B	Fires that involve flammable liquids, greases, oils, tars, oil-base paints, lacquers, and similar materials, where smothering or exclusion of air and interrupting the chemical chain reaction are most effective. This class also includes flammable gases.
Class C	Fires in or near live electrical equipment, where the use of a non-conductive extinguishing agent is of first importance. The material that is burning is, however, either Class A or Class B in nature.
Class D	Fires that occur in combustible metals, such as magnesium, lithium, and sodium. Special extinguishing agents and techniques are needed for fires of this type.

2. *Marking Extinguishers*—The recommendations in NFPA 10 are a guide for marking extinguishers and/or extinguisher locations. Extinguishers suitable for more than one class of fire may be identified by multiple symbols. Decals or paint can be used to ensure legibility and durability. Extinguishers should be marked on the front, in a size and form that can be easily read at a distance of three feet. An easily recognizable picture–symbol label has been devised by the National Association of Fire Equipment Distributors and is now recommended in NFPA 10.

3. *Extinguisher Ratings*—
   - Additional information on the label refers to the water equivalency rating and the square footage that the extinguisher can handle if operated by someone properly trained. A 4A extinguisher would be equivalent to five gallons of water. The B/C rating indicates square footage, such as 25 B:C.
   - There is no rating for Class C or Class D fires. OSHA requires fire extinguishers to be selected and placed based on the class and size of fires anticipated in the work area. OSHA provides guidance on classes of fires and travel distance to an extinguisher. Local fire codes may be more stringent.

4. *Classes of Fire Extinguishers*—
   - Extinguishers suitable for Class A fires should be identified by a triangle containing the letter A. If colored, the triangle should be green.
   - Extinguishers suitable for Class B fires should be identified by a square containing the letter B. If colored, the square should be red.
   - Extinguishers suitable for Class C fires should be identified by a circle containing the letter C. If colored, the circle should be blue.
   - Extinguishers suitable for fires involving metals should be identified by a five-pointed star containing the letter D.

5. *Travel Distances*—
   - Class A must have a travel distance of 75 feet or less. Facilities may meet this requirement with uniformly spaced standpipe systems or hose stations in-

stalled for emergency use. These systems must meet the requirements of 29 CFR 1910.158 or 1910.159.
- Class B must have a travel distance of 50 feet or less.
- Class C travel distance is based on appropriate A or B hazard.
- Class D travel distance must be 75 feet.

6. *Monthly Inspections*—Fire extinguishers must be visually inspected monthly to ensure that:
   - Extinguishers are located in proper locations
   - Extinguishers are not blocked or hidden
   - Extinguishers are mounted according to NFPA Standard 10
   - Extinguisher pressure gauges show adequate pressure
   - Pins and seals are in place
   - Extinguishers show no evidence of damage and nozzles are free of blockage

7. *Maintenance*—Many people do not understand the term maintenance as related to portable fire extinguishers. Maintenance means the complete disassembly and inspection of each part. Hydrostatic testing must be performed by qualified personnel to ensure that failure due to corrosion or abuse has not occurred. Table 1 in 29 CFR 1910.157 provides test intervals for various types of extinguishers.

M. NFPA 101, LIFE SAFETY CODE

Life Safety—The Joint Commission's Life Safety Code (LSC) requirements reference NFPA 101. Life Safety began in 1963 with the publication of the Building Exits Code by the NFPA. The NFPA published the first edition of the LSC in 1966. The Joint Commission uses LSC guidelines because it is the only nationally accepted and uniformly applied code in the United States. Building codes provide design criteria, but the LSC only addresses the general requirements for fire protection and systems safety necessary to assure the safety of building occupants during a fire. The LSC only specifies minimum hourly fire resistance ratings and does not specify how such ratings are to be achieved.

Life Safety Code—The code contains five basic sections or categories published in 32 chapters:

1. *Chapters 1–4*—These chapters address the management of and ways to administer or use the code.
2. *Chapters 5–7*—These chapters cover exit requirements, fire protection, facility service equipment, and fire protection equipment and apply to all types of occupancies.
3. *Chapters 8–30*—These chapters cover the various occupancies covered under the code and the protection requirements for occupancy-related hazards.
4. *Chapter 31*—This chapter addresses fire prevention plans and procedures. It also contains information on drills, furnishings, and smoking policies.
5. *Chapter 32*—This chapter added to the 1991 edition of the code catalogs the specific NFPA standards referenced in the code.

Life Safety Requirements—The code has different requirements for the same type of occupancy based on whether the construction is considered new or existing. A facility will reference different chapters of the LSC depending on the date of construction. The Joint Commission Life Safety standard is designed to produce a safe environment so that all patients, personnel, and visitors can be protected from fire. The Joint Commission recognizes three basic types of occupancies:

- *Healthcare*—These are facilities in which people cannot ensure their own safety during an emergency situation. Chapters 12 and 13 cover requirements for hospitals, hospices, and nursing homes.
- *Residential*—These facilities include dormitories and rooming houses. Occupants must be able to ensure their own safety during an emergency. Chapters 16, 17, and 20 cover the various requirements for residential occupancies.
- *Outpatient*—These facilities include physicians' offices and surgery centers that do not provide overnight stays. Chapters 26 and 27 of the LSC address outpatient facilities, including ambulatory healthcare occupancies.

Life Safety Considerations—The standard specifies general elements to consider when devising an effective life safety management program:

- Understanding and adhering to the appropriate edition of the LSC
- Determining a facility's occupancy type in order to apply the standard correctly
- Applying the unit concept to all fire safety programs
- Understanding equivalencies and how they can be used for compliance with the Joint Commission standard
- Developing a correction plan to deal with safety deficiencies
- Ensuring safe access for physically disabled individuals
- Maintaining a level of safety during all phases of construction
- Maintaining safety in emergency care areas
- Ensuring grounds safety, including areas such as swimming pools and gymnasiums

Life Safety Program Development Factors—When designing a life safety management program, the following factors should be considered:

- Type and classification of facility
- Public/private ownership
- Facility location (rural or urban)
- Patient access to the grounds
- Need to control public access to the grounds
- Lighting required for parking lots, garages, and sidewalks
- Any other factors that affect public safety

The Unit Concept—This Joint Commission safety standard focuses on fire containment efforts. The objective of the unit concept is to contain fire and the products of combustion through compartmentation. The concept defines consecutive units of defense for each occupancy type to prevent the spread of fire and to provide a safe means of egress. Units of defense vary depending on whether a facility is classified as a healthcare occupancy, residential occupancy, or business occupancy. Healthcare facility fires are dangerous because rapid evacuation is not always possible. Several units of containment are defined for healthcare occupancies:

1. ***Unit One (Room)***—The room is the main unit of defense because it provides the initial barrier against the spread of fire and smoke. Evacuation of patients is often very difficult. The main objective is to isolate and protect patients in their rooms.
   - Doors should be at least 1³/₄-inch solid bonded wood core or the equivalent and be constructed to resist fire for at least 20 minutes.
   - Latches on corridor doors should ensure tight closure because the air pressure differentials that develop during a fire may cause unlatched doors to swing open.
   - Louvers, undercuts, and other designs that encourage air circulation are not allowed except in bathrooms, shower areas, and maintenance closets.
   - Rooms in new construction must withstand the passage of smoke and fire for at least one hour. Corridor wall construction must resist the passage of smoke and fire for at least one-half hour.

2. ***Unit Two (Smoke Compartment)***—Partitions designed to stop smoke must be used to subdivide floors into compartments. These fire or smoke zones provide a secondary area of refuge against smoke. Doors and other openings in the smoke partitions must be protected. Doors and dampers must close when the fire alarm or smoke detector is activated. This ensures that the walls become complete. Smoke partitions run from outside wall to outside wall and therefore segment corridors into limited areas.

3. ***Unit Three (Floor Assembly)***—This is the third level of defense. Floor assemblies create barriers between stories and resist vertical passage of fire and smoke. Vertical penetration of floors should be designed to withstand the passage of fire and smoke. Openings such as those made for pipes must be sealed. In the case of stairways, elevators, and chutes, vertical partitions must provide protection.

4. ***Unit Four (Building)***—The structure of the building itself is the fourth unit of defense. The entire building must be built to remain structurally sound during a fire. The building design must contain the fire within its bounds for a specified length of time. The height of the building determines the time that the building must withstand the fire. Building construction definitions are explained in detail in NFPA 220, which identifies five types of buildings. A Type I (332) fire-resistant building would have the following characteristics:
   - The first number (3) indicates that the exterior bearing walls have a fire resistance time of three hours.
   - The second number (3) indicates that the girders, beams, columns, and trusses have a fire resistance time of three hours.
   - The final number (2) specifies that the floor and ceiling assemblies have a fire resistance time of two hours.

5. ***Unit Five (Exits)***—Facilities must have a minimum of two approved exits that are remote from each other for each section or floor. The following are approved for healthcare occupancies:
    *   Doors that open or lead directly outside
    *   Class A or Class B interior stairs and ramps
    *   Smoke-proof towers
    *   Outside stairs or fire escapes
    *   Horizontal means of egress or exit
    *   Exit passageways

Life Safety Deficiencies—Fire safety deficiencies are determined on a building-by-building basis and are rated according to how deficiency affects overall fire safety. Deficiencies are classified as Level I, II, and III.

1. ***Joint Commission Considerations***—The Joint Commission considers the following when determining level of deficiency:
    *   Deficiencies involving basic design or building structures
    *   Number of deficiencies in the building
    *   Whether deficiencies pose a threat to life in the event of fire

2. ***Deficiency Levels***—The Joint Commission then decides whether a plan of correction will require more than three years to implement.
    *   **Level I**—A deficiency or series of deficiencies that indicate a lack of proper maintenance of building components which play a role in the unit concept.
    *   **Level II**—A deficiency or series of deficiencies involving one or more unit-concept units that pose a threat to life. The scope of the deficiencies is significant in a limited area.
    *   **Level III**—A deficiency or deficiencies indicating pervasive violation of one or more of the unit-concept units. The scope of the deficiencies is such that correction in less than three years is not possible.
    Note—It is essential that all fire safety deficiencies be corrected as soon as possible.

3. ***Plans for Improvement***—Organizations with life safety deficiencies must develop a written plan and address the following:
    *   All deficiencies to be corrected
    *   Specific actions underway or to be taken to correct the problems
    *   Source, availability, and management commitment for funding corrective actions
    *   A schedule or time line for correcting all deficiencies

Interim Life Safety Measures—These are detailed in Table 8.3 in the *1995 Plant Technology Safety Management Handbook*. Facilities must use interim measures for all Level III and some Level II deficiencies.

Fire Safety Evaluation System (FSES)—The Center for Fire Research at the National Bureau of Standards, through the support of the U.S. Department of Health and Human Services, has developed the FSES for determining how combinations of fire safety elements can meet the intent of the LSC. This quantitative evaluation system grades fire safety zone by zone in healthcare facilities. The following areas are scored:

- *Containment*—Safety actions taken to control and contain fire and smoke
- *Extinguishment*—Systems and procedures in place to effectively extinguish a fire
- *Moving Occupants*—Safety features and procedures that allow people to safely leave a fire or smoke zone
- *General Fire Safety*—Safety procedures and policies tnat affect the overall safety of the fire and/or smoke zone

Equivalencies—Chapter 1, Section 5 of the LSC specifies that equivalency is an option for meeting compliance. According to the Joint Commission, equivalency is the documented acknowledgment that the intent of a standard has been met in an alternative manner. An equivalency may be developed using the traditional approach or the FSES described above. The traditional approach must consider the following:

- Identification of the deficiency and the intent of the applicable LSC requirement
- A written description of the alternative methods that will be employed to ensure life safety
- Verification in writing by a qualified person, such as a fire protection engineer, registered architect, or local authority, stating that alternative methods meet the intent of LSC requirements

Fire Warning and Safety—The Joint Commission does not require the use of any specific system. Any system that meets NFPA standards and local requirements is acceptable. Each facility must have a manually operated fire system which is electrically supervised. The system must automatically transmit an alarm to the fire department. The local fire department must also be notified by other means when the alarm has been activated.

1. *Fire Alarms*—NFPA 101 suggests four methods for transmitting fire alarms to the local fire department:
   - Auxiliary Protective Signaling Systems (NFPA 72-1990)
   - Central Station Service (NFPA 71-1989)
   - Proprietary Protective Signaling System (NFPA 72-1990)
   - Remote Station Protective Signaling System (NFPA 72-1990)
   Note—NFPA 72, National Fire Alarm Code should be referenced for specific fire alarm requirements.

2. *General Requirements*—
   - Fire alarms must be received at a central location within the facility.
   - The supervised location must be continuously manned.
   - The supervised location must be adequately protected as a hazardous area in accordance with NFPA 101.
   - The supervised location must have the necessary equipment to receive fire alarm signals.
   - A record of all system status changes must be maintained at the receiving location.

- Signals received must be transmitted at once to the local fire department.
- A copy of the master fire plan and other emergency planning documents must be available at the supervised location.

3. *Manual Alarm Stations*—Manual alarm stations should be located throughout the facility. They should be positioned so that no more than 200 feet must be traveled to reach an alarm station on the same floor.

4. *Alarm Levels*—Audible alarms must be designed to exceed the level of any operational noise. Visual alarms accompanying the audible alarm are recommended.

5. *Electrically Supervised Systems*—System components must be monitored to ensure personnel are aware when a part of the system needs repair. The system should signal trouble when:
   - A break or ground fault prohibits normal system operation
   - The main power source fails
   - A break occurs in the circuit wiring

N. NFPA 99, HEALTHCARE FACILITIES

NFPA 99, along with NFPA 101, Life Safety Code, establish criteria for safeguarding patients/residents and employees from fire, electrical, and explosion hazards. The current NFPA 99 was developed from 12 independent documents and first published in 1987. NFPA 99 addresses fire, explosion, and electrical hazards found in healthcare facilities, including information on performance, maintenance, testing, and safety practices for material, equipment, and appliances. Risk managers and safety officers should pay attention to the following references:

- General Disaster Planning　　　　　　　Annex 1
- Respiratory Therapy Fire Response　　　Appendix C-8.3
- Anesthetizing Locations Fire Response　Appendix C-12.4
- Hyperbaric Chambers Fire Response　　　Appendix C-19.2 and 19.3
- Fire Incidents in Laboratories　　　　　Appendix C-10.1

O. FIRE-EXTINGUISHING SYSTEMS

Automatic Systems—Automatic fire-extinguishing systems should not create situations that could endanger the lives or safety of patients, personnel, or visitors. Systems should be designed to meet extinguishing needs while considering occupancy. The Joint Commission does not require that fire pumps be installed in sprinkler systems or that buildings be protected by standpipes or fire hose systems. If these systems are installed, they must be properly maintained and inspected.

Portable Systems—An adequate number of portable extinguishers, appropriate for the types of areas being protected, are to be installed throughout a facility. Extinguishers should be inspected quarterly and maintained according to manufacturers' requirements. Extinguishers should be located no more than 75 feet apart.

Special Requirements for Cooking Areas—Unique hazards found in cooking areas may require the use of special systems. Approved systems to protect cooking surfaces, exhaust hoods, and ducts for commercial cooking equipment are as follows:

- Automatic carbon dioxide systems
- Automatic dry chemical system
- Automatic foam, water, or wet chemical system
- Automatic sprinkler systems approved in NFPA 13

Inspection—All components of a fire alarm system must undergo regular inspection and testing. All systems should have a visual inspection each quarter. Each automatic system should be tested and/or inspected on an annual basis. Systems should also be a part of the facility's preventive program.

Fire Loads—Facilities must implement management policies to help control the amount of material that will burn. The items listed below should be considered when evaluating fire hazards located in a facility. There should always be a balance between the needs of the facility and fire safety considerations.

- Flame-resistant materials
- Location and placement of equipment, furnishings, and bedding
- Substituting non-combustible materials for combustible materials
- Procedures for maintaining flame-resistant coatings and coverings

SUMMARY

Emergency management and planning involves an entire healthcare organization. This chapter covered the basic concepts and principles of emergency management, life safety, and fire control. Other topics presented included planning for external disasters and weather-related disasters. Information on sabotage, terrorism, and bomb threats was also provided. Additional topics covered included understanding fires, flammable/combustible liquids, compressed gases, and evacuation procedures. The chapter provided readers an overview of NFPA 99, Healthcare Facilities and NFPA 101, Life Safety Code. Understanding the information, concepts, and principles presented in this chapter will help healthcare organizations reduce confusion and continue to provide quality care during emergency situations.

REVIEW QUESTIONS

1 The *main* purpose of an effective emergency action plan is to:
   a. reduce confusion and allow the organization to operate normally
   b. provide detailed requirements for every conceivable contingency

    c.   provide an environment of predictable behavior for all staff

    d.   meet OSHA and Joint Commission requirements

2   Which of the following statements concerning healthcare emergency planning is true?

    a.   the emergency plan should never address environmental issues

    b.   OSHA does not have a standard that covers emergency action plans

    c.   emergency drills should be conducted semi-annually by all facilities

    d.   the Joint Commission does not place much emphasis on emergency planning

3   A successful disaster plan should focus on which of the following planning items?

    a.   facility preparation

    b.   worker and staff readiness

    c.   patient control measures

    d.   all of the above

4   The National Weather Service issues a severe thunderstorm warning when which of the following conditions exists?

    a.   damaging winds of greater than 50 mph

    b.   hail $3/4$ inch or larger

    c.   heavy rain

    d.   a or b

5   The community risk management program that addresses hazardous substance releases/spills must include:

    a.   a hazard assessment study

    b.   a release prevention plan

    c.   emergency response procedures

    d.   all of the above

6   Healthcare facilities must perform _____ inspections for selected fire zones.

    a.   monthly

    b.   quarterly

    c.   semi-annual

    d.   annual

7   According to the NFPA, one-third of hospital fires originate in:

    a.   storage areas

    b.   machinery and boiler areas

    c.   kitchen and laundry areas

    d.   patient areas and staff quarters

8   To be classified as flammable, a liquid must have a flash point:

    a.   above 200°F

    b.   below 200°F

    c.   above 100°F

    d.   below 100°F

9 The minimum temperature at which a liquid gives off vapor in sufficient concentration to form an ignitable mixture is the definition of:
a. flash point
b. flammable limits
c. auto-ignition temperature
d. flammable range

10 Flammable and combustible storage containers must meet approval standards found in:
a. NFPA 13
b. NFPA 30
c. NFPA 90A
d. NFPA 99

11 Compressed gas cylinders should never be exposed to temperatures higher than:
a. 100°F
b. 125°F
c. 150°F
d. 200°F

12 Oxygen and other compressed gas cylinders that are stored inside should be:
a. separated by at least 10 feet
b. separated by at least 20 feet
c. stored with a fire-resistant partition separating the cylinders
d. b or c

13 A fire located in or near energized electrical equipment would be a Class ___ fire.
a. A
b. B
c. C
d. D

14 Class A fire protection and extinguishing systems (extinguishers, hoses, etc.) must have a travel distance of no more than _____ feet.
a. 50
b. 75
c. 100
d. 150

15 The three basic types of life safety occupancies recognized by the Joint Commission are:
a. healthcare, residential, and outpatient
b. nursing home, ambulatory, and acute care
c. long-term, sub-acute, and acute care
d. nursing home, home health, and acute care

Answers: 1–a   2–c   3–d   4–d   5–d   6–b   7–d   8–d   9–a   10–b   11–b
            12–d   13–c   14–b   15–a

CHAPTER **6**

GENERAL AND PHYSICAL PLANT SAFETY

A. FACILITY FIRE PREVENTION

Fire Prevention—The facility must have an active fire prevention and inspection program. The plan must direct at least quarterly fire drills for all shifts. Written report evaluation must be provided for each drill performed. An on-the-scene critique of each drill should emphasize the performance of personnel taking part and assess future training needs. A quarterly fire inspection must be performed for each fire zone. The inspection should emphasize:

- Assessment of all equipment
- Testing of alarms, detectors, and pull stations
- Evaluation of housekeeping practices and security alarms
- Inspection of sprinkler pressure
- Water availability and hydrant operation
- Annual suppression, detection, and activation systems checks
- Coordination with fire marshal and engineering/maintenance department

Note—Fire prevention and life safety are covered more fully in Chapter 5.

B. HEALTHCARE ERGONOMICS

Healthcare workers rank very high in incidence of back-related injuries. Many injuries are the result of long-term wear and tear on the back, which can lead to disc degeneration and nerve damage. Muscles and ligaments can also become damaged through continuous or repetitive pulling, tugging, or pushing.

Healthcare Ergonomics—The word ergonomics has a very broad meaning and has several definitions related to body motion in a work environment. Ergonomics is derived from the Greek words *ergo,* which means work, and *nomos,* which means law. It

159

TABLE 6.1   Ergonomic Recommendations for Healthcare Facilities

Equipment	Important Features	Recommendations
Mechanical lifts	Ease/speed of use Availability Patient comfort Number of slings	Less than two minutes to complete transfer At least one per wing Specific to task and patient condition At least two per applicable patient
Gait belts	Well-designed handles Availability	Four inches long, cylindrical One for each nurse or team of two
Wheelchairs	Easily accessible Easily secured	Drop or swivel arms Swing-away footrests Wheel locks
Geri-chairs	Easily accessible Easily secured	Drop or swivel arms Secures in upright position Wheel locks
Shower/toilet chairs	Easily accessible Easily secured	Drop or swivel arms Swing-away footrests Wheel locks Wall-mounted brackets
Bath lifts	Easily accessible Patient comfort/security	Drop or swivel arm Detachable, mobile chair/stretcher Ability to screen off area Contoured seat with belt
Scales	Easily accessible Part of other equipment	Recessed or ramped surface Drop or swivel arm chair Patient lift, bath lift, tub
Beds	Height adjustability Ease/speed of adjustment Easily accessible	20 to 43 inches high <20 seconds low–high cycle time (electric) Conveniently located controls Fold-down side rails

Adapted from information in "A Perfect Fit," by Michael L. Rienerth, ergonomics consultant, Ohio Division of Safety & Hygiene, printed in *Ohio Monitor*, March/April 1995.

literally means a study of the customs, habits, or laws of work. Table 6.1 provides some basic ergonomic recommendations for healthcare facilities.

The basic premise of ergonomics is that if a person is mismatched with his or her job or environment, then production and/or quality may suffer. Ergonomics was born as a broad discipline during World War II. Teams of specialists in the fields of physiology, psychology, medicine, and engineering provided advice on flying aircraft, surviving arctic waters, and using anti-aircraft weaponry. Some other common definitions include:

- The study of human characteristics for the appropriate design of living and work environments
- The study of the interface between people and equipment in their environments at home, at work, and in places of leisure

- The study of human physiology and behavior to design work processes or stations that meet the capabilities and limitations of human operators
- The science of work that can be described by other terms such as human factors, human engineering, human performance engineering, and human hardware systems

Purpose of Ergonomic Programs—There are several reasons why organizations should consider ergonomics:

- Improve system designs to eliminate a problem
- Focus on structural response and human adaptation to stress through biomechanical analysis
- Prevention to preclude worker development of cumulative trauma disorders
- The Occupational Safety and Health Administration (OSHA) considers cumulative trauma disorders (CTDs) and repetitive motion injuries (RMIs) to be illnesses that develop over time

Cumulative Trauma Disorders—CTDs are being reported more frequently in healthcare facilities. The Bureau of Labor Statistics reports that hospitals experienced more than 4000 cases of CTDs in 1992. In non-industrial environments, only grocery stores reported more cases. Increased publicity and a greater understanding of CTDs probably account for the increase. However, CTDs are a real problem for repetitive-type work. Many of the back-related injuries in healthcare environments could correctly be classified as CTDs. Recently hired workers and those working in new positions are especially vulnerable. Employees transferred to new job positions and those working overtime can also be candidates for CTDs. Workers with increased body weight or those prone to retain fluids seem to have more complaints.

1. *New ANSI Standard*—The American National Standards Institute (ANSI) consensus (draft) standard for CTDs focuses primarily on the upper extremities. The draft standard outlines several important elements for controlling these exposures:
   - Identification of exposed or affected workers
   - Assessment of work-related risk factors
   - Evaluation of tasks to develop intervention procedures
   - Implementation of medical and workplace interventions
   - Identification of job functions for compliance with the Americans with Disabilities Act
   - Evaluation of job factors for return-to-work programs
   - Identification of job functions for post-offer testing
   - Studies to determine body parts subjected to stresses
   - Creation of injury risk profiles for each job
   - Providing information for creating task-specific exercises

2. *World Health Organization Ergonomic-Related Definitions*—
   - Anatomy has two major divisions: anthropometry, or the study of dimensions of the body, and biomechanics, which is concerned with the application of forces.

- Skill psychology is the processing of information and making decisions.
- Occupational psychology is concerned with training, effort, and individual differences.
- Psychological factors include attitude and motivation.
- Somatic factors include age, sex, health, and size.
- Learning factors are training, knowledge, and adaptability.
- Physiological factors include fuel intake and storage and pulmonary and cardiac functions.
- Environment includes heat, cold, noise, vibration, altitude, and pollution.
- Nature of work includes intensity, duration, rhythm, technique, and position.

Managing Ergonomic Hazards—OSHA is considering a broad performance-oriented standard for ergonomics. The standard would cover work-related musculoskeletal disorders of the back and upper and lower extremities. Some elements of a proposed statement would include:

1. *Ergonomic Assessments*—The following situations require an ergonomic assessment:
   - Accomplishing the same motion or motion pattern every few seconds for a period exceeding two hours
   - Fixed or awkward work posture for more than two hours
   - Use of vibrating or impact tools for more than two hours
   - Manual handling of objects weighing more than 25 pounds more than once each work shift
   - Work that is mechanically or electronically paced for more than four hours at a time

2. *Medical Management*—Medical management programs may help eliminate or reduce development of ergonomic-related problems. The goal should be early identification, evaluation, and treatment of problems. Elements of a medical management program include:
   - Accurate recording of occupational illnesses and injuries to permit trends to be identified and problems addressed
   - Establishment of procedures for managing work-related illnesses and injuries, including guidelines for evaluating and responding to employee symptoms
   - Education of employees to ensure early identification and reporting of cumulative trauma
   - Use of annual surveys to determine employee work-related disorders and the location, frequency, and duration of discomfort
   - A surveillance program where workers are given a baseline health assessment against which subsequent periodic health evaluations are compared in order to identify and correct hazards in the workplace
   - A health surveillance program that focuses on musculoskeletal problems relating to the back, the hands or wrists, or other body parts, depending upon the worker's exposure, to help employees identify changes in health status

3. *Employee Education*—Training programs can go a long way toward increasing safety awareness among both managers and employees.

- Training and education can ensure employees are sufficiently informed about workplace hazards.
- Suggestions from workers who are educated about ergonomic hazards can be helpful in designing improved work practices.
- A good ergonomics training program can teach employees how to properly use equipment, tools, and machine controls.

Workstation Evaluations—Workstation evaluations should assess prolonged work in any posture may result in CTD or RMI. Healthcare organizations should assess offices, computer usage areas, and nursing stations. Basic considerations include evaluating force, duration, position, frequency, and metabolic expenditure. Lighting is also a key ergonomic complaint.

- The most significant factor in the ergonomic equation is usually position.
- Work area dimensions should take into consideration whether a tall person has enough room and whether an extremely short person can reach everything to accomplish the work.
- Work should be within normal arm and/or leg reaches. Workers should be provided with good chairs that have arm/leg rests if required.
- Workers should be at workstations where their posture can be varied. Sufficient space for knees and feet must be provided.
- The distance between the eyes and the work should not cause eye strain or an unusual work position (posture).

Symptoms of Ergonomic Problems—Signs of ergonomic problems include reporting pain, tingling, numbness, swelling, and other body discomforts. The affected areas could be the back, shoulders, and neck, although hand, wrist, and arm problems are the most common ergonomic complaint.

1. *Analyzing Problems and Complaints—*
   - Employers should analyze trends, absenteeism, and turnover rates.
   - A change in a workstation and a decrease in quality is also a good indicator that there may be a problem.
   - Repetitive tasks that require awkward positions or excessive grasping, reaching, or pinching can lead to worker complaints.
   - Extreme temperatures and vibrating tools are strong contributing factors.
   - Material handling tasks including lifting from level, lifting more than 30 pounds repeatedly, twisting the body to lift, lifting above shoulder height, and pulling/pushing heavy loads are all important factors when evaluating ergonomic problems.
   - Lighting considerations are often overlooked during evaluations. Lighting should be adequate (not excessive) for the task. Employees should be able to read and accomplish work easily.
   - Additional light should be added to work areas where workers complain. Lighting should always be direct or reflect down.
   - Computer video screens and display terminals with tilt monitors that minimize glare should be installed. Screen and work distances should be adjustable to individual workers.

- Proper height, adjustable keyboards, and correct seating are important for all operators. Evaluations should focus on worker position.

2. *Neutral Position*—The body is relaxed with arms loose, wrists straight, elbows close to the body, with neck and spine straight.
   - A standing workstation should have an anti-fatigue mat, work surface below the elbows, and a footrest so the worker can elevate one foot.
   - A sitting station should have a surface 18 inches wide and rounded in the front. The upholstery should be firm and made of woven fabric.
   - Chairs should allow unrestricted movement, allow for the knees to bend, be adjustable, have support for the lower back, and be equipped with footrests.

Types of CTDs—The most common disorder is tenosynovitis, which is inflammation of the tendons and their sheaths. It often occurs at the wrist and is associated with extreme wrist movement from side to side. Other conditions caused by repetitive tasks include:

- *DeQuervain's Disease*—The tendon sheath of both the long and the short abductor muscles of the thumb narrows, causing wrist deviation. It is common among workers who perform repetitive manual tasks that involve inward hand motion and a firm grip.
- *Trigger Finger*—A condition caused by any finger being frequently flexed against resistance.
- *Tendinitis*—A condition where the muscle–tendon junction becomes inflamed due to repeated abduction of a body member away from the member to which it is attached.
- *Tennis Elbow*—This form of tendinitis is an inflammatory reaction of tissues in the elbow region caused by palm-upward hand motion against resistance such as the violent upward extension of the wrist with the palm down.
- *Carpal Tunnel Syndrome*—A common affliction caused by the compression of the median nerve in the carpal tunnel. It is often characterized by tingling, pain, or numbness in the thumb and first three fingers. It is often associated with repeated wrist flexion.
- *Raynaud's Syndrome*—A condition where the blood vessels in the hand constrict from cold temperature, vibration, emotion, or unknown causes. It is easily confused with the one-sided numbness of carpal tunnel syndrome.

C. OFFICE SAFETY

Office areas are frequently overlooked during health and safety inspections in hospitals and other healthcare facilities.

1. *General Safety Rules*—The following guidelines should be included in health and safety programs for officer workers:
   - Desks and countertops should be free of sharp, square corners.
   - Material should be evenly distributed in file cabinets so that the weight of the

upper drawers does not cause the cabinet to fall over. Only one drawer should be opened at a time, and each drawer should be closed immediately after use.
- Papers and other office materials should be properly stored and not stacked on top of filing cabinets.
- Aisles and passageways should be sufficiently wide for easy movement and should be kept clear at all times. Temporary electrical cords and telephone cables that cross aisles should be taped to the floor or covered with material designed to anchor them.
- Electrical equipment should be properly grounded, and the use of extension cords should be discouraged.
- Carpets that bulge or become bunched should be relaid or stretched to prevent tripping hazards.
- Heavy materials should not be stored on high shelves.

2. *Office Hazards*—Electrical cords may cause employees to trip and fall. Use of extension cords should be avoided. Whenever possible, any other cords should be taped to the floor. Anti-slip surfaces should be utilized for all uncarpeted floors. Adequate lighting must be provided. The heating and air conditioning system should be inspected regularly with particular attention to symptoms of sick building syndrome. This condition may lead to chronic health problems, such as colds or other illnesses, among employees. Safety guidelines to consider include:
   - Sound electrical safety measures should be followed.
   - Aisles should be wide enough to permit flow of personnel.
   - Some copying machines use chemicals in toner and other fluids that can endanger employee health.
   - For sedentary or monotonous jobs such as data entry, ergonomic chairs that have casters designed for carpet or tile should be provided.

Video Display Terminal (VDT) Safety—One of the main sources of worker complaints is related to VDT usage. Eyestrain, back pain, neck strain, fatigue, headaches, blurred vision, CTDs, and other arm/hand discomfort have been associated with VDT work.

1. *General VDT Safety Tips*—
   - Workers should take vision breaks about each hour to allow the eye muscles to relax. Glancing at an object at least 20 feet away can be of great help. Some workers get relief by rolling, blinking, or shutting their eyes for a few seconds.
   - The workstation should have a padded keyboard, adjustable table, and tilt screen. This allows the worker to experiment and find a position that is comfortable.
   - Light should be adequate to help alleviate eyestrain and glare. Glare control devices should be used as necessary.
   - Workers should take a minimum of one break for every hour of continuous work to allow the hands, neck, and arms to relax.

- Chairs are individual and should meet an individual worker's needs.
- Chair height is normally correct when the entire sole of the foot can rest on the floor or a footrest and the back of the knee is slightly higher than the seat of the chair.
- Some believe that a seat pan design should be slightly concave with a rounded or waterfall edge. Seat design preference will vary with individual workers.
- Seat backrests should support the entire back. Chairs should be adjustable to fit individual needs.
- Workers should do conditioning exercises for the hands, neck, and shoulders.
- Printers and other accessories should be arranged to prevent the worker from twisting and turning.
- Workers should have a vision test before being assigned to a VDT position. Periodic vision testing is also highly recommended.

2. *Correct Posture—*
   - Workers should be trained to keep the head straight and balanced when looking forward at the screen.
   - Elbows should be maintained in the bent position at 90 degrees when the hands are on the keyboard.
   - The wrists should be kept in the neutral position. Wrist rests can also help alleviate discomfort.
   - Feet should be positioned flat on the floor or a footrest should be provided.
   - A back support for the lumbar region should be provided if necessary.

3. *Proper Lighting—*
   - Drapes, blinds, or shades should be adjusted to reduce glare on the work area and screen.
   - Indoor lighting and lamps should be positioned to reduce reflection on the screen.
   - Low-watt lights provide better lighting than single, more powerful sources.
   - Light should be at a at 90-degree angle, and overhead lighting should not be used if possible.
   - Walls should be painted a dark color or a color that does not produce a glare.
   - Indirect or shielded light sources should be used whenever possible.

4. *Screen Adjustment—*
   - Contrast or screen brightness should be set at a comfortable level and adjusted as necessary throughout the day.
   - The screen should be positioned so as to reduce glare and reflections from other light sources.
   - Windows should not be directly behind or in front of the operator.
   - The screen should be positioned just below eye level when sitting normally at the keyboard.
   - An anti-glare filter should be provided for screens where light reflection or glare cannot be avoided.
   - Anti-glare devices can affect image clarity and should be used only when absolutely necessary.

5. *Chair Placement—*
   - Chair height should be adjusted to keep the thighs horizontal, feet resting flat, and arms/hands comfortably positioned at the keyboard.
   - Footrests should be provided for locations where the chair is high and cannot be properly adjusted.
   - Armrests should be padded and adjustable for the up/down and in/out positions.
   - Seat cushions should be firm but comfortable.
   - Mats should be used to increase mobility on carpeted surfaces.

6. *Practice Smart Work Habits—*
   - The document holder should be close to the monitor and at the same distance and level from the eye as the screen.
   - The document holder should be switched to the opposite side of the screen on a regular basis.
   - Workers should be encouraged to change positions and move around for a short time when they become fatigued.
   - Workers should be encouraged to use a soft keyboard touch and keep fingers, wrists, and hands as relaxed as possible.
   - Workers should remember that incorrect positioning is the main culprit in CTDs.
   - Workers should be aware that many other office tasks can also contribute to RMI.

D. HEALTHCARE LIFTING

Any approach to preventing back injury involves reducing manual lifting and other load-handling tasks that are biomechanically stressful. Emphasis must be on teaching workers how to perform stressful tasks while minimizing the biomechanical forces on their backs, to maintain flexibility, and to strengthen the back and abdominal muscles.

1. *Healthcare Back Injuries—*According to a study conducted by the National Council on Compensation Insurance, over 30% of workers' compensation lost-time cases are due to back injuries. Over 50% of back injuries result from lifting.
   - Slips, falls, pushing, and pulling were also listed as frequent causes of back injuries.
   - Sprains and strains were the result in over 90% of the back claims.
   - Lower back injuries were found to be involved in 90% of all back injury claims, while 7% involved the upper back.
   - The study also reported that back injury claims last longer than other types of claims. Employees of nursing homes, public health nursing associations, and taxicab companies reported the highest percentage of back injury claims.

2. *Mechanical Lifting Devices—*Mechanical lifting devices should be used for heavy or unusual loads whenever practical. Manual lifting tasks should be evaluated to ensure that extremely heavy loads are not lifted. The direction of the work

motion should be correct in relation to the force required. Torques around the axis of the body should be avoided if possible.

3. *Basic Lifting Techniques—*
   - **Get a Firm Footing**—Keep feet apart for a stable base; point toes out.
   - **Bend Your Knees**—Do not bend at the waist. Keep the principles of leverage in mind at all times. Do not do more work than necessary.
   - **Tighten Stomach Muscles**—Abdominal muscles support the spine when lifting, offsetting the force of the load. Train muscle groups to work together.
   - **Lift with Your Legs**—Let the powerful leg muscles do the work of lifting instead of the weaker back muscles.
   - **Keep Load Close**—Do not hold the load away from the body. The closer it is to the spine, the less force it exerts on the back. Never carry anything above waist level.
   - **Keep Your Back Upright**—Whether lifting or putting down the load, do not add body weight to the load.
   - **Avoid Twisting**—Turn by moving the feet, not by twisting the back.

4. *Understanding Causes of Back Pain—*
   - **Back Muscles**—Back curves are held in place and supported by the muscles of the back and abdomen. These muscles must be strong and healthy so they can keep the back curves in their normal, balanced position.
   - **Poor Posture**—Slouching, rounded shoulders, one hip higher than the other, and a swayback look with too much forward curve in the lower back are examples of poor posture. It looks bad and feels even worse and causes muscle tension, stiffness, backaches, neck aches, and fatigue.
   - **Stand Easy**—Stand tall, with the back flat and knees relaxed. If standing for a prolonged period, one foot should rest on a low stool to support the lower back.
   - **Walk Tall**—Keep head up and chest lifted.
   - **Sit Smart**—Pick a chair that supports the lower back and is not too high. Tuck the buttocks and keep feet flat on the floor or on a stool so that knees are bent. If nothing is available for the feet, cross legs (change legs every so often). While driving, seat should be close to the steering wheel and pedals. (Lap belt and shoulder harness should always be worn.)
   - **Sleep Tight**—Sleep on back, with a small pillow under knees, or sleep on side with knees bent. Never sleep on stomach, nor on back with legs straight out. Mattress should be firm with no sag in the middle. A bed board should be placed between the mattress and box spring if necessary.
   - **Lack of Exercise**—The muscles that support the back need regular exercise to stretch them and keep them strong. A program of aerobic exercise and back exercises should be established. Just 20 minutes at least three times a week can keep the back fit. Each individual should choose an exercise program that feels right.

5. *Risk Factors for Back Disorders—*
   - Excessive repetition, prolonged activity, and forceful exertion, especially of the hands

- Accomplishing tasks that require a pinch grip
- Prolonged static posture of the body, trunk, or extremities while either sitting or standing
- Awkward posture of the upper body, including reaching above the shoulders or behind the back
- Excessive bending or twisting of the wrist and continued elevation of the elbow
- Continuous contact with work surfaces, including contact with edges of machines
- Exposure to temperature extremes
- Inappropriate or inadequate hand tools and vibration from power tools
- Restrictive workstations and inadequate clearances
- Improper body mechanics such as continued bending at the waist, continued lifting below the knuckles or above the shoulders, or twisting at the waist while lifting
- Lifting heavy objects or objects of abnormal size
- Lack of adjustable chair, footrest, body support, or work surface or slippery footing.

6. *Back Pain Complaints—*
   - The most common causes of back pain are poor physical condition, being unaccustomed to the task, poor posture, and lifting weight that approaches the limit of a worker's strength.
   - Contributing factors are understaffing, inadequate training, poor body mechanics, and inadequate safety precautions.
   - Food service workers hurt themselves by pushing/pulling carts, lifting heavy food trays, moving dishes, and storing supplies.
   - Housekeepers bend repeatedly while cleaning, lifting supplies and equipment, operating floor cleaning equipment, and using brooms/mops.
   - Clerical workers use chairs that are not designed for desk work and do not provide the proper support.
   - Maintenance workers hurt themselves by lifting, moving, and handling large packages, boxes, or equipment.
   - Patient care providers hurt themselves by moving, lifting, and transporting patients.

7. *Basic Back-Injury Prevention Tips*—The following are some of the most important elements in a program to prevent back injuries among healthcare workers:
   - Use of mechanical devices for lifting and transferring patients or heavy objects
   - Use of wheels and other devices for transporting heavy, non-portable equipment
   - Adequate staffing to prevent workers from lifting heavy patients or equipment alone; lifting teams and/or two-person lift requirements are important in reducing injury
   - Close supervision for newly trained workers to assure that proper lifting practices have been learned

- In-service education for both new and experienced staff on the proper measures to avoid back injuries
- Pre-placement evaluation of workers; workers with significant pre-existing back disorders should not be assigned jobs that require lifting
- Routine lower back (lumbar) X-rays are not recommended for pre-placement evaluations because studies indicate they do not predict which workers will suffer future back injuries

Back Support Belts—The National Institute of Occupational Safety and Health (NIOSH) has concluded that the use of lumbar support belts to reduce the risk of injury remains unproven. NIOSH based its conclusion on a two-year study that consisted of only reviewing previously published laboratory-based research. NIOSH concluded that the supports do not reduce spinal compression during heavy lifting tasks.

1. *NIOSH Concerns*—NIOSH expressed concern that the belts might give workers a false sense of security and result in persons lifting more than normal. Belt manufacturers and some other groups claimed that the study was incomplete and flawed because no new research was conducted.

2. *Criticism of the Study*—NIOSH admits that the study looked at lumbar orthoses and weight-lifting supports. The verdict is still out concerning the effectiveness of industrial-type lumbar belts. The NIOSH project which reviewed data from other studies has also drawn some strong criticism. Some associations and self-insurance funds claim that the use of support belts has resulted in a significant reduction in workers' compensation costs. *Note*—A copy of the study, "Workplace Use of Backbelts," can be obtained from NIOSH.

3. *Back Belt Usage Tips*—Many healthcare facilities require workers to wear support belts. The following are common-sense safety tips regarding the use of lumbar supports:
   - Workers should be medically screened before being issued belts.
   - The back belt should be properly fitted to the worker by a qualified person.
   - Workers should not wear back belt supports fastened unless involved in planned lifting activities.
   - Continuously wearing a fastened belt could weaken stomach and back muscles.
   - Back belts do not prevent all back injuries or enable a worker to lift more weight.
   - Lumbar supports should not be issued unless they are part of a total back management and education program.
   - Regular training should be provided for all workers involved in lifting activities, and participation in an off-duty exercise program should be encouraged.
   - Back supports should only be used as part of a formal back care management program.

E. PREVENTING SLIPS AND FALLS

Slips, trips, and falls are some of the most common patient-related occurrences in healthcare facilities. Many healthcare workers are also involved in slip, trip, and fall incidents each year. Fall prevention is also addressed in Chapter 9, Safety in Patient Care Areas.

Fall Prevention Program—An effective program should seek to:

- Identify, evaluate, and correct safety hazards that could contribute to employees, patients, or visitors slipping, tripping and/or falling
- Communicate to all staff members the physical hazards and behavioral aspects of slip, trip, and fall prevention
- Provide all employees with training related to slip, trip, and fall prevention
- Establish procedures to better manage and analyze the trends and problems within the facility

Fall Prevention Management—The safety director should manage the program under the guidance and direction of the safety committee. Trends and problem areas should be determined by analyzing data from the risk management and quality improvement areas. Workers' compensation data should also be reviewed to determine worker fall problem areas. During 1992, falls were the fourth leading cause of accidental death on the job. Falls happen because people and conditions do not remain static. Slip, trip, and fall prevention programs must be continuously monitored to be effective in preventing and reducing loss.

Management Elements—The following must be considered crucial to the program:

- Hazard surveillance program for physical hazards
- Regular inspection of all areas, especially patient and visitor areas
- Special emphasis on high-risk areas as determined by a statistical analysis of incident and accident data
- Analysis of accidents/incidents to determine trends
- Involvement of safety committee in approving and/or reviewing all policies
- Awareness training for employees, staff, and students
- Training for those responsible for facility maintenance and housekeeping
- Quarterly assessment of program effectiveness
- Annual audit

F. HEAT-RELATED HAZARDS

The laundry, boiler room, and kitchen are hot environments; other areas may also be hot during the summer.

Heat Stroke—

1. Heat stoke is a serious condition which results from the body's failure or inability to cool itself.
2. The condition is characterized by hot dry skin, dizziness, headache, thirst, nausea, cramps, mental confusion, and even loss of consciousness.
3. A victim's body temperature can exceed 105°F.

4. Quick action is required by:
   - Removing the victim to a cool area
   - Attempting to lower body temperature by soaking the victim with water and fanning vigorously
   - A physician treating the victim immediately

Heat Exhaustion—This condition occurs when a person becomes dehydrated. The symptoms are similar to heat stroke but are much milder. The victim should be moved to a cool area and given large amounts of liquids. Severe cases may require the attention of a physician.

Other Conditions—Some other heat-related conditions include heat cramps, fainting, and heat rashes.

NIOSH Guidelines—In 1986, NIOSH published guidelines to protect workers exposed to hot environments. These guidelines consider both acclimatized and unacclimatized workers and also factor in the effects of clothing.

WBGT Index—One of the most common methods of measuring heat exposure is the wet bulb globe temperature (WBGT) index. This method combines the effects of radiant heat and humidity with the dry bulb temperature.

A Guide to Heat Stress in Agriculture—An excellent source for information about heat stress is the OSHA/EPA booklet "A Guide to Heat Stress in Agriculture," which can be useful to other industries with heat-related risks. The publication is available from the U.S. Government Printing Office.

Heat Control Tips—Some basic heat control suggestions include:
- Schedule heavy work early in the day and allow workers frequent breaks in cool areas.
- Isolate, enclose, and/or insulate hot equipment.
- Remove heat from work areas by mechanical means.
- Install reflective shielding materials where appropriate.
- Provide fans in hot areas to promote sweat evaporation.
- Make cool water readily available to workers.
- Ensure break areas and lunchrooms are cool.
- Train workers to recognize heat-related symptoms.

G. MAINTENANCE DEPARTMENT SAFETY

The maintenance or plant services department performs a variety of functions that support hospital operations. The size of the department depends on the size of the facility and type of medical services provided. The department of a medium-size healthcare facility could be responsible for the following activities:

- Normal and preventive maintenance of facilities and equipment
- Equipment procurement and installation
- Heating, ventilation, and air conditioning systems
- Utility and energy systems
- Interior maintenance, including painting and floor replacement
- Maintenance of facility grounds
- Plant, technology, and safety management compliance
- Monitoring contractors and coordinating renovation activities
- Fire protection, disaster planning, and emergency power
- Handling, storing, and managing facility waste
- Parking, communications, and security

General Safety—Standards pertinent to healthcare facilities and maintenance functions can be found in 29 CFR 1910, National Fire Protection Association (NFPA) codes, and state/local laws and regulations. Crews should dress correctly for the specific job. Rings, watch chains, wristwatches, and other jewelry should not be worn. Tools should be carried in a tool bag or attached to a special belt fitted with tool carriers. Gloves should be supplied as required for different jobs (e.g., rubber gloves for electrical installations and chemical-resistant gloves for handling acids or hazardous materials). Eye protection should be provided to protect workers from flying objects, injurious heat, light rays, dust, wind, and chemicals. In all cases, when a job involves unusual risk, another person should be stationed nearby. Workers should wear protective clothing and equipment when exposed to hazards that require such protection.

Hazard Control Measures—Hazard control measures should be implemented to reduce hazards such as wet floors, stairway obstructions, and faulty ladders. Other general rules for maintenance areas include:

- Drive belts must be guarded. Gears, shafting, chains, and sprockets must be properly enclosed as required by 29 CFR 1910, Subpart O.
- Tool rests, adjustable tongue guards, and spindle guards on grinders must be installed and kept properly adjusted in accordance with 29 CFR 1910, Subpart O.
- Blade guards must be installed on table saws, band saws, and radial arm saws. When used for ripping, saws must have anti-kickback devices installed according to 29 CFR 1910, Subpart O.
- Electrical equipment must be properly grounded or double insulated per 29 CFR 1910.304.
- Extension cords must be of three-wire type and have sufficient capacity to safely carry the current drawn by any devices operated from them.
- Extension cords may be used only in temporary situations and may not be substituted for fixed wiring.
- Electrical switches on circuit boards should be marked with danger tags and physically locked to prevent circuit activation when machinery is being repaired. Circuits should be de-energized before repair work begins using appropriate lock-out procedures.
- Battery-charging areas should be adequately ventilated to prevent a buildup of hydrogen gas. These areas should be designated as no smoking areas.

- Gasoline- and diesel-powered equipment should be properly maintained and operated in well-ventilated areas only. Data obtained recently from animal studies indicate that diesel exhaust is a potential carcinogen.
- Paints, solvents, and other flammable materials must be stored in cabinets or rooms that meet the requirements outlined in NFPA 30.
- Hand tools should be properly maintained and stored.
- Fuel and cylinders of flammable gas must be stored separately from cylinders of oxidizing gas. Cylinders must be kept away from heat sources such as radiators, steam pipes, and direct sunlight.
- Cylinders must be stored in the upright position and chained or secured to prevent falling.
- Trash compactors should never operated in the open position and should have guarding devices such as two-hand controls, electric eyes, and emergency shutoff bars.
- Pipes in laboratories where sodium azide is used for the automatic counting of blood cells must be flushed before plumbing repairs. A buildup of sodium azide in the pipes can result in a violent explosion. A sodium-azide decontamination procedure is available from NIOSH.
- The use of compressed air for cleaning surfaces should be avoided.

Material Handling—Handling and storing material involves operations that can include moving bags, boxes, and materials.

1. *Material Handling Hazards*—
   - Strains and back injuries are common among material handlers in all industries.
   - Fractures and bruises result from being struck by materials or caught in pinch points.
   - Cuts and bruises are also caused by falling materials or by incorrectly cutting ties or other securing devices.

2. *Basic Safe Material Handling Tips*—
   - Get help when the load is too bulky or heavy.
   - When placing blocks under a raised load, ensure the load is not released until hands are clear.
   - Handles or other holders should be attached to materials if possible.
   - Proper protective equipment, including eye protection and gloves, should be used as required by the task.
   - Storage areas should be kept free from materials that could cause tripping, fires, or explosions.
   - Bound material should be stacked, placed on racks, blocked, and interlocked to prevent sliding, falling, or collapsing.
   - Lumber should never be stacked more than 16 feet in height.
   - The maximum height when using a forklift is 20 feet.
   - Drums and barrels must be stacked symmetrically, and when stored on the side, the bottom tier must be blocked.
   - Walkways must be kept clear of all obstructions.

- Flammable and combustible materials should be stored according to their fire characteristics.
- Aisles and walkways should be appropriately marked.
- Clearance signs should be placed to warn of overhead hazards.

Ladder Safety—OSHA standards for ladders are found in 29 CFR 1910.25 and 1926.1053. Ladders should also meet the requirements in ANSI-14.5. The following safety items should be considered:

- Construction of all ladders should conform to the applicable ladder or safety code of the locality or state.
- Portable ladders should be equipped with non-slip bases.
- Ladders should be visually inspected to ensure compliance with applicable codes.
- All ladders should be inspected every three months and documented inspection records maintained.
- Damaged and unsafe ladders should be removed from service. Ladders identified for repair must be tagged DANGEROUS—DO NOT USE.
- Ladders should never be used in a horizontal position as runways or scaffolds.
- A ladder should never be placed in front of a door that opens toward the ladder unless the door is locked, blocked, or guarded.
- A ladder should never be leaned up against unsecured backing such as loose boxes or against a window.
- A ladder should never be placed close to live electrical wiring, against operational piping, or near a sprinkler system where damage could result.
- Metal ladders should be marked CAUTION—DO NOT USE NEAR ELECTRICAL EQUIPMENT.
- Ladder rungs, cleats, and steps must be parallel, level, and uniformly spaced.
- Rungs, cleats, and steps should not be spaced less than 10 inches or more than 14 inches apart along the ladder side rails (does not apply to stepladders or the base section of extension ladders).
- Rungs, cleats, and steps of stepstools must not be less than 8 or more than 12 inches apart between center lines of the rungs.
- The rung spacing of an extension ladder must not be less than 6 inches or more than 12 inches.
- Wood ladders may not be covered with any opaque covering except for warning and identification labels.
- When portable ladders are used for access to an upper landing, the side rails must extend a least three feet above the upper landing surface.

Tool Safety—Employees must be trained in the proper use of all tools. Workers must recognize the hazards associated with different types of tools and take safety precautions to prevent accidents.

1. *Basic Rules*—
   - Keep all tools in good condition with regular maintenance.
   - Use the right tool for the job.

- Examine each tool for defects before use.
- Operate according to the manufacturer's instructions.
- Ensure workers use the appropriate protective equipment.

2. *Power Tools*—Type of power tool is determined by the power source. Types include electric, pneumatic, liquid fuel, hydraulic, and powder-actuated tools. Power tool safety measures include:
   - Never carry a tool by the cord or hose.
   - Never pull the cord or hose when disconnecting from a receptacle.
   - Keep cords and hoses away from heat, oil, and sharp edges.
   - Disconnect tools when not in use and when changing blades, bits, and cutters.
   - Secure bulky work with a clamp or vise and ensure both hands are free to operate the tool.
   - Avoid accidental starting and never hold fingers on the switch button while carrying a plugged-in tool.
   - Avoid wearing loose clothing, ties, and jewelry that can become caught in moving parts.
   - Remove all damaged electric tools from use immediately and tag them until repaired.
   - Radial saws must have the direction of the saw rotation marked on the hood. A permanent label with the following warning must be affixed to the rear guard: DANGER—DO NOT RIP OR PLOUGH FROM THIS END.

3. *Safeguarding*—The hazardous moving parts of power tools should be safeguarded. Belts, gears, shafts, pulleys, sprockets, spindles, drums, flywheels, chains, and other moving parts must be guarded if workers can come into contact with them. Safety guards must never be removed when a tool is being used. Guards protect the operator and others from:
   - Point of operation hazards
   - In-running nip points
   - Rotating parts
   - Flying chips and sparks

4. *Electric Tools*—Electric tool hazards include the possibility of electrocution. Tools must have either a three-wire cord with ground or be powered by a low-voltage transformer. Double insulation provides added safety. Some general safety practices include:
   - Operate tools within the design limitations.
   - Use gloves and safety footwear when using electric tools.
   - Store tools in a dry place.
   - Never use tools in a wet or damp location.
   - Keep work area well lighted.
   - Ensure power cords do not create tripping hazards.

5. *Powered Abrasive Wheel Tools*—Before mounting an abrasive wheel, it should be inspected closely and tested to ensure the wheel is free of cracks.
   - To prevent the wheel from cracking, the user should be sure it fits freely on the spindle.

- Ensure the spindle does not exceed the abrasive wheel specifications because wheels can disintegrate or explode when started up.
- Never stand directly in front of the wheel as it accelerates to operating speed.
- The wheel must be equipped with proper guards.
- Always use eye protection and always turn off the power when not in use.
- Never clamp a handheld grinder in a vise.

6. *Grinder Safety*—The rules for safe grinding are as follows:
- Never grind the side of the wheel unless it is designed for that purpose.
- Never jam work onto the wheel.
- Never stand in front of the wheel when the grinder is started.
- Hold material with hands away from the wheel and work rest.
- Wear eye protection and/or a face shield.
- Never wear gloves or hold material with a rag.
- Use jigs or fixtures to hold frequently ground pieces.
- Use vise grips/channel locks to hold pieces not frequently ground.
- Use a file or portable grinder to dress very small pieces.

7. *Pedestal Grinder Checklist*—
- Glass shields must be in place, undamaged, and clear.
- All guards should be attached correctly and securely.
- Wheel face should be clean with no foreign material, chips, or nicks.
- Grinder should not have excessive vibration.
- Adjustable tongue must be $1/4$ inch or less from the wheel and secure.
- Work rest must be $1/8$ inch or less from the wheel.
- Work rest should be securely attached at the center of the wheel or above.

H. CONSTRUCTION SAFETY

OSHA Construction Standards (29 CFR 1926)—Construction projects at healthcare facilities are not covered under the general industry standards. Construction safety and health standards cover many hazards encountered during construction and renovation projects. Healthcare hazard control managers and engineering department heads must be familiar with the standards found in 29 CFR Part 1926. The following are some key subjects of the construction safety and health standards:

- Concrete and Masonry Construction (1926.701)
- Cranes and Derricks (1926.550)
- Electrical Installations (1926.402–408, 416, 417)
- Excavating and Trenching (1926.651)
- Fall Protection (1926.501)
- Hoists, Material and Personnel (1926.552)
- Joiners and Saws (1926.304)
- Railings (1926.502)
- Scaffolds (1926.451)
- Steel Erection (1926.750)

- Wall Openings (1926.501)
- Welding and Cutting (1926.350 and 351)
- Wire Ropes and Chains (1926.251)

I. WARNING SIGNS

Color Codes for Safety (29 CFR 1910.144 and ANSI Z53.1)—

- *Red: Danger—*
  1. Red is used for fire protection equipment and fire apparatus.
  2. Red is also used to indicate STOP.
  3. Safety cans for flammable liquids with a flash point of 80°F or below must be painted red with other visible markings.
  4. Red lights should be provided at barricades and temporary obstructions as specified in ANSI Standard A10.2-1944.
- *Yellow*—Caution. Designates physical hazards and used for marking physical hazards such as stumbling, falling, and tripping.
- *Blue*—Used to warn against moving, starting, or using equipment being repaired. Blue is also used for informational signs and handicap access.
- *Green*—Used for safety and first-aid equipment.
- *Magenta/Yellow*—Used to designate alpha, beta, gamma, neutron, proton, and X-ray radiation hazards.
- *Black, White, or Combination of the Two*—Used for traffic, boundaries, and housekeeping markings.
- *Orange*—Used to mark dangerous parts of machines or equipment. Fluorescent orange and orange-red are used to designate biohazardous materials.

Accident Prevention Signs and Tags (29 CFR 1910.145)—

- Wording should be easy to read and concise.
- The wording should make a positive rather than a negative suggestion.
- Accident prevention tags are used to identify hazardous conditions and provide a message to employees regarding the condition.
- Tags must contain a signal word and a major message.
- Danger signs/tags are required to warn workers of immediate danger and that special precautions apply. Standard colors are red, black, and white.
- Caution signs/tags warn against potential hazards or unsafe work practices. Standard color is yellow with black lettering.
- General safety signs provide information about safe working practices. Standard color is green with black or white letters or a white background with black or green letters.
- Warning signs should be orange or predominantly orange, with letters or symbols in a contrasting color.
- Fire and emergency information is designated with white letters on a red background.
- Informational signs are a white background with blue letters.

ANSI Standards on Color Codes and Signs—

- Z353.1 Color Codes for Safety Signs
- Z353.2 Environmental and Facility Safety Signs
- Z353.3 Safety Symbols
- Z353.4 Product Safety Signs and Labels
- Z353.5 Temporary Hazard Signs

Identification of Piping Systems (ANSI A13.1)—

- *Safety Red*—Fire protection
- *Safety Yellow*—Dangerous
- *Safety Green*—Safe
- *Safety Blue*—Protective materials

J. ELECTRICAL SAFETY

Electrical Hazards—Handling electricity safely requires some knowledge about how it works and what hazards it presents. Electrical current travels through electrical conductors. Its pressure is measured in volts. Resistance to the flow of electricity is measured in ohms and can vary widely. Resistance is determined by the nature of the substance itself, the length and area of the substance, and the temperature of the substance.

Conductive Materials—Some materials, like metal, offer very little resistance and become conductors very easily. Other substances, such as porcelain and dry wood, offer high resistance. Materials that prevent the flow of electricity are called insulators. Water that contains impurities such as salts and acids makes a ready conductor. Electricity travels in closed circuits and its normal route is through a conductor.

Electrical Shock—Electrical shock occurs when the body becomes part of the circuit. Shock normally occurs when a person:

- Contacts both wires of an electrical circuit
- Contacts one wire of an energized circuit and the ground
- Contacts a "hot" metallic part that has become energized, while in contact with the ground

Shock Severity—The severity of shock is affected by several factors:

- The amount of current (amperes) flowing through the body
- The path of the current through the body
- The length of time the person is in the circuit
- The phase of the heart cycle when the shock occurs
- The general health of the person involved

Electrical Burns and Other Injuries—Severe shock can cause falls, cuts, burns, and broken bones. Three types of burns can result from shocks:

- Electrical burns result from current flowing through tissue or bone. The damage is from the intense heat which damages tissues.
- Thermal burns occur when the skin comes into contact with the hot surfaces of overheated conductors or other energized parts.
- Arc burns are caused by high temperatures near the body and are produced by an electrical arc or explosion.

Protecting Workers—

- Proper insulation protects workers from electrically energized wires and parts. Insulation should always be checked before working with electrical equipment.
- Insulation requirements are regulated by 29 CFR 1910, Subpart S. The standard requires insulation to be suitable for the voltage and existing conditions.
- Conductors and cable are marked by the manufacturer to show maximum voltage, American Wire Gage size, letter of the insulation, and manufacturer's name or trademark.
- Insulation is often color coded. Grounding conductors are green or green with yellow stripes. Grounded conductors that complete a circuit are usually white or natural gray.
- Hot wires are colors other than those above, often black or red. Live parts of electrical equipment operating at 50 volts or higher must be guarded against accidental contact.

Guarding—Guarding can be accomplished by:

- Location in a room, vault, or enclosure that has limited access
- Use of permanent substantial partitions or screens
- Location on a suitable balcony or platform elevated and arranged to limit access
- Elevation of eight feet or greater above the floor
- Use of warning signs to mark entrances to locations with exposed live parts
- Containing indoor electric installations over 600 volts in metal enclosures or in vaults or areas controlled by locks
- Marking high-voltage equipment with caution signs

Grounding—

- Grounding refers to a conductive connection, usually with the earth. This low-resistance path prevents the buildup of voltage that results in shock.
- 29 CFR 1910, Subpart S covers two types of grounds:
  1. *Neutral Conductor*—The neutral conductor is grounded and is normally the white or gray wire at the generator or transformer and again at the service entrance to a building. This ground protects machines, tools, and insulation against damage.
  2. *Equipment Ground*—This additional ground offers enhanced protection for the worker by providing another path from the machine or tool through which the current flows into the ground. This protects the worker should the metal frame of the tool become accidentally energized. The resulting heavy surge of current will activate the circuit protection devices and open the circuit.

Circuit Protection Devices—

- Circuit protection devices are designed to limit or shut off the flow of electricity in the event of a ground-fault overload or short circuit in the wiring system.
- Fuses and circuit breakers are over current devices that automatically open or break when the amount of current becomes excessive.
- Fuses and circuit breakers primarily protect equipment and conductors.
- Ground-fault circuit interrupters are designed to shut off electrical power immediately by comparing the amount of current going to the equipment and the amount returning along the circuit conductors. They should be used in wet locations and construction areas.

Safe Work Practices—Electrical safety-related work practice requirements are contained in 29 CFR 1910.331–335.

- Employers must ensure workers are trained in safety-related work practices.
- Maintenance employees should be qualified electricians who also have been instructed in lockout/tagout procedures.
- Workers whose jobs require them to work constantly and directly with electricity must use required personal protective equipment.
- Equipment may consist of rubber insulating gloves, hood, sleeves, line hose, and protective helmet. Workers should always use tools that are designed to withstand voltage and stresses of electricity.

Electrical Equipment Safety—Electrical malfunction is the second leading cause (after matches and smoking) of fires in hospitals. Violations of standards governing the use of electrical equipment are the most frequently cited causes of fires.

1. Hospital personnel use a wide variety of electric equipment in all areas, including general patient care, intensive care units, emergency rooms, maintenance, housekeeping service, food preparation, and research.
2. Thorough electrical maintenance records should be kept, and considerable effort should be devoted to electrical safety, particularly in areas where patient care is involved.
3. Equipment and appliances that are frequently ungrounded or incorrectly grounded include:
   - Three-wire plugs attached to two-wire cords
   - Grounding prongs that are bent or cut off
   - Ungrounded appliances resting on metal surfaces
   - Extension cords or improper grounding
   - Cords molded into plugs that are not properly wired
   - Ungrounded multiple-plug "spiders" often found in office areas and nurses' stations
4. Personal electrical appliances, such as radios, coffeepots, fans, power tools, and electric heaters, that are not grounded, have frayed cords or poor insulation, or are otherwise in poor repair.

National Electrical Code (ANSI/NFPA 70)—OSHA has adopted the National Electrical Code (NEC) in NFPA 70 as a national consensus standard.

1. The NEC is designed to safeguard persons and property from the hazards of using electricity.
2. Article 517 of NFPA 70 and NFPA 76A and 76B contain special electrical requirements for healthcare facilities. In addition, state and local laws and regulations may be applicable.
3. Electricians and maintenance personnel should consult OSHA's electrical safety standards found in 29 CFR 1910.301–399 and refer to NFPA 99 for information on special hazards.
4. Each circuit breaker or fuse box to be disconnected should be legibly marked to indicate its purpose unless the purpose is evident.
5. Frames of electrical motors should be grounded regardless of voltage.
6. Exposed non-current-carrying metal parts of fixed equipment which may become energized under abnormal conditions should be grounded under any of the following circumstances:
   • If the equipment is in a wet or damp location
   • If the equipment is operated in excess of 150 volts
   • If the equipment is in a hazardous location
   • If the equipment is near grounded metal objects and subject to contact by workers
   • If the equipment is in contact with metal
   • If the equipment is supplied by metal-clad, metal-sheathed, or grounded metal raceway wiring
7. Exposed non-current-carrying metal parts of plug-connected equipment that may become energized should be grounded under any of the following circumstances:
   • If the equipment is a portable, handheld lamp or motor-operated tool
   • If the equipment is an appliance, sump pump, hedge clipper, lawn mower, snow blower, wet scrubber, or portable X-ray equipment
   • If the equipment is operated in excess of 150 volts
   • If the equipment is in a hazardous location
   • If the equipment is used in a wet or damp location
   • If the equipment is used by workers standing on the ground or on metal floors
8. Outlets, switches, junction boxes, etc. should be covered
9. Flexible cords should not be used as a substitute for fixed wiring.
10. Flexible cords should be connected without any tension on joints or terminal screws.
11. Frayed cords or cords with deteriorated insulation should be replaced.
12. Splices in flexible cords should be brazed, welded, soldered, or joined with suitable splicing devices. Splices, joints, or free ends of conductors must be properly insulated.

Healthcare Electrical Safety Requirements—Healthcare facilities contain many damp or wet areas which make electrical safety requirements important.

1. A switch or circuit breaker in a wet area or outside a building should be protected by a weatherproof enclosure.
2. Cabinets and surface-type cutout boxes in damp or wet areas should be weath-

erproofed and located to prevent moisture from entering and accumulating in the cabinet or box.

3. The boxes should be mounted with at least 0.25 inches of air space between the enclosure and the wall or supporting surface.
4. Non-metallic-sheathed cable and boxes made of non-conductive material are recommended.
5. In areas where walls are washed frequently or where surfaces consist of absorbent materials, the entire wiring system including all boxes, fittings, conduit, and cable should be mounted with at least 0.25 inches of air space between the electrical device and the wall or support surface.
6. Specific NEC recommendations apply in areas where flammable materials are stored or handled, in operating rooms, and in patient care areas.
7. Orientation and continuing in-service training programs are necessary to maintain worker awareness of electrical hazards.
8. The following work practices can also help prevent shocks to hospital workers:
   - Develop a policy for using extension cords.
   - Use a sign-out system to list the number and location of all extension cords currently in use.
   - Do not work near electrical equipment or outlets when hands, counters, floors, or equipment are wet.
   - Consider defective any device that blows a fuse or trips a circuit breaker, and prohibit its use until it has been inspected.
   - Do not use any electrical equipment, appliance, or receptacle that appears to be damaged or in poor repair.
   - Report all shocks immediately; even small tingles may indicate trouble and precede major shocks. Do not use the equipment again until it has been inspected and repaired if necessary.

Electrical Safety Rules—

- Under no circumstance should an electrically operated apparatus be inside of any canopy when oxygen is flowing.
- Electrically powered hair dryers used in the general patient floor areas should be checked for safety by the biomedical engineering department.
- Portable electric heaters or heating pads should not be used unless approved in advance.
- Patients and/or visitors should not use personal appliances unless an exception has been granted by the administration and biomedical engineering department in writing.
- Unapproved appliances should be removed immediately and placed with security until further notice.
- Extension cords are prohibited except in temporary emergency situations. Extension should be at least 16 gauge consisting of three copper wires with maximum length of 25 feet.
- Multiple-outlet strips are prohibited; the only exception should be office environments to support personal computers.

- Frayed or broken cords should not be used and should be reported to the maintenance department for repair.
- Adapters which convert three prongs to two prongs have no place in the healthcare environment.
- Broken or cracked receptacles should be reported immediately and not used until repaired.
- Only one piece of medical electronic equipment should be plugged in to a receptacle.
- Outlets that emit smoke or odor should be reported to the maintenance department immediately.

Control of Hazardous Energy Sources—

1. *Lockout/Tagout*—This safety standard covers servicing and maintenance of machines and equipment in which unexpected start-up of the machine or power source could cause injury. The standard in 29 CFR 1910.147 places four basic requirements on employers whose workers are engaged in service and/or maintenance functions:
   - Written procedures for lockout/tagout
   - Training of employees
   - Accountability of engaged employees
   - Administrative controls

2. *Definitions*—
   - Maintenance is the act of maintaining or the state of being maintained.
   - To maintain means keeping in an existing state or preserving from failure.
   - To service is to repair or provide maintenance.
   - To refurbish means to brighten up, freshen up, or renovate.
   - To renovate is to restore to a former better state.
   - To modify is to make fundamental change or to give a new orientation.

3. *Maintenance*—The definition for servicing and/or maintenance requires further clarification concerning the unexpected release of hazardous energy.
   - Production equipment and machines are safeguarded under the requirements of 29 CFR, Subpart O.
   - The OSHA lockout/tagout standard applies only during those associated functions when the normal machine or process equipment safeguards are bypassed or are rendered ineffective.
   - OSHA requires the employer to conduct a periodic inspection to ensure the procedures and requirements are being followed. This periodic inspection includes a review of each authorized worker's responsibilities under the energy control program.
   - The inspections and the reviews are intended to be a representative sample of compliance with the requirements of the standard and not a 100% inspection.

4. *Purpose of Lockout/Tagout*—The purpose of any lockout procedure is to render inoperative electrical systems, pumps, pipelines, valves, and any other systems that could be energized while employees are working.

- Lockout systems are normally administered by the maintenance department and/or contractor personnel.
- All energy sources locked out should be tagged as follows: DO NOT OPERATE. The tag should also indicate who installed the lock, their craft, and the reason the system was locked out.
- Electrical systems should be locked out until they are released for service.
- When either temporary or permanent repairs or modifications are made to electrical systems, the systems must be locked out. Locks should be applied to the main disconnect switch whenever possible.
- Pipelines, valves, and other such sources that could be inadvertently activated must be locked out, blanked off, or otherwise secured to prevent accidental activation.
- Lines, valves, and similar systems being tested pneumatically or with other gases must be tagged and/or locked out to prevent an accidental discharge of the pressure within the line.
- A lock inadvertently left in service may be removed only after a visual check of the system has been made.

5. *General Lockout Guidelines*—
   - Open the main switch, which is usually a side arm switch located on the main electrical panel, and pull down.
   - Insert the lock into the slotted hole, which should prevent the side arm switch from moving upward to close the main switch.
   - Insert a danger tag through the slotted hole next to the lock.
   - Lockout procedures are for everyone's safety, and the procedures must be used when working on machines and equipment.
   - In multi-problem situations, it may be necessary for more than one technician to attach tags and locks to the disconnect switch.
   - Never trade assigned lock, tags, or duplicate keys with other employees.
   - Tagout devices should include information such as the following: DO NOT START, DO NOT OPEN, DO NOT CLOSE, DO NOT OPERATE, DO NOT ENERGIZE.
   - When maintenance is complete and normal operation is ready to begin, areas around the equipment should be checked to ensure that no one is exposed.
   - When all tools have been removed from the equipment, guards have been reinstalled, and employees are in the clear, all lockout or tagout devices should be removed.
   - Operate the energy-isolating devices to restore energy to the machine or equipment.

K. CONFINED SPACES

Working in Confined Spaces (29 CFR 1910.146)—OSHA recognizes that employers are diverse in their activities, resources, and safety concerns. Accordingly, some flexibility is being allowed in compliance.

OSHA Compliance—OSHA considers three compliance approaches to be appropriate:

- Preparation of a written permit at the time entry is authorized, which contains all of the information needed to document compliance with the proposed standard.
- Preparation of a written permit that identifies the place, date, and time of the entry and the personnel who are involved. A checklist must be developed that details the hazards potentially present and the precautions which have been taken to protect entrants.
- Direct supervision of the entry by the person authorizing entry using a checklist-type permit which specifies the hazards potentially present and precautions taken to protect entrants.

Confined Space Characteristics—

- The size and shape allow for a person's entry
- Limited openings for individuals to enter and exit
- Not designed for continuous occupancy
- Unfavorable natural ventilation

Permit Required Spaces—A "permit required" confined space possesses at least one of the following characteristics:

- Contains or has the potential to contain a hazardous atmosphere
- Contains a material that could engulf a person
- Has an internal configuration that could cause a person to become trapped or asphyxiated by converging walls or a downward sloping floor
- Carries a recognized serious health or safety hazard

Potential Hazards—Evaluation of a confined space should consider the following potential hazards:

- The atmosphere in a confined space may be extremely hazardous due to the lack of natural air movement. This can result in an oxygen-deficient atmosphere of less than 19.5% available oxygen (O_2) and require an approved self-contained breathing apparatus (SCBA).
- The oxygen level in a confined space can decrease due to work activities such as welding, cutting, or brazing.
- Oxygen levels can also be decreased by certain chemical reactions (rusting) or through bacterial action (fermentation).
- Oxygen can be displaced by other gases such as carbon dioxide or nitrogen.

Flammable Atmospheres—

- An oxygen-enriched atmosphere (above 21%) will cause flammable materials such as clothing and hair to burn violently when ignited.
- Pure oxygen should never be used to ventilate a confined space; always ventilate with normal air.
- Different gases have different flammable ranges; if a source of ignition is introduced into a space containing a flammable atmosphere, an explosion will result.

Hazardous Substances—Most liquids, vapors, gases, mists, solid materials, and dusts should be considered hazardous in a confined space. Toxic substances can come from the following:

- Toxic materials can be absorbed into the walls and give off toxic gases when removed; this includes the cleaned-out residue of a stored product.
- Toxic atmospheres can be generated in various processes. Cleaning solvent vapors can be very toxic in a confined space.
- Toxic materials produced by work in the area of confined spaces can accumulate and become a hazard to workers.

Other Hazards—

- Extremely cold temperatures can present problems.
- Loose, granular material stored in bins and hoppers, such as grain, sand, coal, or similar material, can engulf and suffocate a worker.
- Noise within a confined space can be amplified because of the design and acoustic properties of the space. Excessive noise can damage hearing and affect communication.
- Slips and falls can occur on a wet surface, resulting in injury or death to workers.
- Wet surfaces increase the chance of electric shock in areas where electrical circuits, equipment, and tools are used.
- Workers should be mindful of falling objects, particularly in spaces that have topside openings or where work is being done above the worker.

Controlling Hazards—

- Identify all confined spaces in the workplace.
- Develop employee awareness through training.
- Place appropriate signs and prevent unauthorized entry.
- Implement a written program.
- Provide workers with appropriate protective equipment.

Entry Permits—All permits should be thoroughly reviewed by the entry supervisor to ensure all safety measures are in place. Each permit must include:

- Specific confined-space identification
- Purpose, date, and expected duration of entry
- Names of those authorized to enter confined space
- Actual hazards of the specified confined space
- Control and isolation methods to be used
- Results of atmospheric tests
- Rescue, communication, and emergency procedures

Preparation for Entry—

- Notify all departments affected
- Erect signs and barriers as required
- Lock out hazardous energy

- Empty the space of hazardous materials
- Ensure ventilation and air supply are adequate

Atmospheric Requirements—

- Oxygen content to be between 19.5 and 23.5%
- Flammable gas concentration less than 10% of lower flammable limit
- Toxic materials not to exceed permissible exposure limit
- Provisions to monitor air and heat-stress conditions
- Ventilation by a blower or fan to remove harmful gases and vapors

Entry Equipment—

- Appropriate personal protective equipment for the space
- Equipment must be listed and available at the site
- Determine need for and type of respirators required
- Test communication system and review any special procedures
- List any special lighting or tool requirements

Emergency Procedures—

- Self-rescue is the best option at the first sign of trouble.
- A trained standby person should be assigned to remain on the outside of the confined space and be in constant contact with the workers inside.
- The standby person should not have any other duties and should know who to notify in case of emergency.
- Standby personnel should not enter a confined space until help arrives, and then only with proper protective equipment, lifelines, and respirators.

L. HEARING CONSERVATION

Controlling Noise Hazards (29 CFR 1910.95)—Noise is any unwanted sound. It is created by sound waves, which are rapid vibrations in the air.

1. *Noise Terms*—
   - Frequency or pitch is measured in cycles per second or Hertz (Hz).
   - Amplitude or intensity is measured in decibels (dB).
   - The decibel scale is a logarithmic measure of intensity. An increase of 10 dB is 10 times as intense but is perceived as being twice as loud.
   - Perceived loudness is a subjective perception and therefore cannot be measured by an instrument.

2. *Noise Exposure*—Exposure to high levels of noise in the workplace is a common job hazard, even in healthcare facilities. A 1979 survey of noise levels in hospitals indicated five work areas with noise levels high enough to reduce productivity: the food service department, the laboratory, the engineering department, the business office, and the medical records section.

- The ear changes air pressure waves into impulses that the brain interprets as sound. Hair cells in the inner ear stimulate nerves that carry the message to the brain.
- Loud noise damages these nerves and decreases hearing acuity. Noise may also trigger changes in cardiovascular, endocrine, neurologic, and other physiologic functions.
- Noise also hinders communication among workers.

3. *Protecting Workers*—Healthcare facilities must protect all workers from occupational noise exposure that exceeds an 8-hour time-weighted average (TWA) of 90 decibels (dBA). The 6-hour level is 92 dB, 4-hour exposure is 95 dB, 3-hour exposure is 97 dB, and the 2-hour exposure level is 100 dB. To protect workers, the employer must:
   - Monitor noise exposure
   - Institute control measures
   - Implement a hearing conservation program (HCP) when occupational noise exposure exceeds an 8-hour TWA of 85 dBA

4. *Noise Controls*—The employer must institute engineering and/or administrative controls whenever possible. If these controls fail to reduce employee noise exposure to an 8-hour TWA of 90 dBA or less, then the employer must provide and enforce the use of hearing protectors that attenuate employee exposure to at least an 8-hour TWA of 90 dBA.

   - **Engineering Controls**—
     1. Use technology to reduce noise levels.
     2. Keep machinery in good maintenance repair to minimize noise.
     3. Erect total or partial barriers to confine noise.

   - **Administrative Controls**—
     1. Limit employees' scheduled work time in a noisy area.
     2. Limit noisy operations and activities per shift.

5. *Hearing Protectors*—Employers must provide at no cost to the employee a selection of hearing protection appropriate for noise levels in the environment. The employer must also provide training on the selection, fitting, use, and care of hearing protectors and ensure that protectors are worn. Monitoring requirements include:
   - Use only measuring instruments that meet ANSI specifications.
   - Use a sampling strategy that will pick up all continuous, intermittent, and impulsive sound levels from 80 to 130 dBA, and include all of these sound levels in the total noise measurement.
   - Permit employees or their representatives to observe monitoring.
   - Notify employees of noise exposure at or above an 8-hour TWA of 85 dBA.

6. *Audiometric Testing Requirements*—
   - Employers must provide testing free of cost to employees with noise exposure equal to or above an 8-hour TWA of 85 dBA.

- Audiometers must be calibrated to meet ANSI standards.
- Only a licensed or certified audiologist, otolaryngologist, other physician, or a technician who is certified by the Council of Accreditation in Occupational Hearing Conservation or who has demonstrated competence in performing audiometric testing can perform such testing.
- Baseline testing must be preceded by at least 14 hours without workplace noise exposure.
- The use of hearing protectors during work hours may substitute for the 14-hour requirement.
- A baseline must be established within six months of first exposure or within one year if using a mobile van to test. Hearing protection must be worn from the sixth month until testing is performed.
- An audiogram must be obtained annually from the baseline date.

7. *Audiograms—*
   - Compare subsequent audiograms to the baseline audiogram to determine if there is a change in hearing threshold of 10 dBA or greater in either ear at 2000, 3000, and 4000 Hz (known as a standard threshold shift [STS]).
   - If an STS exists, the employer may retest the employee within 30 days and use the test results as the annual audiogram.
   - Employees not already using hearing protectors must be fitted with hearing protectors, trained in their use and care, and required to use them.
   - Employees already using protectors must be refitted, retrained in their use, and provided with hearing protectors that offer greater attenuation if necessary.
   - An employee should be referred for a clinical audiological evaluation or otological examination, as appropriate, if additional testing is necessary or if the employer suspects that a medical pathology of the ear has been caused or aggravated by the wearing of hearing protectors.
   - An employee should be informed of the need for an otological examination if a medical pathology of the ear that is unrelated to the use of hearing protection is suspected.

8. *Training and Documentation—*
   - A training and education program should be implemented for those employees whose noise exposure equals or exceeds 85 dBA.
   - The training/education program should be repeated annually for employees included in the HCP.
   - The training program should include:
     1. The effects of noise on hearing and the purpose of hearing protectors
     2. Advantages, disadvantages, and attenuation of various hearing protectors and instructions on how to select, fit, use, and care for them
     3. Purpose of audiometric testing and an explanation of the testing procedure
   - Audiometric test records must include name and job classification of the employee, date of the test, examiner's name, date of the last acoustic or exhaustive calibration of the audiometer, and the employee's most recent noise exposure assessment.

- Audiometric test records should be retained for the duration of the affected employee's employment.
- Noise exposure measurement records should be retained for two years.
- Test room background noise measurements should be recorded and maintained.
- Access to audiometric test records and noise exposure measurement records should be provided upon request to the employee, former employees, the employee's designated representative, or the Assistant Secretary of Labor for Occupational Safety and Health.

9. ***Reducing Noise in the Workplace***—To be successful, an occupational HCP must be conducted using basic management concepts, beginning with the active support of management. It is management's responsibility to provide and enforce the use of hearing protection, whether that protection is wearing personal protective equipment or engineering controls. All efforts taken to initiate the program should be documented. All engineering control efforts should be recorded and used in determining the feasibility of control.

10. ***Abatement Suggestions***—
   - Mount tabletop equipment on rubber feet or pads.
   - Install sound-absorbent floor tiles.
   - Use acoustical ceiling tiles and wall hangings where possible.
   - Install mufflers where possible on generators, air compressors, etc.
   - Decrease volume of intercom speakers, televisions, and radios.
   - Keep wheels, hinges, and latches lubricated.
   - Adjust door closing mechanisms to prevent slamming.
   - Use sound-absorbent materials wherever possible.
   - Enclose noisy equipment and reduce metal-to-metal contact.
   - Limit worker exposure by implementing administrative controls.

11. ***Hearing Protection Evaluation***—Ear protection provided for employees is effective in substantially reducing noise exposure. Such protection may be provided as either earplugs, earmuffs, or both. Employers are required by the standard to evaluate the sound attenuation provided by ear protectors for the specific environment in which the protector will be used. Evaluation methods that must be used, according to 29 CFR 1910.95(j)(1), are described in Appendix B of the OSHA standard.

Ultrasound Exposures—Ultrasound is the mechanical vibration of an elastic medium that is produced in the form of alternating compressions and expansions. The vibrations may be produced by a continuous or impulse sound in the form of a sequel of interrupted vibrations.

- Exposure to audible high-frequency radiation above 10 kHz can result in nausea, headaches, ringing in the ears, dizziness, and fatigue.
- Temporary hearing loss and threshold shifts are possible from high-frequency ultrasound radiation.

- Low-frequency ultrasound radiation may produce effects if a person touches parts of the materials being processed by the ultrasound.
- Exposure to powerful sources may damage nervous and vascular structures at the point of contact.
- Airborne ultrasound may affect the central nervous system and organs through the ear. Exposures can be reduced by the use of enclosures and shields.
- Workers should be provided with appropriate personal protective equipment for the task being performed. Protection should also be provided for exposure to radiation above 10 kHz or when in contact with low-frequency sources.

M. PLANT MANAGEMENT AREAS

Utilities Management—Building utilities and utility systems are critical to the daily operation of healthcare facilities.

1. *Joint Commission Requirements*—The Joint Commission on Accreditation of Healthcare Organizations stresses the importance of several key components in the utilities management program:
   - A current inventory must be kept for all systems that support patient care areas. This inventory must detail systems that contribute to the effectiveness of life support, infection control, equipment support, and environmental support.
   - Facilities must have procedures to ensure that planned and preventive utility maintenance is conducted as required to ensure each system is properly identified, operated, and maintained.
   - Each component of a system must be evaluated to determine testing procedures, inspections, calibration, and parts servicing. Preventive maintenance programs should stress a thorough training program for operating and maintenance personnel.
   - Facilities should develop a contingency plan that outlines procedures to ensure prompt repair of failed systems. The plan should also address the facility's policies for dealing with total failures and subsequent evacuation of patients.
   - Facilities must maintain records and documentation for all inspections, tests, and repairs. Policy and procedure manuals should be available and used in servicing utility systems.

2. *Emergency Procedures*—Utility plans must:
   - Address specific procedures to be followed during disruptions and failures
   - Identify alternate utility sources
   - Discuss operating procedures for shutoff controls
   - Provide for notifying staff in affected areas
   - Provide guidance on how to obtain repair services
   - Cover the requirements and methods for emergency clinical interventions

3. *Electrical Power Distribution*—A program of preventive maintenance and inspection helps ensure that the electrical distribution system and its components operate safely and reliably. Components include panels, switch gears, transformers, electrical closets, receptacles, and elements of the emergency power system.

4. ***The National Electrical Code (NFPA 70 and NFPA 99)***—These codes establish standards and practices for maintaining electrical distribution systems.
   - Policies and procedures should detail the action to be taken in the event of an electrical distribution system failure.
   - It is important that facilities develop policies that define the actions to be taken during a failure of essential electrical systems, equipment, or sources of electrical supply.
   - An electrical distribution system is composed of numerous elements. It is important that safety procedures relating to each element be part of the total safety program.
   - A ungrounded electrical system isolates the area served from the grounding network of the building's electrical system or a grounded three-wire system. Each type offers protection from electric shock by shunting fault currents to ground.

5. ***Emergency Power System Requirements***—The facility must maintain a reliable system to supply electricity during interruption of normal service. Refer to NFPA 99 and NFPA 110, Standard for Emergency and Standby Power Systems for additional information. Areas to be supplied include:
   - Blood, bone, and tissue storage units
   - Emergency care areas
   - Medical air compressors
   - Surgical vacuum systems
   - Newborn nurseries
   - Surgery and delivery rooms
   - Recovery and special care units
   - Alarm systems, exit signs, and egress illumination
   - Elevators and emergency communication systems

Vertical/Horizontal Transport—Multi-story buildings use transport systems to move people, supplies, food, medicine, and messages. Vertical and horizontal transport systems include passenger and freight elevators, escalators, moving walkways, conveyors, and pneumatic tube systems. The location of a facility's elevators is crucial to the successful movement of people and materials throughout the building.

1. The daily operation of any multi-story facility can be considerably hampered if one or more elevators become inoperable. The loss of any transport system could cause modification of facility operations.
2. It is important to establish a comprehensive management/maintenance program to address the following:
   - Maintenance and inspection procedures
   - System components
   - Permits and licenses
   - System failures
   - Training and education programs

Plumbing—Plumbing systems are composed of equipment used for the supply, distribution, and disposal of potable and non-potable water. Water is often processed at the facility for purposes such as softening, sterilization, deionization, distillation, and treatment for boiler feed water.

1. Pumps, valves, fittings, traps, drains, vacuum breakers, and back-flow prevention devices are among the parts of a typical plumbing system. At the user end are the fixtures, including lavatories, urinals, water closets, showers, bathtubs, sinks, water fountains, and utility sinks.
2. The plumbing system also provides for the removal of wastewater and stormwater, usually through discharge into a municipal or rural sewer system.
3. Care must be taken in dumping certain chemicals into this system. Unless they are handled and disposed of properly, the various waste systems can be damaged.
4. Design and use of a plumbing system usually require strict adherence to state or local codes and environmental regulations.
5. Maintenance personnel in healthcare facilities must be trained in the hazards of exposure to bloodborne pathogens and explosives in laboratory systems.
6. Plumbing cross-connections that merge with potable and non-potable water sources can be candidates for waterborne diseases. The prevention of back pressure can be attained by removing the cross-connections or installing connection control devices. Consider the following possible scenarios:
   - Hose submerged in a laboratory sink
   - A hazardous materials tank with a submerged inlet
   - Water supply to dishwasher without a vacuum breaker
   - Valve connection between potable and non-potable water supplies

Boiler/Steam Systems—The American Society of Mechanical Engineers (ASME) Boiler and Inspection Code covers the design, fabrication, and inspection of boilers during construction. The National Board of Boiler and Pressure Vessel Inspectors has published an inspection code to be used after installation.

- Boiler systems should be maintained according to the instructions provided by the manufacturer. A boiler is simply a closed vessel in which water is heated by a heat source to form steam or hot water under pressure. Most boilers are either water-tube or fire-tube boilers. Boilers utilize a number of different fuels, depending on the design of each particular system.
- An effective management/maintenance program stresses several elements, including water treatment, maintenance and inspection, system components, permits and licenses, system failure, emergency shutdown, and training and education programs. The water that goes into and through a boiler eventually supplies hot water or steam throughout a healthcare facility; the treatment of the water is of prime importance.
- In an effective maintenance program, water used for boiler operations is treated to protect the boilers and related equipment and system lines from corrosion buildup within the boiler or lines. A corrosion buildup can result in malfunction, inefficiency, and eventual breakdown. In addition, the steam provided needs to be suitable for its intended end use.

- A program of preventive maintenance and periodic inspection will help assure that the boiler and steam systems operate safely and reliably. Periodic inspections of the systems will permit an immediate check of integrity and safety between preventive maintenance checks and tests.
- An effective management program provides for the proper maintenance of boiler controls and system safety controls or devices. NFPA 85A–E, Boiler Furnace Standards also detail minimum design and installation requirements for high-pressure boilers. Other standards that provide guidance are the ASME Boiler and Pressure Vessel Code (1990) and ANSI/ASME PVHO-1A-1990, Safety Standard for Pressure Vessels for Human Occupancy.

Medical Gas Systems—These systems provide oxygen, nitrous oxide, and compressed air throughout a healthcare facility. When such systems are used, heavy and bulky bottles or tanks do not have to be physically transported throughout a building. Refer to CGA Pamphlet P-2, Characteristics and Safe Handling of Medical Gases.

1. A major concern in the use of a medical gas system is that the right gas is connected to the right system lines via special connectors.
2. Also of concern are the purity of the gas provided and the maintenance of adequate line pressure.
3. An effective management/maintenance program for medical gas systems addresses the following concerns:
   - Maintenance and inspection of system components and failures
   - Responsibility for the effectiveness and safety of the system
   - Testing for proper configuration
   - Training and education programs
4. A program of preventive maintenance and periodic inspection helps assure that the medical gas systems operate safely and reliably.
5. As part of an effective management/maintenance program, inspections and corrective actions should be documented, and any faulty fittings should be repaired or replaced immediately. In the event of a system outage, it is important to notify nursing personnel and medical staff.
6. In addition, labeling shutoff controls and providing signs at outlet locations, as well as identifying piping, will help assure safety in a facility. System piping is also to be kept free from contamination.
7. It is essential that cylinders be protected from the extremes of weather, that empty and full gas cylinders be kept separate from each other, and that empty cylinders be marked.
8. NFPA 99 contains technical information on flammable agents and non-flammable gases.
9. Bulk medical gas systems involving oxygen and nitrous oxide should meet requirements of CGA Pamphlet 8.1, Standard for the Installation of Nitrous Oxide Systems or NFPA 50, Standard for Bulk Oxygen Systems at Consumer Sites.

Medical/Surgical Vacuums—Many healthcare facilities have medical/surgical vacuum systems which are used instead of portable units. Medical/surgical vacuum systems are used for aspiration during patient treatment and certain surgical procedures and for the evacuation or removal of non-flammable anesthetic waste gas.

- Laboratories should have a separate vacuum system to prevent contamination of and damage to the vacuum system used for patients.
- An effective program for medical/surgical vacuum systems addresses maintenance and inspection programs, system failures, testing for correct piping network configuration, and training and education programs.
- A program of preventive maintenance and periodic inspection will help assure that medical/surgical vacuum systems operate safely and reliably.
- Components to be inspected include vacuum pumps, receivers, valves, inlets, terminals, gauges, meters, switches, system vacuum alarms, and protective devices.
- As part of an effective safety program, major valves, inlets, and piping should be tagged or identified and emergency shutoff controls labeled throughout each system.

Communication Systems—Communication systems play a useful and vital role in healthcare facilities by assisting patients and staff in obtaining the help they need to deal with both routine and emergency situations as expeditiously as possible.

- Communication systems include such basic components as the telephone system, internal and external paging systems, the nurse call system, and data exchange systems (computers).
- As equipment has grown more complex, the Joint Commission has been concerned that the programs developed to monitor the equipment reflect this growing complexity so that the basic channels of communication of patient safety will not be lost.
- To address this concern, it is essential that communication systems operate effectively and reliably and that they are designed in accordance with the services provided by the organization. Refer to NFPA 101 and NFPA 99 for information on life safety and emergency communications.

Medical Equipment Maintenance—Maintaining medical equipment is costly and has resulted in many healthcare organizations establishing in-house clinical engineering departments to maintain the growing inventory of medical and patient care equipment. Many highly specialized items such as ventilators, pulse oximeters, EKG monitors, defibrillators, nebulizers, ultrasound equipment, lasers, and X-ray machines must be maintained by the department or outside contractors. The department's main goal should be to maintain equipment in safe operating condition.

1. *Departmental Functions*—Primary functions include:
   - Provide fast and efficient repair services to all departments
   - Manage an effective preventive maintenance program
   - Conduct operator training and safe-use instruction
   - Maintain accurate records on all medical equipment
   - Provide cost analysis and management assistance

2. ***Equipment Maintenance***—Maintenance requirements address the need for preventative maintenance. Equipment that is used frequently, is used in a harsh environment, or is wholly or partially a mechanical assembly is likely to require preventive maintenance. Equipment that employs solid-state technology often has no requirement for preventive maintenance.
   - Equipment incident history involves measuring the reliability and actual performance of equipment. Through an ongoing process, equipment with a history of breakdown, misuse, or malfunction must be identified.
   - An equipment management program that identifies, maintains, and improves performance/safety will help detect damage caused by abuse or neglect.

3. ***The Joint Commission Four-Point Process***—The Joint Commission views equipment management as a four-step process:
   - Develop a policy to meet the organization's goals and objectives.
   - Establish equipment performance monitoring procedures.
   - Collect and document equipment performance data.
   - Analyze data to identify problems and ways to strengthen the program.
   Note—Each type of patient care equipment has special maintenance requirements.

4. ***Management Procedures***—Procedures covering patient care policies, testing requirements, and operating performance standards should be developed. Sources of information include codes, standards, manufacturers' requirements, and experience.
   - The Joint Commission has traditionally specified a six-month interval for testing of equipment. The risks associated with a piece or type of equipment will often suggest a higher or lower level of management intensity.
   - The Joint Commission standards give organizations an opportunity to select intervals appropriate for identified risks.
   - Effective programs allow a variety of intervals, reflecting the identified risk levels; the only constraint is a specified one-year maximum interval for any equipment in a test program.
   - Training and education processes begin with the initial decision to purchase equipment.
   - A pre-purchase evaluation that involves medical, nursing, technical, and administrative personnel will set the course for choosing safe, effective, and serviceable equipment.
   - Involving employees in the pre-purchase evaluation will permit user and service training programs to be developed.

5. ***Electrically Powered Equipment Standards***—Previous Joint Commission standards on electrically powered equipment required facilities to perform a basic safety check on all non-clinical equipment. It is the responsibility of a facility to design a program that identifies areas or uses that pose a significant risk of injury from electric shock and to manage that risk through periodic preventive maintenance.
   - Electrical safety is the goal of this portion of an organization's equipment management program.

- Keeping electrically powered equipment clean internally and externally and tracking the deterioration of wearing components will help ensure that equipment is replaced when necessary and will help avoid costly breakdowns.
- An effective program also addresses personally owned electrical devices.
- A facility should look at three important factors:
    1. Impact of the equipment on the quality of care
    2. Cost of each piece of equipment
    3. Ways in which advanced technology affects traditional preventive maintenance
- The standard emphasizes the identification of conditions or situations that pose a significantly high risk to patients. As part of a planned program of risk management, an organization should establish a system to manage the risk posed by electrical devices and to record and document all findings.
- The standard provides the opportunity to significantly reduce the amount of electrically powered equipment that is to be inspected. By placing an emphasis on areas that pose the greatest risk, a safer environment for patients and employees will be created.
- Refer to NFPA 99, Chapter 7 for additional information.

6. ***Summary of Joint Commission Equipment Management Requirements***—The Joint Commission has established requirements that permit organizations to evaluate the risks associated with biomedical and electrical equipment.
    - An effective management program becomes part of a network vital to effective quality assurance, risk management, and safety management.
    - Equipment management must be evaluated in the performance measurement of all departments or services affected.
    - The sophistication of equipment requires that healthcare engineers and clinical staffs work together. The success of an integrated program depends on the communication links between the two groups.
    - Equipment functioning addresses what role a type of equipment plays in patient care. The equipment may, for example, provide life support or it may measure and present information used in diagnosing or monitoring the physiologic condition of a patient.
    - Clinical application involves evaluation of possible electrical, mechanical, or physiologic risks associated with use of equipment. Much biomedical technology, such as infusion pumps, respirators, and dialysis equipment, requires invasive connection to the patient's body and presents risk of injury.
    - Research devices are often prototypes and can pose unusual risks due to incomplete designs or "rough" construction and packaging.

7. ***Investigating Problems and Incidents***—The Safe Medical Device Act (SMDA) of 1990 requires all healthcare facilities to report all medical device-related serious injuries and deaths to the manufacturer. The Food and Drug Administration (FDA) must be notified of all incidents that result in death. The SMDA gives FDA investigators access to the facility and to organizational equipment records. Healthcare facilities should:
    - Develop a medical device management program

- Establish formal reporting and investigation procedures
- Train staff members how to respond to potentially serious medical device incidents
- Never release the device or accessories to any outside party until independent testing has been conducted and documented

N. HEATING, VENTILATION, AND AIR CONDITIONING SYSTEM OPERATION

NFPA 90A, Installation of Air Conditioning and Ventilating Systems details specific suggestions for maintaining systems. Proper temperature, humidity, and air flow will provide a comfortable environment inside a building regardless of climatic conditions outside. Ventilation is in accordance with the guidelines established in the American Society of Heating, Refrigerating, and Air Conditioning Engineers (ASHRAE) Standard 62-1989. The initial design of a facility's heating, ventilating, and air conditioning (HVAC) system is the first step in providing a comfortable environment. During design stages, system loads and building capacities are estimated and calculated according to the intended occupant load of the building. Future expansion possibilities and potential load requirements should be considered at that time.

1. *General Requirements*—
   - To maintain safety in the event of an emergency shutdown, a current and complete set of documents indicating the distribution of the HVAC system and controls for partial or complete shutdown should be maintained.
   - Systems normally are not sufficiently marked, and maintenance personnel are seldom able to maintain HVAC systems without plans showing the system components and branch outlay.
   - Equipment failures often result in system outages which may result in the loss of heating or cooling, special environmental support such as hood exhaust systems, and temperature conditions required to maintain the operating status of equipment.
   - Training programs help provide maintenance and operating personnel with appropriate information on the HVAC systems and equipment.
   - Training should focus on the technical aspects of maintaining the systems and on the role the systems play in specific areas of buildings.

2. *Correctly Designed and Functioning System*—
   - Provides thermal comfort to all building occupants. ASHRAE Standard 55-1981 describes temperature and humidity ranges for building occupants.
   - Distributes and blends the proper amounts of outdoor and recirculated air to meet ventilation requirements in ASHRAE 62-1989.
   - Isolates and removes odors/contaminants through pressure control, filtration, and exhaust systems. Positive-pressure rooms have more air supplied than is exhausted, whereas negative-pressure rooms have less air than is exhausted.

3. *Maintenance*—A maintenance plan should describe the equipment covered, maintenance procedures, and frequency schedule. HVAC systems must be operated

in accordance with their original design or that applied when the facility was constructed or renovated, whichever is the most recent.

- Employers must implement controls for specific contaminants and their sources within indoor work environments.
- Employees must be notified at least 24 hours prior to the use of any cleaning or other chemical substance in the facility.
- Buildings that are not already smoke-free must provide a separate enclosed room that is vented directly outside as a smoking area.
- The room must be maintained under negative pressure. Smoking is not permitted during the cleaning of the room, and employees should not have to enter under normal working conditions.
- Signs must be posted that clearly inform all persons entering the workplace about the smoking restrictions.

4. *Worker Training*—Maintenance and operation workers must be trained in:
- The use of personal protective equipment
- How to maintain adequate ventilation during cleaning and maintenance
- How to minimize adverse effects during use and disposal of chemicals
- How to detect and report building-related illness situations

Indoor Air Quality—The quality of air inside healthcare buildings should be monitored because many infectious agents can be spread through the air and ventilation systems. Healthcare facilities also use a number of chemical substances that can contaminate the air. The poor ventilation in many modern buildings can be attributed to ventilation systems that bring in an insufficient volume of outside air. ASHRAE increased outside air requirements by four times in its revised 1989 standard.

1. *Factors Affecting Indoor Air Quality (IAQ)—*
- Source of contamination or discomfort
- HVAC system not able to control air contaminants
- HVAC system not able to maintain occupant comfort
- Pollutant pathways connect the source to occupants
- Occupants in the building with the contaminants

2. *Factors Contributing to Good IAQ—*
- Proper ventilation
- Temperature, humidity, and air movement
- Maintenance of equipment and building surfaces
- Isolation of emission sources from occupied spaces
- Major contamination sources properly controlled
- Controlling maintenance and construction contaminants

3. *Types and Symptoms of IAQ Complaints—*
- Headache
- Sneezing
- Fatigue
- Eye, nose, and throat irritation
- Shortness of breath

- Skin irritation
- Sinus congestion
- Dizziness
- Cough
- Nausea

4. ***Indoor Airborne Contaminants***—Hospitals can be home to a variety of airborne contaminants. IAQ is concerned with the safety of air in non-industrial environments such as offices and healthcare facilities. OSHA has proposed amending 29 CFR 1910.19 to address IAQ issues.

5. ***Chemical Contaminants***—Most air contaminants refer to substances contained in vapors from paint, cleaning substances, pesticides, solvents, particulate materials, outdoor air pollutants, and other airborne substances.
   - Three of the most common healthcare air contaminants, ethylene oxide, glutaraldehyde, and formaldehyde, are discussed in Chapter 9.
   - Healthcare facilities must be aware that new carpet and particleboard can release volatile organic compounds such as formaldehyde.
   - Healthcare facilities must also contend with a number of other potential contaminants such as antibiotics and antineoplastic drugs.

6. ***Microbial Contamination***—Healthcare facilities must also be concerned about microbiological contamination. A microbe is a small living organism that can contaminate ventilation systems.
   - Microbes such as Legionnaires' bacilli have been discovered in the cooling towers of healthcare facilities.
   - Wet, moist, and damp areas can be breeding grounds for microbes that can become airborne and cause problems for workers.
   - Healthcare organizations are also becoming concerned about tuberculosis.

7. ***Sources of Indoor Air Contaminants***—
   - **Outdoor Sources**—Outdoor air contaminants include pollen, dust, fungal spores, industrial pollutants, and vehicle emissions; emissions from nearby sources such as vehicle exhaust from roadways or parking garages; odors from dumpsters; pollutants drawn in from the building itself or other sources; unsanitary conditions or debris near the outdoor air intake; soil gases such as radon; leakage from underground storage tanks; contaminants from landfills; and pesticide usage.
   - **Moisture Sources**—Moisture or standing water promotes excess microbial growth from cooling towers, flat roofs, or in crawlspaces. HVAC systems can contain contaminants in ducts or microbiological growth in drip pans, humidifiers, and coils.
   - **Other Sources**—Sources include improper use of biocides, sealants, and cleaning materials; improper venting of combustion products; refrigerant leaks; emissions from office equipment, especially volatile organic compounds; and elevator motors and other mechanical systems. Carpets, curtains, and other furnishings can also contain or emit pollutants.

8. *IAQ Terms*—There are three basic IAQ terms that should be understood:
   - **Sick Building Syndrome (SBS)**—Describes a situation in which a building is believed to be contaminated but no specific illness can be identified.
   - **Building-Related Illness (BRI)**—Refers to an illness brought on by exposure to contaminants in a building. Legionnaires' disease and hypersensitivity pneumonia are two examples of BRIs that can have serious consequences.
   - **Multiple Chemical Sensitivity (MCS)**—A controversial term that is not recognized by many medical organizations. MCS is simply being exposed to a number of substances at small concentrations and experiencing health problems as a result.

Preventing IAQ Problems—Healthcare facilities should consider establishing a written IAQ program to help manage the ventilation system. The written program should cover a number of areas, including maintenance, testing, monitoring, and training. The program should have established protocols to deal with complaints. Some basic things can help prevent IAQ problems:

- *Preventive Maintenance*—Effective building care, HVAC maintenance, and intelligently designed renovations.
- *Analyze Ventilation System*—The building's ventilation system should be analyzed for comfort, ventilation, and sanitation. This can be accomplished by inspecting accessible areas for obvious problems, poor design, or signs of contamination. The inspection should determine airflow, temperature, humidity, carbon dioxide concentration, and air pressure differentials.
- *HVAC System Inspections*—The HVAC system must be periodically inspected and serviced. Many buildings have been designed to be "energy efficient," and the result is an inadequate or contaminated supply of outside air.
- *Effective Filtration System*—A workplace could be contaminated with pollen if the air filtration system does not work properly. An excellent reference source is the NIOSH/EPA publication entitled "Building Air Quality—A Guide for Building Owners and Facility Managers."

Building-Related Illnesses—BRIs refer to specific medical conditions which can be determined by physical signs and laboratory findings. BRI includes any medical condition that prompts an evaluation of the building to determine whether HVAC systems are operating correctly. The following can be classified as BRIs caused by indoor contaminants:

- Respiratory allergies
- Nosocomial infections
- Asthma
- Humidifier fever
- Hypersensitivity pneumonitis
- Legionnaires' disease (*Legionella* bacteria)
- Carbon monoxide poisoning
- Formaldehyde poisoning
- Pesticide exposure

BRI Assessment—
- Designate a person responsible for implementing the program.
- Prepare a written narrative description of the building systems.
- Refer to schematic drawings that describe the building's systems.
- Establish normal operating procedures for the building's systems.
- Describe the building's function, activity, and systems.
- Document known releases of air contaminants.
- Establish a preventive maintenance program covering all standards and codes.
- Prepare a checklist for visual inspection of all systems.

Investigating IAQ Complaints—Investigators must follow safety precautions during routine inspections of buildings with suspected IAQ problems.

1. *Potential Safety Hazards—*
   - Electrical hazards
   - Injury from contacting fans, belts, and dampers
   - Burns from steam or water lines
   - Falls from ventilation shafts, ladders, or roofs

2. *Microbiological—*Use extreme care when serious illnesses are under investigation. People prone to allergy problems should be cautious. Minimize exposure to air ducts or when contaminants are suspected to be growing. Use respiratory protection if visible contamination exists.

3. *Chemical—*Use caution if severe chemical contamination is suspected. Follow precautions for exposure to the suspected substance(s). A pesticide spill in a confined space would require appropriate protective equipment, including a respirator.

4. *Asbestos—*When conducting an IAQ investigation, contact with asbestos is likely. Take appropriate precautions, including wearing overalls and a respirator.

5. *HVAC Systems—*Compare the original design to the current system. Evaluate changes in ventilation and temperature control zones. Note changes in HVAC equipment. Also review operating procedures for unoccupied and occupied periods. Evaluate the condition of the system and look for unsanitary conditions such as moisture, standing water, debris, mold growth, or excessive dust. Consider the following HVAC malfunctions:
   - Equipment breakdown
   - Obstructed grilles or diffusers
   - Air distribution and mixing problems
   - Condition of air bypass filters
   - Leaks in the distribution system
   - Adjustments in ventilation-affected areas

6. *Pollutant Pathways—*Attempt to identify all pathways that could allow contaminants to enter and consider:
   - Doors and operable windows
   - Stairways

- Elevator shafts
- Utility chases
- Ductwork and plenums

Ventilation—OSHA standards cover ventilation requirements for a variety of operations including abrasive blasting, grinding/polishing operations, and spray finishing operations.

1. *Mechanical Ventilation*—Mechanical ventilation systems should be designed to bring in outside air and mix it with a percentage of return (inside) air. The air is then cooled, heated, or humidified and distributed.
   - Over 50% of all IAQ problems are the result of insufficient or ineffective ventilation.
   - ASHRAE has set limits on the amount of outside air that must be brought in.
   - General office space requires 20 cubic feet of intake air per minute for each occupant.
   - Carbon dioxide is a good indicator of insufficient intake air. OSHA's proposed IAQ standard specifies that levels be below 800 ppm. When levels reach 1000 ppm, ventilation is not adequate. *Note*—OSHA's permissible exposure limit for carbon monoxide is 5000 ppm.
   - ASHRAE 62-1989, Ventilation for Acceptable Indoor Air Quality specifies minimum ventilation rates and IAQ needed to prevent adverse health effects to occupants.

2. *Engineering Standards*—Standards also cover some engineering requirements for various operations. Healthcare facilities should refer to NFPA 90A, Standard for the Installation of Air Conditioning and Ventilation Systems.
   - Local exhaust ventilation systems must conform to the construction, installation, and maintenance requirements found in ANSI Fundamentals Governing the Design and Operation of Local Exhaust Systems, Z9.2-1960 and ANSI Z33.1-1961.
   - Information on local exhaust duct system, independent exhaust, and room intakes is provided in 29 CFR 1910.107d.
   - Ventilation requirements for inside storage rooms with flammable materials are provided in 29 CFR 1910.106.
   - OSHA also covers a number of air contaminants in 29 CFR, Subpart Z. Laboratory ventilation is referenced in 29 CFR 1910.1450.
   - Hospitals must also be concerned about ventilation and respiratory protection areas that have hazardous materials such as anesthetic gases and formaldehyde. Refer to NFPA 99 for additional guidance.

3. *Engineering Controls*—Two basic engineering controls for ensuring adequate ventilation in a work area are defined below:
   - Local exhaust refers to the method designed to capture airborne contaminants near the point of generation or release. The system should draw the contaminant away from a person's breathing zone.
   - Mechanical exhaust involves the continuous introduction of fresh air into the workplace to dilute the contaminated air and lower the concentration of the

hazardous substance. The effectiveness depends on the number of air changes per hour.

4. ***Fume Hoods and Ventilation***—A fume hood or fume removal system is a device used to capture hazardous air contaminants such as vapors, dusts, mists, gases, and metal fumes. A fume removal system consists of:
   - A blower, which is the major component that must be carefully selected. Blowers must be sufficient to remove the contaminant. They are sized or rated by the amount of cubic feet of air moved at a given resistance. This resistance or static pressure can be influenced by type of collection mechanism and length of ducting.
   - Blower flywheels are available in a number of materials, depending on the contaminant being removed. Some are designed to be explosion-proof or non-sparking.
   - Ducting materials are normally galvanized steel, stainless steel, or PVC-type materials.
   - Air purification devices include chemical adsorption and mechanical filters which remove particulate matter.
   - Collection devices include:
     a. Cabinet hoods are often used in laboratories because they are effective against a variety of chemicals. This three-sided enclosure is normally made of chemically resistant materials. Air is pulled through the front and away from the worker.
     b. Canopy hoods are mounted on walls or hung from ceilings over the work area. They are effective for contaminants that rise.
     c. Local collection hoods are directly attached to the duct. They are designed for operations where a contaminant is generated at a specific place.
   Note—NFPA 45, Standard on Fire Protection for Laboratories Using Chemicals has information on fume hood requirements.

5. ***Definitions***—
   - An anemometer measures air velocity, normally in feet per minute.
   - Capture velocity refers to the velocity of air produced by a hood to capture contaminants outside the hood area.
   - Dilution ventilation is an exposure control method that uses an air purification device and returns the exhaust to work area air.
   - A manometer measures pressure differences, usually in inches of a water gauge.
   - Static pressure is developed in a duct by a fan.

ASHRAE Standards and Guidelines—

1. ***Standard 62: Ventilation for Acceptable Air Quality***—This standard can assist professionals in the proper design of building ventilation systems. Important aspects of the standard include:
   - A definition of acceptable air quality
   - Information of ventilation effectiveness
   - Recommendation on using source control through isolation and local exhaust

- Information on the use of heat recovery ventilation
- A guideline for allowable carbon dioxide levels
- Appendices listing guidelines to control common indoor pollutants

2. *Standard 55: Thermal Environmental Conditions for Human Occupancy*—This standard covers several areas, including temperature, humidity, and air movement. Important aspects of the standard include:
   - A definition of acceptable thermal comfort
   - Information on environmental parameters that must be considered
   - Recommendations for summer and winter comfort zones for humidity and temperature
   - Guidelines for making measurements

3. *Standard 52: Method of Testing Air-Cleaning Devices in General Ventilation for Removing Particulate Matter*—This standard can assist professionals in the evaluation of air-cleaning systems for particle removal. Highlights of the standard include:
   - Definitions of arrestance and efficiency
   - Information about the uniform comparative testing procedure
   - Establishment of a standard reporting method for performance
   - Methods of assessing resistance to airflow and dust-holding capacity

4. *Guideline 1: Guideline for the Commissioning of HVAC Systems*—This guideline can assist professionals by providing methods and procedures for documenting the performance of HVAC systems. Important features include:
   - Information on the commissioning process
   - Sample forms and specification logs
   - Recommended corrective measures
   - Operator training requirements
   - Information on periodic maintenance and recommissioning

5. *Standard 52-76: Method of Testing Air Cleaning Devices Used in General Ventilation for Removing Particulate Matter*

Refrigerant Recycling—In May 1993, the U.S. Environmental Protection Agency established a comprehensive recycling plan under the Clean Air Act, Section 608 for ozone-depleting refrigerants. The plan addresses the service and disposal of air conditioning and refrigeration equipment. The regulations require persons servicing such equipment to follow certain practices to reduce emissions. The rules also address equipment/reclamation requirements and technician certification requirements.

- The final rules established leak repair requirements for equipment holding 50 pounds or more of refrigerant.
- Ozone-depleting compounds contained in appliances must also be removed prior to final disposal of the appliance.
- The federal standard for purity when reprocessing refrigerants has been updated to incorporate the 1993 edition of the Air Conditioning and Refrigeration Institute's standard for purity.

- The 1993 rules established a certification requirement for recycling and recovery equipment to verify that all recycling or recovery equipment and reclaimed refrigerant sold is of known acceptable quality to avoid failure of equipment by contaminated refrigerant.
- All persons involved in maintenance, service, and repair of refrigerant equipment must have at least one piece of certified, self-contained recovery equipment on the premises.

O. PRINT SHOP SAFETY

Print Shop Safety Guidelines—Facility print shops pose a hazard in a number of ways. Consider the following suggestions:

- Material Safety Data Sheets should be available for all hazardous substances used in the print shop.
- Smoking should be prohibited because highly flammable inks and solvents are used in the print shop.
- Water-based inks should be used whenever possible.
- Safety cans should be used to store all flammable liquids.
- Ink-cleaning chemicals should be dispensed from plunger-type safety cans.
- All rags soaked with solvent or solvent-based ink should be disposed of in covered metal containers that are emptied at least daily.
- Ventilation should be provided as needed to control airborne concentrations of solvents and other toxic substances used in the print shop.
- The cutting edge of guillotine paper cutters should be guarded; two-hand controls are an effective method of reducing this hazard.
- All gears, belts, pulleys, and pinch points should be guarded.
- Because printing equipment produces a noisy environment, control measures should be implemented to reduce noise to the lowest possible level. Adequate hearing protection should be provided, and surveys of noise level should be conducted routinely.

SUMMARY

This chapter covered a wide variety of topics, hazards, and management programs that should be addressed by healthcare organizations on a continuing basis. Some of the subjects are addressed in Joint Commission standards, while others are important OSHA compliance issues. The chapter provided a brief review of fire prevention principles that were covered more fully in Chapter 5. Information on topics such as ergonomics, office hazards, and video display terminal safety was presented. The chapter briefly covered some hazards that should receive management attention throughout the facility, including material handling, indoor air quality, and preventing slips, trips, and falls. A large portion of the chapter presented information on hazards found in healthcare maintenance departments, including sections on correct tool use, ladder safety, electrical safety, hearing conservation, and working in confined spaces. The importance of effective equipment and utility management was also briefly touched upon. In addition, many hazards that can be easily overlooked or not adequately covered by the safety management program were presented.

REVIEW QUESTIONS

1 Which of the following areas is likely to experience the most trauma complaints?
   a. office areas and computer stations
   b. nursing stations
   c. kitchen and food preparation areas
   d. emergency rooms

2 A commonly reported affliction often associated with tingling, pain, or numbness in the thumb and the first three fingers is referred to as:
   a. carpal tunnel syndrome
   b. trigger finger
   c. Raynaud's syndrome
   d. DeQuervain's disease

3 Which of the following is true concerning video display terminal lighting?
   a. lighting should be bright and located directly overhead
   b. lighting should be behind the terminal and come in at 90 degrees
   c. walls should be a light color
   d. wall glare normally does not affect video display terminal lighting

4 A serious condition characterized by hot dry skin, dizziness, headache, thirst, nausea, and mental confusion is known as:
   a. heat exhaustion
   b. heat stroke
   c. heat cramps
   d. heat prostration

5 Standards that apply to healthcare maintenance and engineering departments can be found in:
   a. 29 CFR Part 1910
   b. 29 CFR Part 1926
   c. NFPA 99 and 101
   d. all of the above

6 Live parts of electrical equipment operating at ____ volts or higher must be guarded against accidental contact.
   a. 12
   b. 24
   c. 50
   d. 100

7   Devices used to shut off electrical power immediately by comparing the amount of current going to the equipment and the amount returning along the circuit conductors are:
a. grounding devices
b. fuses
c. circuit breakers
d. ground-fault circuit interrupters

8   The OSHA lockout/tagout standard applies to which of the following types of energy?
a. electrical
b. hydraulic
c. pneumatic
d. all of the above

9   The OSHA confined space standard requires that a space be designated as "permit required" when:
a. it contains a hazardous atmosphere
b. it contains material that could engulf a person
c. it is configured in such a manner that a person could become trapped
d. all of the above

10 Occupational noise exceeding an 8-hour time-weighted average of 85 dBA requires the:
a. use of hearing protectors by all workers
b. employer to provide baseline audiograms for exposed employees
c. employer to implement a hearing conservation program
d. all of the above

11 Which NFPA publications establish standards for maintaining electrical distribution systems?
a. NFPA 30 and 45
b. NFPA 70 and 99
c. NFPA 70 and 101
d. NFPA 101 and 130

12 Healthcare ventilation guidelines can be found in which of the following publication(s)?
a. ASHRAE 62-1989 and NFPA 70
b. ASHRAE 62-1989 and NFPA 99
c. ASHRAE 55 and 29 CFR 1910
d. ASHRAE 55 and NFPA 101

Answers: 1–a 2–a 3–a 4–b 5–d 6–c 7–d 8–d 9–d 10–d 11–b 12–b

CHAPTER 7

MANAGING HAZARDOUS MATERIALS

A. CHARACTERISTICS OF HAZARDOUS CHEMICALS

Each chemical substance is endowed with a unique set of properties that differentiates it from other substances. Properties and characteristics of hazardous materials include:

- *Corrosivity*—The ability to degrade the structure or integrity of another substance, object, or material (e.g., acids and alkalis)
- *Ignitability*—The ability to readily burn or ignite (some chemicals can auto-ignite upon contact with the air)
- *Reactivity*—The ability to readily combine with other chemicals, often with the sudden or violent release of heat or energy (e.g., chlorine gas)
- *Toxicity*—The ability to cause illness or death in man, animals, fish, plants, etc.

Hazardous Materials—Hazard control managers should obtain as much information as possible about each hazardous substance found in the facility. Some excellent information can be obtained from organizations such as the Occupational Safety and Health Administration (OSHA), the Centers for Disease Control and Prevention (CDC), the Environmental Protection Agency (EPA), the Nuclear Regulatory Commission, and the American Society of Hospital Pharmacists.

1. *Effects of Hazardous Materials*—Chemicals and hazardous materials can cause worker stress when inhaled, absorbed, or ingested through the mouth. The effects on humans depend on concentration, duration of exposure, route of exposure, physical properties, and chemical properties. The effects exerted by a hazardous substance can be influenced by other chemicals and physical agents.

2. *Identify Hazardous Substances*—Identify all chemical, biological, and radiation hazardous materials, including:
   - Quantity used
   - Quantity on hand and where stored

- Currency of Material Safety Data Sheets (MSDSs)
- Availability of a less hazardous substitute

3. **Classes of Hazardous Materials**—
   - Chemicals and disinfectants
   - Medical waste and sharps
   - Chemotherapeutic substances
   - Radioactive materials

4. **OSHA Hazardous Material Regulations**—Exposure to chemical hazards should be a top concern for hospital or healthcare hazard control managers. OSHA requires employers to provide workers with information about hazardous materials found in the workplace. OSHA has specific standards on the following toxic or hazardous substances in 29 CFR 1910:
   - Asbestos   1910.1001
   - Benzene   1910.1028
   - Bloodborne Pathogens 1910.1030
   - Ethylene Oxide   1910.1047
   - Formaldehyde   1910.1048

5. **Hazardous Chemical Determination**—Substances are considered to be hazardous if:
   - Regulated by OSHA in 29 CFR Part 1910, Subpart Z, Toxic and Hazardous Substances
   - Included in the American Conference of Governmental Industrial Hygienists' (ACGIH) latest edition of *Threshold Limit Values for Chemical Substances and Physical Agents in the Work Environment*
   - Found to be confirmed or suspected carcinogens by the National Toxicology Program in the latest edition of the *Annual Report on Carcinogens*
   - Listed by the International Agency on Research on Cancer in the latest edition of *IARC Monographs*

6. **OSHA Revision of PELs**—OSHA has started updating the permissible exposure limits (PELs) found in 29 CFR 1910.1000, Subpart Z. A 1992 decision by the 11th Circuit Court of Appeals vacated the 1988 revision to PELs for air contaminants. OSHA has enforced PELs set in the early 1970s and has indicated that it plans to address risk assessment methodology for non-carcinogens and carcinogens.

7. **NIOSH Policy Documents**—The National Institute of Occupational Safety and Health (NIOSH) has published policy documents on many hazardous substances. Documents are available on the following substances found in healthcare facilities:
   - Ammonia
   - Asbestos
   - Benzene
   - Benzidine
   - Dioxane
   - Ethylene oxide
   - Formaldehyde
   - Isopropyl alcohol
   - Phenol
   - Toluene
   - Waste anesthetic gases
   - Xylene

8. *Labeling*—Hazardous material labels should follow OSHA, EPA, National Fire Protection Association (NFPA), Department of Transportation, and Joint Commission on Accreditation of Healthcare Organizations guidelines. All labels should include:
   - Name of substance
   - Chemical name if appropriate
   - Hazard classification
   - Safeguards to use when handling the substance
   - Organs that could be affected by the substance

9. *Handling and Disposal*—Organizations must provide adequate handling and storage facilities. Workers should be trained how to safely handle, store, use, and segregate hazardous materials and waste products.

10. *Training*—Joint Commission standards outline training requirements for workers exposed to hazardous materials. The OSHA hazard communication standard found at 29 CFR 1910.1200 specifies training for users of hazardous chemicals. Biohazardous material handlers should conform to standards published in the OSHA bloodborne pathogens standard (29 CFR 1910.1030). The CDC publishes guidelines concerning infection control, isolation protocols, and tuberculosis prevention. Hazardous waste contingency training requirements are found in OSHA 29 CFR 1910.120 and EPA regulations.

B. HAZARD CATEGORIES

Toxic Materials—Toxic substances are poisonous. They can also have other qualities such as being flammable, combustible, corrosive, or irritants.

1. *Toxic Exposure Considerations*—When evaluating the toxicity of a substance, the following factors should be considered:
   - Concentration of the chemical
   - Duration of exposure
   - Available ventilation
   - Temperature of the chemical
   - Temperature of the surrounding air

2. *Toxic Substances Entrance Routes*—
   - Absorption through the skin and/or breaks in the skin (cuts, sores, or dermatitis)
   - Inhalation of contaminated air through the nose and mouth
   - Ingesting the substance through eating or drinking

3. *Precautions When Working Around Toxic Materials*—
   - Wear required personal protective equipment including respirators when necessary.
   - Never eat, drink, or smoke around toxic materials.
   - Cover cuts and other breaks in the skin carefully.
   - Know the exposure limits.
   - Follow ventilation instructions.
   - Wash thoroughly after exposure.
   - Take personal responsibility for self-protection.

TABLE 7.1   Hazardous Exposure Limits and Terms

Exposure Limit	Agency	Definition
TLV[a]	ACGIH	**Threshold limit value**—The airborne concentration of a substance to which most workers can be exposed on a daily basis without adverse health effects.
TWA	ACGIH	**Time-weighted average**—The average concentration of a substance for a normal 8-hour workday and 40-hour work week.
STEL	ACGIH OSHA	**Short-term exposure limit**—A 15-minute TWA that should not be exceeded at any time during the workday. There should be 60 minutes between each 15-minute exposure, up to four times a day.
CEILING (TLV-C)	ACGIH OSHA	**Ceiling limit**—The airborne concentration that is not to be exceeded at any time during the workday.
PEL[b]	OSHA	**Permissible exposure limit**—Limit based on 8-hour TWAs. Exposures below the PEL do not require respiratory protection.
REL	NIOSH	**Recommended exposure limit**—A TWA for up to a 10-hour workday during a 40-hour work week.
IDLH	NIOSH	**Immediately dangerous to life or health**—Levels may cause severe health effects which may impair a person or prevent the ability to escape from a dangerous situation.

Acute Exposure	Single exposure or several short-term exposures to a toxic substance.
Chronic Exposure	Long-term exposure at a rate at which the body cannot get rid of the toxic substance.
Toxicity	A hazardous substance that poses a poison hazard to human health. • Immediate toxicity occurs rapidly after a single exposure episode. • Delayed toxicity occurs after a lapse of time. • Systematic toxicity is characterized by effects at a place other than the point of entry. • Local toxicity occurs when effects arise at the point of entry.

[a] TLVs are given in parts per million (ppm) or milligrams per cubic meter of air (mg/m^3).
[b] PEL values can be found in 29 CFR 1910.1000.

Reactives, Oxidizers, and Corrosives—

- *Reactives*—Reactive substances can burn, explode, or give off hazardous vapors when mixed with other chemicals or when exposed to air or water. Reactive substances can self-ignite and the chemical reaction itself creates the hazard.
- *Oxidizers*—Oxidizers easily release oxygen which can fuel fires when stored near flammable substances. Even though most oxidizers will not burn by themselves, they can cause other materials to burn. Oxidizers should be stored separately from flammable/combustible materials. Storage areas should be kept away

from heat sources because warming causes oxygen release, which can create the perfect environment for a fire.

- *Corrosives*—Corrosive chemicals are materials that can eat through other materials, including human skin. Irritants such as ammonia are corrosive substances that attack the mucous membranes of the nose and mouth.

Flammable/Combustible Liquids—Flammable liquids have a flash point less than 100°F (38°C), whereas combustible liquids have a flash point of 100°F (38°C) or more.

- *Vapor*—Vapor is the gaseous state of material. Sometimes vapors can be smelled; other times their presence is not readily detected. When vapors from some materials combine with the oxygen in the air, they form a mixture that will ignite easily and burn rapidly, often with explosive force.
- *Flash Point*—The flash point of a liquid is the lowest temperature at which it gives off enough vapor to form an ignitable mixture with the air around it. Combustible liquids have a flash point of at least 100°F (38°C). Flammable liquids have a flash point less than 100°F (38°C). The lower the flash point, the more easily the liquid will burn and the more carefully it has to be handled.
- *Ignition Source*—An ignition source is anything that causes something to start burning. Common ignition sources include sparks from tools and equipment; open flames such as torches, smoking materials, and pilot lights; hot particles and embers generated while grinding or welding; and hot surfaces such as electric coils and overheated bearings.
- *Grounding*—Static electricity can be generated by the flowing of liquids. Grounding ensures that an electrical charge goes to the ground rather than building up on the drum of flammable or combustible material.
- *Bonding*—This process equalizes the electrical charge between the drum and the transfer container so that there is no buildup of electrical charges in either container.
- *Vapor Density*—Vapor density is the ratio of the weight of a volume of vapor or gas to the weight of an equal volume of clean but dry air. Vapor density is listed in the MSDS for a chemical. Knowing the vapor density can help predict how a vapor will act. If the MSDS indicates the vapor density is less than 1.0, the vapor will tend to rise and spread out. This means it is less likely to be a hazard. If the vapor density is 1.0 or more, then the vapor is heavier than air and will tend to sink to the lowest point on the ground. These vapors can then travel along the ground, sometimes for long distances, and find ignition sources. This makes chemicals with high vapor densities particularly dangerous.
- *Ignition Temperature*—Ignition temperature is the minimum temperature at which a chemical will burn and continue burning without the need for an ignition source.
- *Explosive*—The main difference between a flammable and an explosive is the rate of combustion, or the speed at which a material burns. A fire is a rapid release of energy, and its rate of combustion is fast. An explosion is the instantaneous release of energy and involves extremely rapid rate of combustion.

Note—Flammable and combustible liquids are covered more fully in Chapter 5.

C. EXPOSURE TO HAZARDS

Hazardous Airborne Substances—Substances that can be inhaled include:

- **Dust**—Airborne solid particles 0.1 to 25 μm (micrometer) in size. Dust particles as small as 50 μm can be detected with the naked eye.
- **Fumes**—The result of volatile solids condensing in cool air; in some cases, the solid particles react with air to form an oxide. Gases and vapors are often incorrectly called fumes.
- **Smoke**—Carbon or soot particles less than 0.1 μm in size formed because of incomplete combustion of carbon-based materials.
- **Aerosols**—Liquid droplets or solid fine particles that remain airborne for a period of time.
- **Mists**—Suspended liquid droplets created by a material condensing from a gaseous to a liquid state. Mists also form when liquids break into a dispersed state.
- **Gases**—Formless liquids occupying space that can be changed to a liquid or solid state through increased temperature and pressure.
- **Vapors**—Gaseous form of a substance that appears solid or liquid at room pressure and temperature. Vapors are water soluble and can irritate the eyes and respiratory system.
- **Fibers**—Solid particles such as asbestos whose length is several times larger than their diameter.

Airborne Exposures—The concentration (exposure) is the per unit volume of air to which a worker is exposed. Airborne concentrations are usually expressed in milligrams of substance per cubic meter of air (mg/m^3) or parts of substance per million parts of air (ppm).

- **Measuring Fibers**—Asbestos is expressed as fibers per cubic centimeter (f/cc) or fibers per cubic meter (f/m^3) of air.
- **Dose**—The amount of substance that actually enters the body during the period of exposure.
- **Elimination of Substances**—Substances ingested are present in the body until metabolized or eliminated. Some chemicals are rapidly metabolized while others are not and may be excreted unchanged or stored in fatty tissue.
- **Interactions**—Possible interactions may occur as a result of the multiple exposures that exist in a hospital environment. These interactions may increase exposures to chemicals through an individual's use of tobacco, alcohol, or drugs.
- **Other Considerations**—The physiological or psychological state of the worker can also affect exposure potential.

Exposure Routes—Toxic substances can enter the body through the skin, respiratory system, mouth, and eyes. Some substances can also damage the skin or eyes directly without being absorbed. Inorganic lead can be inhaled or swallowed, but it does not penetrate the skin. Sometimes a chemical substance can enter through more than one route.

1. *Chemical and Physical Properties*—The physical properties of a chemical substance include vapor pressure, solubility in water, boiling point, melting point, molecular weight, and specific gravity. Chemical properties describe the reactivity of a substance with other chemicals.

2. *Olfactory Warning Properties*—Some hazardous materials can be perceived by smell, which serves as a warning of exposure. The warning depends on a person's ability to detect the odor of a substance. Hazardous substances provide either good or poor warning of their presence. The lowest concentration at which the odor of a chemical can be detected is called the odor threshold.
   - Some hazardous substances have no odor and therefore provide no warning.
   - Concentrations high enough to give odor warnings can also cause adverse health effects at the same concentrations.
   - Exposure to some substances causes olfactory fatigue, which prevents workers from detecting odors.
   - Some people might detect the odor of chlorine at 0.02 ppm, while others cannot detect the odor until the concentration is much higher.
   - Never rely solely on sense of smell to detect the presence of hazardous substances.

3. *Absorption*—Hazardous substances can quickly enter the body through broken skin. Examples of substances that can be absorbed through skin contact are:
   - Phenolic compounds
   - Mercury
   - Arsenic
   - Nitrobenzene

Chemical Sensitization—Some people exposed to a chemical become sensitized to the substance; some may develop allergic dermatitis, while others do not.

1. *Common Allergens*—
   - Metallic salts of nickel, chrome, cobalt, gold, and mercury
   - Plastic resins such as epoxies, phenolics, and acrylics
   - Rubber accelerators and antioxidants
   - Organic dyes such as those in color developing solutions
   - First-aid preparations such as neomycin and benzocaine
   - Laboratory agents such as formaldehyde and phenol

2. *Exposure by Type of Work*—
   - Food service workers are exposed to heat, moisture, yeast, bacteria, detergents, acids, spices, soaps, and fruit juices.
   - Environmental service workers come into contact with bacteria, disinfecting chemicals, house plants, soaps, solvents, synthetic gloves, polishes, bowl cleaners, floor strippers, and waxes.
   - Laundry workers are exposed to alkalis, bleaches, enzymes, bacteria, fungicides, heat, moisture, biohazards, and soaps.
   - Nurses and nursing assistants are exposed to local anesthetics, antibiotics,

antiseptics, bacteria, biohazards, ethylene oxide, synthetic gloves, soaps, drugs, fungi, and moisture.

3. *Skin Disorders*—Skin disorders and diseases account for a large portion of occupational illnesses each year. Skin injuries in the form of cuts, lacerations, punctures, abrasions, and burns are also prevalent in the healthcare environment. Chemicals can irritate the skin and/or cause an allergic reaction. Dermatitis is the most common and often the most preventable of all job-related problems. The skin is the natural defense system of the body. It has a rough, waxy coating called keratin. This layer of protein and the outer layer of dead cells help prevent chemicals from penetrating the tissues.

4. *Contact Dermatitis*—Many substances cause irritation on contact with the skin. This condition is called contact dermatitis because the substance dissolves the protective fats and causes the skin to dehydrate, thereby destroying the skin cells.

5. *Skin-Related Workers' Compensation Claims*—Exposure to the following agents is responsible for the greatest number of workers' compensation claims:
   * Soaps, detergents, and cleaning agents
   * Solvents
   * Hard and particulate dusts
   * Food products
   * Plastics and resins

Chemical Storage Requirements—Proper storage information can be found on the MSDS for the substance. The container label will often have storage information.

1. *Segregation of Chemicals*—Acids should never be stored with bases. Oxidizers should be stored away from organic materials or reducing agents. Corrosives and acids will corrode most metal surfaces, including storage shelves or cabinets. Flammable or combustible liquids should be stored in a proper fire-safe area or cabinet.

2. *Storage Requirements*—
   * Temperature
   * Ignition control
   * Ventilation requirements
   * Reactive properties
   * Type and size of container
   * Chemical substance

D. THE OSHA HAZARD COMMUNICATION PROGRAM

OSHA Hazard Communication Standard (29 CFR 1910.1200)—The standard requires employers to develop, implement, and maintain at the workplace a written, comprehensive hazard communication program that includes provisions for container labeling, collection and availability of MSDSs, and an employee training program. Steps to ensure compliance are as follows.

1. ***Written Program***—A written program must be available to employees on all work shifts. Hazard information must be communicated to all affected or exposed employees. Chemical manufacturers and distributors must provide hazard information on their products to customers by means of a container label and MSDS.

2. ***Hazardous Materials Listing***—A listing of the hazardous chemicals in the workplace should be compiled by:
    *   Conducting a thorough workplace inventory to determine all chemicals currently in use and hazards created, such as welding fumes
    *   Coordinating the list with the purchasing department to ensure all purchased substances are listed
    *   Developing procedures to keep the chemical list current

3. ***Hazard Evaluations***—The quality of any hazard communication program relies on the accuracy of the hazard assessment.
    *   Chemical manufacturers and distributors are required to review available scientific evidence concerning the hazards of the chemicals produced and report this information to those who use the products.
    *   Most employers rely on the evaluations performed by the chemical manufacturers to determine the hazards.
    *   Each chemical must be evaluated for its potential to cause adverse health effects and its potential to pose physical hazards such as flammability.
    *   Criteria used to determine whether a chemical is hazardous can be found in 29 CFR 1910.1200, Appendix B.

4. ***Hazard Rating Information***—Hazard rating information can be obtained from such publications as the *Fire Protection Guide to Hazardous Materials* and the *National Fire Rating Guide.* Under OSHA regulation, labels are not required on portable transfer containers of less than 10 gallons as long as the chemical is used or returned to its original container by the end of the work shift. All mixed or diluted solutions should be labeled as full-strength solutions.

5. ***Non-Routine Tasks***—Employers must inform employees of the hazards of non-routine tasks and the hazards associated with chemicals in unlabeled pipes.

6. ***Hazard Communication Training Requirements***—Employers must establish a training and information program for employees who may be exposed to hazardous chemicals in their work area at the time of initial assignment and whenever a new hazard is introduced in their work area. Training may be done on each individual substance or by categories of hazards such as flammability or carcinogenicity. The training must include the following topics:
    *   The existence of the hazard communication standard and the requirements of the standard
    *   The components of the hazard communication program in the workplace
    *   Operations in work areas where hazardous chemicals are present
    *   The location of the written hazard evaluation procedures, communications program, lists of hazardous chemicals, and the required MSDS forms

7. **Reviewing Training Programs**—The following items should be considered when reviewing or evaluating training and informational programs:
   - Designation of persons responsible for conducting the training
   - Effectiveness of the format used to present the material
   - Training program content consistent with provision of paragraph (h) of the OSHA standard
   - Procedures used to train new employees upon initial assignment
   - Procedures used to train employees when a new chemical hazard is introduced into the workplace

Material Safety Data Sheets—Chemical manufacturers, distributors, and importers are required to obtain or develop an MSDS for each hazardous chemical they produce, import, or distribute.

1. **OSHA Requirements**—OSHA specifies minimum information to be included on an MSDS but does not prescribe the format. The MSDS must be provided with the shipment of a hazardous chemical to the user. Distributors must ensure that employers are similarly provided an MSDS. Workplaces may rely on the information received from their suppliers.

2. **MSDS Information**—Each MSDS must be in English and must contain the following information:
   - Section I provides information regarding the specific chemical identity of the hazardous chemical(s) involved and its common name(s).
   - Section II must provide information about hazardous ingredients.
   - Section III covers the physical and chemical characteristics of the hazardous chemical, known acute and chronic health effects, and related health information.
   - Section IV contains fire and explosion hazard data.
   - Section V describes reactivity information and chemical interactions.
   - Section VI provides information on health hazards. The MSDS also contains information on exposure limits, whether the chemical is considered to be a carcinogen, precautionary measures, emergency and first-aid procedures, and the identification of the preparing organization.
   - Section VII outlines the precautions for safe handling and use.
   - Section VIII covers general control information including the use of personal protective equipment and engineering controls such as ventilation.
   Note—MSDS information must be located close to workers and readily available to them during each work shift.

3. **ANSI-Proposed MSDS Standard**—The Chemical Manufacturers Association and the American National Standards Institute (ANSI) recently developed Standard Z400.1-1993 which prescribes a uniform format for the MSDS. This proposed format contains 16 sections including information on first-aid measures, accidental release measures, detailed regulatory information, information on toxicology, and transport guidelines. The standard is voluntary and OSHA has not indicated that the format will be required under 29 CFR 1910.1200. The sections of the new standard are:

- Section 1—Chemical Product and Company Identification
- Section 2—Composition Information or Ingredients
- Section 3—Hazard Identification
- Section 4—First-Aid Measures
- Section 5—Fire-Fighting Measures
- Section 6—Accidental Release Measures
- Section 7—Handling and Storage
- Section 8—Exposure Control and Personal Protection
- Section 9—Physical and Chemical Properties
- Section 10—Stability and Reactivity
- Section 11—Toxicological Information
- Section 12—Ecological Information
- Section 13—Disposal Considerations
- Section 14—Transport Information
- Section 15—Regulatory Information
- Section 16—Other Information

E. HAZARDOUS MATERIAL LABELING

General Labeling Requirements—

- All containers should be properly labeled when received.
- Incoming shipments should be checked to ensure that all information is contained on the label.
- Labels must contain the name of the chemical and appropriate hazard warnings.
- Workers should be instructed not to deface existing labels.
- Small containers used by more than one worker must also be labeled with chemical name and hazard warnings.
- Minimum labeling requirements specify that all labels must be in English, contain the chemical identification name, and identify any appropriate hazard.

Right-to-Know (RTK)—This versatile system can be used for labels, signs, pipes, valves, and tags. A new improved system is now available that contains information on target organs. RTK labels provide complete information and a chart is not necessary. Labels in the RTK system contain:

- Chemical name
- Common name (if any)
- Signal word (warning/caution)
- Statement of hazards
- Precautionary measures
- Instruction in case of contact or exposure
- CAS number

National Fire Rating (NFR) System—This system, designated in NFPA 704, uses a hazard rating diamond with colors and codes to rank the health risks, flammability, and reactivity of hazardous materials.

TABLE 7.2   Numerical Hazard Ratings

HEALTH HAZARDS

4   **Deadly**—The slightest exposure to this substance could be life threatening. Only specialized protective clothing should be worn when working with this substance.

3   **Extremely Dangerous**—Serious injury could result from exposure to this substance. Do not expose any body surface to this material. Full protective measures should be taken.

2   **Dangerous**—Exposure to this substance would be hazardous to health. Protective measures are indicated.

1   **Slight Health Hazard**—Irritation or minor injury would occur from exposure to this substance. Protective measures are indicated.

FLAMMABILITY

4   **Flash Point Below 73°F**—Substance is very volatile, explosive, or flammable. Use extreme caution when handling or storing the substance.

3   **Flash Point Below 100°F**—Flammable, explosive, or volatile under most normal temperature conditions. Exercise great caution in using and storing the substance.

2   **Flash Point Below 200°F**—Moderately heated conditions may ignite this substance. Use caution in handling or storage.

1   **Flash Point Above 200°F**—This substance must be pre-heated to ignite.

REACTIVITY

4   **May Detonate**—Capable of explosion at normal temperatures. Evacuate area if exposed to heat or fire.

3   **Explosive**—Substance is capable of explosion by strong initiating source such as heat, shock, or water.

2   **Unstable**—Subject to violent chemical changes at normal or elevated temperatures. Potential violent explosive reaction may occur if exposed to water.

1   **Normally Stable**—May become unstable at elevated temperatures.

- *Blue (Health Hazard)*—May cause health problems if acute exposure occurs by ingestion, inhalation, or physical contact.
- *Red (Flammability)*—Evaluates the risk of materials to fireburst based on factors relative to the substance and surrounding environment.
- *Yellow (Reactivity)*—Advises that a substance may react violently under certain conditions or exposures.
- *White (Specific Hazard)*—Refers to substances with specific hazards or properties such as oxidizers.

Note—A number system is used to rate the hazards, where 0 is the lowest and 4 is the highest risk. This system is used throughout healthcare facilities to designate hazards. Refer to NFPA 99.

Hazard Material Identification Guide (HMIG) System—This system, which is similar to the NFR system, is composed of four colors:

- *Blue*—Health
- *Red*—Flammability
- *Yellow*—Reactivity
- *White*—Protective equipment

Note—System references explain hazard ratings, and the label is in a bar format with rectangles instead of diamonds. Personal protective equipment is covered by visual representations on some charts and tags.

Hazardous Materials Identification System (HMIS)—This system, developed by the National Paint and Coatings Association, is similar to the HMIG system except that personal protective equipment is designated by an alphabetic code.

DOT HM-181 Hazard Classes—The Performance Oriented Packaging Regulation became effective in 1991 and is administered by the U.S. Department of Transportation (DOT) system for materials shipped through transportation systems within the United States. The law is based on the United Nations Recommendations on the Transport of Dangerous Materials. A diamond-shaped label specifies the kind of hazard by listing a class number and recognizable hazard symbol, such as a "flame." OSHA recently published a requirement in 29 CFR 1910.1201 that original DOT labels must remain on vehicles, tanks, and containers until the substances are removed or transferred to other containers. Effective October 1, 1993, secondary labels are required to identify the secondary hazard of a package. Information can be found in 49 CFR 172.101. The DOT system, which has nine classes, is now required on more than 1400 hazardous materials divided into the following nine classes:

- Class 1—Explosive and Blasting Agents
- Class 2—Oxygen, Flammable, and Poison Gases
- Class 3—Flammable Liquids, Combustible Liquids, and Fuel Oils
- Class 4—Flammable Solids, Dangerous Combustibles
- Class 5—Oxidizers and Organic Peroxide
- Class 6—Poisons and Infectious Substances
- Class 7—Radioactive Materials
- Class 8—Corrosive Materials
- Class 9—Miscellaneous

F. DISINFECTING SUBSTANCES

Isopropyl Alcohol—Isopropyl alcohol, a widely used antiseptic and disinfectant, is used to disinfect thermometers, needles, anesthesia equipment, and other instruments.

- *Exposure Considerations*—The odor of isopropyl alcohol can be detected at concentrations of 40 to 200 ppm. Exposure to isopropyl alcohol can cause irritation of the eyes and mucous membranes. Contact with the liquid may also cause skin rashes. The OSHA PEL is 400 ppm for an 8-hour TWA (29 CFR 1910.1000).
- *Personal Protection*—Workers should use appropriate protective clothing such as gloves and face shields to prevent repeated or prolonged skin contact with iso-

propyl alcohol. Splash-proof safety goggles should also be provided and required for use where isopropyl alcohol may come in contact with the eyes.

Ethyl Alcohol—Most hospitals utilize 70% ethyl alcohol as a topical application in local skin disinfection.

1. *Exposure/Hazard Information*—Ethyl alcohol, according to NFPA 325M, is flammable in all dilutions where vapor may come in contact with an ignition source. The flash point of a 70% solution is approximately 70°F; therefore, it is considered a possible fire hazard.
   - When used topically, ethyl alcohol enhances drying of the skin. Care should be exercised in its use to avoid dermatitis.
   - Because ethyl alcohol is highly miscible in water, it should be disposed of through dilution with water in an area where adequate ventilation is provided.

2. *Handling and Storage*—Healthcare facilities must utilize reasonable care in the use, handling, and disposal of ethyl alcohol.
   - A designated pharmacy area should be the only place where 100% ethyl alcohol is stored. A flammable storage area should be provided in these locations.
   - Ethyl alcohol at a volume over 70% should be kept in a flammable storage cabinet away from patient care areas.
   - A type BC fire extinguisher should be used to extinguish ethyl alcohol flames.

Ammonia—Ammonia is used as a liquid cleaning agent and as a refrigerant gas. Concentrated solutions of ammonia can cause severe burns. Workers should avoid skin contact by wearing protective clothing.

1. *Handling/Storage Requirements*—
   - If skin or eye contact occurs, the affected area should be washed and rinsed promptly.
   - Workers handling concentrated solutions should wear rubber gloves and face protection (goggles or face shields).
   - Adequate ventilation should be provided in areas where ammonia gas is released from concentrated solutions.
   - Ammonia should never be stored with deodorizing chemicals because the reaction can produce harmful by-products such as chlorine gas.

2. *Exposure Information*—The NIOSH REL for ammonia is 50 ppm (35 mg/m^3) as a 5-minute ceiling. The OSHA PEL is 50 ppm as an 8-hour TWA. ACGIH recommends a TLV of 25 ppm (18 mg/m^3) as an 8-hour TWA. The STEL is 35 ppm (27 mg/m^3).

3. *Quaternary Ammonium Compounds*—These substances are widely used as disinfectants. Many are ineffective against tuberculosis and Gram-negative bacteria. Quaternary ammonium compounds are used in central supply, housekeeping, patient care, and surgical services areas.
   - Ammonium compounds can cause contact dermatitis but seem to be less irritating to hands than other substances. They can also cause nasal irritation.

- No OSHA PEL, NIOSH REL, or ACGIH TLV exists for quaternary ammonium compounds.

Sodium Hypochlorite—Also known as household bleach, sodium hypochlorite is commonly used as a disinfecting solution (mixed $1/4$ cup to a gallon of water). Chlorine should be mixed fresh daily and used to disinfect non-critical surfaces such as water tanks and bathrooms. It also is used as a laundry additive, a sanitizer for dishwashing, and a disinfectant for floors.

1. *Hazards*—Chlorine-based substances should not be mixed with materials that contain ammonia because the reaction produces a toxic gas. Chlorine is released slowly from cleaning and bleaching solutions as they are used. Repeated exposures may result in runny nose, coughing, wheezing, and other respiratory problems.

2. *Exposure Information*—Mild irritation of the mucous membranes can occur at exposure concentrations of 0.5 ppm. The OSHA PEL for chlorine is a ceiling of 1 ppm according to 29 CFR 1910.1000, Table Z-1. Chlorine has an odor threshold between 0.02 and 0.2 ppm, but a person's sense of smell is dulled by continued exposure.
   - Odor alone does not provide adequate warning.
   - Splash-proof safety goggles should be used in instances where there is a possibility of contact with the eyes.
   - Workers should use appropriate personal protective equipment such as gloves, face shields, and respirators as required.
   - Primary control measures include good exhaust ventilation or using another substance.

Iodine—Iodine is used as a general disinfectant and can be mixed with alcohol for use as a skin antiseptic or with other substances for general disinfecting purposes.

1. *Exposure Information*—Exposure can include irritation of the eyes and mucous membranes, headaches, and breathing difficulties. Crystalline iodine or strong solutions of iodine may cause severe skin irritation because they are not easily removed and may cause burns.
   - The OSHA PEL for iodine is a ceiling of 0.1 ppm according to 29 CFR 1901.1000, Table Z-1.
   - The ACGIH recommends a TLV of 0.1 ppm as a ceiling.

2. *Personal Protection*—Workers should use personal protective equipment such as gloves, face shields, and any other appropriate protective clothing deemed necessary. Contaminated clothing should be changed before leaving the work area.

3. *Iodophors*—These substances are available in concentrated preparations and contain a surfactant. Some iodophors that claim to control tuberculocidal activity may be used as intermediate disinfectants for certain surfaces and equipment.
   - Iodophors are used on countertops, chairs, and amalgams; they can be used on impressions and prosthetics as a "holding solution."

- Iodophors also work well on certain items that are not sterilized, such as dental handpieces.
- Contact with skin should be avoided, especially by individuals who are allergic to iodine.

Phenol-Based Disinfectants—These were among the first disinfectants used in hospitals. They are generally used for a wide range of bacteria, and some may also be used for intermediate-level disinfection if tuberculocidal. Phenolics are available as sprays and liquid. They can be used for surfaces, equipment, prosthetics, bite registrations, partial dentures, handpieces, etc. and can also be used as a holding solution.

- *Hazards*—Contact with skin or mucous membrane should be avoided. Phenol may be detected by odor at a concentration of about 0.05 ppm. Serious health effects may follow exposure to phenol through skin adsorption, inhalation, or ingestion. These effects may include local tissue irritation and necrosis, severe burns of the eyes and skin, irregular pulse, difficulty breathing, darkened urine, convulsions, coma, and even death.
- *Protection*—The OSHA PEL for phenol is 5 ppm for an 8-hour TWA skin exposure. Workers should use protective clothing, gloves, face shields, splash-proof safety goggles, or other protective clothing as necessary to protect against skin or eye contact. Workers should wash their hands thoroughly before eating, smoking, or using toilet facilities.

G. GLUTARALDEHYDE/FORMALDEHYDE

Glutaraldehyde—Solutions with glutaraldehyde contain surfactant substances to promote wetting and rinsing of surfaces. Most contain sodium nitrite to inhibit corrosion, peppermint oil as an odorant, and yellow and blue dyes to indicate activation of the solution. Buffered glutaraldehyde solutions are stable for less than two weeks. Solutions must be dated when mixed.

1. *Hazards*—Solutions can be absorbed by inhalation, and extensive skin contact may result in allergic eczema. The nervous system can also be affected. Skin contact and vapors should be avoided. Glutaraldehyde has an odor threshold of about 0.04 ppm and is irritating to the skin/mucous membranes at a concentration of about 0.3 ppm.

2. *Exposure*—Glutaraldehyde-based solutions are registered by the EPA as disinfectants or sterilants. They may be used for either sterilization or high-level disinfection. Items treated with glutaraldehyde must be thoroughly cleaned before and rinsed after immersion.
   - The ACGIH-recommended ceiling limit for glutaraldehyde is 0.2 ppm.
   - OSHA does not have a PEL for glutaraldehyde.
   - Workers should avoid breathing the vapors and use splash-proof safety goggles to protect the eyes.

Formaldehyde (29 CFR 1910.1048)—Formaldehyde is one of the most common chemicals in use today. It is used in manufacturing urea, phenol, and melamine resins. Studies indicate that formaldehyde is a potential carcinogen. Airborne concentrations above 0.1 ppm can irritate the eyes, nose, and throat. NIOSH also regards formaldehyde as a potential carcinogen. In healthcare settings it is used primarily for cold sterilization of instruments. It is not used as a general disinfectant because of its caustic properties. Formaldehyde is found in the laboratory as a tissue preservative and in central supply and dialysis units as a sterilant. It is often combined with methanol and water to make formalin.

1. *Exposure Information*—The odor of formaldehyde can be detected in the air at about 0.8 ppm. Formalin solutions splashed in the eyes may cause severe injury and corneal damage. Low ambient concentrations of formaldehyde (0.1 to 5 ppm) may cause burning and tearing of the eyes and irritation of the upper respiratory tract.
   - Higher concentrations (10 to 20 ppm) may cause coughing, chest tightness, increased heart rate, and a sensation of pressure in the head.
   - Exposures of 50 to 100 ppm may cause pulmonary edema. Repeated exposure to formaldehyde may cause some persons to become sensitized.
   - Sensitized individuals will experience eye or upper respiratory irritation or an asthmatic reaction at levels too low to cause symptoms in most people. Dermatitis is a common problem with formaldehyde exposure.

2. *Exposure Limits*—
   - The 1992 OSHA standard lowered the PEL for formaldehyde to 0.75 ppm as an 8-hour TWA.
   - The ceiling concentration is 2.0 ppm for a 15-minute STEL.
   - The action level is 0.5 ppm for an 8-hour period.

3. *ACGIH Exposure Recommendations*—The ACGIH has designated formaldehyde as a suspected human carcinogen and has recommended a TLV of 1 ppm with a STEL of 2.0 ppm. Odor is not a reliable warning for the presence of formaldehyde.

4. *Personal Protective Equipment*—Skin and eye contact with formaldehyde should be avoided. Goggles, face shields, aprons, NIOSH-certified positive-pressure air-supplied respirators, and boots should be used when formaldehyde spills and splashes are likely. Appropriate protective gloves should be used whenever hand contact is probable.

5. *Medical Surveillance*—Pre-employment baseline data should document the respiratory tract, liver, and skin condition of individuals who will be exposed to formaldehyde. Periodic monitoring should be conducted to detect symptoms of pulmonary or skin sensitization or effects on the liver. Employers must implement a medical surveillance program for employees exposed at or above the action level or above the STEL. The program should consist of medical and physical examinations to determine overexposure. A medical disease questionnaire must be completed by workers prior to assignment to potential exposure areas.

6. ***Warning Signs***—Employers are required to establish regulated areas where the concentration exceeds either the TWA or STEL and post all entrances and doorways with signs that read: DANGER—FORMALDEHYDE—IRRITANT AND POTENTIAL CANCER HAZARD—AUTHORIZED PERSONNEL ONLY.

7. ***Labeling***—Specific hazard labeling is required for all forms of formaldehyde, including all mixtures and solutions composed of 0.1 ppm formaldehyde or greater. Labels must also be placed on materials capable of releasing formaldehyde in quantities of 0.1 ppm or greater. Warning labels must state that formaldehyde presents a potential cancer hazard in all situations where the level may potentially exceed 0.5 ppm.

8. ***Contaminated Clothing***—Containers for contaminated clothing and equipment must be labeled as follows: DANGER—FORMALDEHYDE-CONTAMINATED CLOTHING/EQUIPMENT—AVOID INHALATION AND SKIN CONTACT. *Note*—Employers must ensure that hazard warning labels comply with the requirements in 29 CFR 1910.1200.

9. ***Exposure Monitoring***—Employers are required to conduct initial monitoring to identify all employees who are exposed at or above the action level and to determine the exposure of each employee.
   • When the exposure level is maintained below the STEL and the action level, employers may discontinue monitoring if two consecutive sampling periods taken seven days apart reveal concentrations below the STEL and action level.
   • Employers must monitor exposure when reports of formaldehyde-related symptoms are received.
   • If monitoring reveals that the concentration exceeds the action level, monitoring must be repeated every six months.
   • If the level is at or above the STEL, annual monitoring is required.

10. ***Work Practice and Engineering Controls***—Employers must institute necessary controls to reduce or maintain exposure at or below the TWA or STEL. Controls must be supplemented with respirators where and when necessary to protect workers.

11. ***Training***—Training is required at least annually for all workers exposed to concentrations of 0.1 ppm or greater. The training should make the employees aware of the specific hazards and the controls in place to reduce the chance of exposure. Employees must be informed about the signs and symptoms of formaldehyde exposure and how to report them to the employer.

Formalin—Formalin, a 37% solution of the formaldehyde gas, is used as a preservative in surgical pathology and histopathology laboratories, in autopsy rooms, and for cold sterilization of equipment in central supply and dialysis units.

   • ***Hazards***—Formalin is also a contact hazard. It can cause contact dermatitis and severe eye injury if accidentally splashed. A 10% formalin solution is widely used in hospitals. Many hospitals dilute their own from stock 37% formalin. Unless the process is automated, it should be done in a fume cabinet or hood.

* *Protection*—In situations where splashes or spills are possible, goggles, face shields, aprons, boots, and heavy rubber gloves should be used to prevent eye injury or contact dermatitis. In case of spills, which can result at high exposure, full-face respirators and special bags to absorb spills should be available.

Procedures for Overexposure to Disinfectants—If inhaled, the exposed individual should be moved into fresh air, supplement with humidified 100% oxygen if necessary.

* *Skin*—Decontaminate the area involved with water or saline for at least 15 minutes. Wash the affected area with soap and water. Use topical steroids or antibiotics as required.
* *Eyes*—Irrigate with water for at least 15 minutes and refer for medical evaluation.
* *Ingestion*—Induce emesis unless the victim is convulsing or comatose. Use ipecac syrup.
* *General Precautions*—Monitor heart and vital signs.
* *Allergic Reaction*—Administer prednisolone, dexamethasone, or antihistamines.

H. ANTINEOPLASTIC DRUGS (CHEMOTHERAPEUTIC)

Many chemotherapeutic agents have been reported to cause cancer in animals and thus can be considered to be potential human carcinogens. Antineoplastic drugs derive their name from the fact that they interfere with or prevent the growth and development of malignant cells and neoplasms. They may also be called cytotoxic or cytostatic because they have the ability to prevent the growth and proliferation of cells.

Exposure Information—Primary routes of worker exposure to antineoplastic drugs are inhalation and dermal absorption. Exposure by inhalation can occur during drug preparation or administration.

* Aerosols can be generated when inserting needles into or withdrawing them from vials and when expelling air from syringes before injection.
* Skin absorption may occur when antineoplastic drugs are spilled during their preparation or administration. Skin exposure may also occur as a result of contact with the urine of patients being treated with antineoplastic drugs.

Worker Evaluations—Workers exposed to antineoplastic drugs should receive preplacement and periodic medical evaluations that include gathering complete work and medical histories. The examination should emphasize the hematopoietic, reproductive, and nervous systems.

Safe Handling Procedures—Workers should be thoroughly trained in safe handling procedures. Exposure can occur during administration (through accidental needlesticks) and disposal of these drugs. OSHA currently has no standard for exposures. The use of class II vertical-flow biological safety cabinets (BSCs) in the preparation of cytotoxic drugs, rather than the horizontal laminar-flow hoods still found in many pharmacies, has been shown to significantly reduce mutagenicity in pharmacy workers. Face masks, polyvinyl chloride gloves, goggles, and disposable gowns and masks should be used when preparing or administering these drugs.

Mixing and Storing—All cytotoxic substances should be stored in accordance with the manufacturers' recommendations and in such a manner to assure the protection of the environment and all personnel from potential exposure and contamination.

- All mixing and reconstitution of cytotoxic agents should be performed in a class II vertical-flow BSC on a disposable, plastic-backed, spill-proof preparation mat.
- Before and after use, the hood should be thoroughly cleaned and wiped down with 70% alcohol and decontaminated monthly with detergent followed by thorough rinsing.
- The blower of the hood must be left on 24 hours a day, 7 days a week.
- A 0.22-micron hydrophobic-filtered vented set can be used, or a negative pressure must be maintained when reconstituting and preparing doses to ensure that aerosolized/airborne particles do not come in contact with the preparer.

Worker Protection—Methods for preventing exposure to antineoplastic drugs are outlined in OSHA Publication 8-1.1 These guidelines address drug preparation, drug administration, waste disposal, spills, medical surveillance, storage and transport, training, and dissemination of information. Professional recommendations have also been issued by the National Institutes of Health and the American Society of Hospital Pharmacists.

- Latex gloves must be worn by all personnel reconstituting and preparing doses of hazardous cytotoxic agents.
- A protective disposable gown made of lint-free low-permeability fabric with a closed front, long sleeves, with cuffs tucked under the gloves must be worn by personnel using the hood to protect the skin from possible exposure.
- Used gloves and gowns should be disposed of according to the hazardous waste procedures.
- Approved aseptic techniques should be observed in the preparation of all antineoplastic agents.

Dispensing Guidelines—

- All hazardous chemotherapeutic agents dispensed by the pharmacy must be issued in a plastic bag and labeled to identify the contents as being potentially dangerous and requiring special handling when administering or disposing of used materials and equipment.
- All supplies and equipment that come in contact with hazardous cytotoxic agents during preparation and administration should be placed in appropriate containers for disposal as biohazardous material. This includes all empty vials, syringes, needles, intravenous bottles and tubing, gloves, disposable gowns, and any other items that may be contaminated.

Spill Procedures—Spill control procedures must be posted in preparation areas. In the event of an accidental spill or the breaking of a container of a cytotoxic agent, the pharmacy should be contacted immediately. Instructions for the proper handling of individuals that may have been exposed or contaminated should be in accordance with policies and procedures developed by the safety and the infection control committees.

- Care should be exercised by nursing personnel to avoid direct contact with antineoplastic agents during administration.
- An alcohol swab should be used to prevent exposure when removing bubbles from syringes or when connecting intravenous tubing.
- All syringes and needles should be returned to the hematology lab or oncology clinic for disposal as biohazardous waste.
- In case of skin contact with an antineoplastic drug, the affected area should be washed thoroughly and the affected eye(s) flushed with water.
- The employee health nurse should visually examine the affected area and make appropriate recommendations.
- Exposure must be reported to the hospital safety officer as well as the risk manager by the employee health nurse.
- Baseline CBC and differential should be drawn at the time of exposure and followed up with a repeat CBC and differential in six months. Both sets of lab work should be reviewed by the occupational health department and any differences noted should be referred to the treating physician.

Reproductive Hazards—Exposures to anesthetic waste gases, solvents, ethylene oxide, and chemotherapeutic agents have the most potentially harmful effects on reproductive health. Each employee should be trained so that the proper decision regarding his or her reproductive health can be made. Training for males should include information on the increasing number of spontaneous abortions for wives of male anesthesiologists and pharmacists who handle chemotherapeutic agents. The mutagenic and teratogenic properties of ionizing radiation should be stressed.

1. *Counseling*—Employees working in areas with reproductive hazards should seek counseling prior to planned pregnancy. Female reproductive hazard training should stress reproductive health and cover the hazards of radiation, drugs, chemicals, and infectious agents. Information on the effects of smoking and alcohol on fetal development should be included. Pre-pregnancy counseling should be encouraged and specific training provided. Factors to consider in pre-pregnancy counseling include:
   - Days and hours worked, shift or schedule changes, overtime, breaks, and work flow
   - Physical requirements, including sitting, standing, walking, bending, climbing, lifting, balance, and coordination
   - Climate and temperature, noise, vibration, radiation, biologic agents, airborne agents, and other hazardous materials
   - Home tasks, environment, stress, habits (smoking, drinking), age, weight, and nutrition

2. *Pregnant Workers*—Because of the unique nature of the occupational responsibilities in healthcare facilities, policies should provide options for transfers and/or leaves of absence for workers exposed to hazards that could affect their pregnancy.

I. SOLVENTS

Solvents such as methyl ethyl ketone, acetone, and Stoddard solvent may be used to clean parts in maintenance shops. Recommended personal protective equipment should be worn by workers who come into contact with solvents. Many solvents remove the natural fats and oils from the skin and may be absorbed through the skin. Organic solvents are flammable and should be stored in approved safety containers. Cleaning tanks should be kept closed when not in use. Solvents are also found in a number of chemicals used in medical laboratories. Some are widely used as cleaning agents in housekeeping and maintenance, and some are present in inks and cleaning agents in print shops.

1. *Exposure Hazards*—Many solvents act as central nervous system depressants, causing headaches, dizziness, weakness, nausea, and other symptoms. Solvents may also irritate the eyes, skin, and upper respiratory tract. Prolonged contact may result in defatting and dehydration of the skin. The hazard control manager should develop an inventory of solvents in use and consult 29 CFR 1910.1000 for the pertinent OSHA PEL.

2. *Protection*—Local exhaust ventilation and enclosure of solvent vapor sources are the preferred methods of controlling exposures to solvents in laboratories. When selecting engineering and other controls, the toxicity of the solvent, as well as its flammability and explosion potential, must be considered.
   - Protective gloves help prevent absorption of solvents through the skin. Respirators, rubber aprons, goggles, and boots may be required during certain procedures or during cleanup of spills.
   - Workers should be thoroughly trained to recognize the symptoms of solvent exposure, to avoid eating in potentially contaminated areas, to work only under exhaust hoods when handling solvents, and to follow those work practices recommended for specific solvents.

3. *Solvent Hazard Severity*—The hazard severity of solvents depends on several factors:
   - Type of chemical and how the substance is used
   - Length and type of worker exposure
   - Effectiveness of ventilation
   - Evaporation rate and concentration of vapor
   - Housekeeping and safety practices
   - Engineering controls and protective equipment

Acetone—Many hospitals use acetone to remove adhesives from topical skin areas on patients. NFPA 325M indicates that acetone presents only a slight health risk when used in patient applications. Concern is noted due to its flammability. Acetone may be ignited under almost all conditions, and water may not be effective in extinguishing it. A policy similar to the following should be implemented in all areas where acetone is used.

1. ***Exposure and Hazard Information***—Acetone is a flammable liquid and as such is covered under general guidelines established in the hazardous materials policy of this hospital.
   - Acetone is to be used as a topical application only and not for internal use. Should internal exposure occur, the victim should be immediately treated under accepted guidelines.
   - In the unlikely event of ignition, a type BC fire extinguisher should be used to extinguish the flames.

2. ***Protection***—Acetone should not be stocked in the unit in quantities greater than necessary for use in a one-week period.
   - Precaution should be taken to eliminate breakage of acetone containers.
   - The general waste receptacles should not be utilized for disposal of acetone. The maintenance department should be contacted for removal.
   - Topical use on patients should be limited to small areas of the skin, as acetone enhances the drying effect of the skin and may result in dermatitis.
   - Further questions concerning contraindications of acetone usage are available through the Poison Control Center.

Benzene (29 CFR 1910.1028)—Benzene is used as a solvent in laboratories and maintenance departments, often to clean up after painting.

1. ***Exposure Information***—High levels of exposure can cause acute central nervous system depression, with such symptoms as headache, dizziness, and nausea. Exposure can also cause eye, skin, and upper respiratory irritation. Effects of long-term exposure to benzene include liver and kidney damage, aplastic and hypoplastic anemia, and leukemia.
   - The OSHA standard is 10 ppm, and the NIOSH standard is 1.0 ppm.
   - Whenever possible, a less toxic solvent, such as toluene, cyclohexane, or mineral spirits, should be substituted for benzene.
   - If there is no substitute, benzene should be used with local exhaust ventilation, such as under an exhaust hood.

2. ***Protection***—Workers should wear gloves. Respirators, goggles, and other protective equipment and clothing should be worn for cleaning up spills.
   - Warning signs must be posted at entrances to regulated areas and must contain the following information: DANGER—BENZENE—CANCER HAZARD—FLAMMABLE—NO SMOKING—AUTHORIZED PERSONNEL ONLY—RESPIRATOR REQUIRED.
   - Benzene containers must be labeled as follows: DANGER—CONTAINS BENZENE—CANCER HAZARD.

Dioxane—Dioxane can be found in laboratories and is used as a cleaning agent by housekeeping and maintenance departments. It is found in ink and is used as a cleaning agent in print shops; therefore, it can also be found in hospitals that do their own printing.

1. ***Exposure Information***—High levels of exposure can cause acute central nervous system depression, with such symptoms as headache, dizziness, and nausea.
   - It can also cause eye, skin, and upper respiratory irritation. Long-term effects may include liver and kidney damage. Dioxane should be considered a suspect carcinogen.
   - The OSHA standard is 100 ppm, and the NIOSH standard is 100 ppm/30 minutes. Whenever possible, a less toxic solvent, such as an alcohol- or ketone-based substance, should be substituted for dioxane. If dioxane must be used, it should be used with local exhaust ventilation, such as under an exhaust hood.

2. ***Protection***—Workers should wear gloves. Respirators, goggles, and other protective equipment and clothing should be worn for cleaning up spills.

Perchloroethylene—This solvent is found in adhesive tape remover. It is used mostly in the central supply area to remove residual gum from medical instruments. Exposure may cause skin irritation and can affect the liver and central nervous system. Repeated skin contact should be avoided.

Xylene—This solvent is found in histopathology and cytology laboratories and in some maintenance departments.

1. ***Exposure Information***—High levels of exposure can cause acute central nervous system depression, with such symptoms as headache, dizziness, and nausea. It can also cause eye, skin, and upper respiratory irritation. Cardiovascular and reproductive effects have been noted with long-term exposure. OSHA and NIOSH exposure limits are 100 ppm.

2. ***Protection***—Whenever possible, a less toxic solvent, such as toluene, cyclohexane, or acetate, should be substituted for xylene. If there is no substitute, it should be used with local exhaust ventilation, such as under an exhaust hood.
   - Care should be taken during routine laboratory procedures, such as pouring off xylene solution from automatic tissue processors and staining line trays, which may result in short-term elevation of exposure.
   - Respirators, goggles, and other protective equipment and clothing should be worn when cleaning up spills.

J. OTHER HAZARDOUS SUBSTANCES

Methyl Methacrylate—This substance is an acrylic cement used in operating rooms to secure surgical prostheses to bone, especially in total hip replacements. The compound is also used in dental prostheses. The two components, a liquid and a powder, are mixed immediately before use. Methyl methacrylate has been reported to have an odor threshold of about 0.08 ppm. At concentrations in excess of 400 ppm, methyl methacrylate affects the central nervous system.

- *Exposure Information*—Methyl methacrylate is an eye, skin, and mucous membrane irritant in concentrations at or above 170 to 250 ppm. Patients exposed to this compound have suffered acute episodes of hypotension (low blood pressure) and cardiac arrest. The OSHA PEL, as well as the ACGIH TLV, for methyl methacrylate is 100 ppm as an 8-hour TWA (29 CFR 1910.1000, Table Z-1).
- *Protection*—A local exhaust hood should be used to conduct exhaust fumes from the area in which methyl methacrylate is mixed. A tent hood may be used unless mixing can be done in a separately ventilated area. Workers who handle methyl methacrylate should wear personal protective equipment including gloves, goggles, face shields, and respirators. Portable hoods should be available for operating room use.

Peracetic Acid—Peracetic acid is used to sterilize the surface of medical instruments and may be found in laboratories, central supply, and patient care units. Peracetic acid or peroxyacetic acid is a strong skin, eye, and mucous membrane irritant.

- *Exposure Information*—Currently no standards exist for regulating exposures to peracetic acid, and no recommendations have been made by NIOSH, ACGIH, or ANSI.
- *Protection*—Use of an isolation chamber should eliminate major exposure to peracetic acid vapors in hospitals. This chamber should be checked frequently for defects. Peracetic acid should never be used outside this chamber. Protective clothing such as waterproof gloves is recommended.

Mercury—Elemental mercury is a metallic element that is liquid at room temperature. Mercury is used in many types of hospital equipment, such as thermometers, Coulter counters, Van Slyke apparatus, Miller-Abbot and Cantor tubes, and sphygmomanometers. Mercury is also used in dental amalgams and is found in paint, fluorescent light bulbs, and batteries.

1. *Exposure Information*—Most exposures are the result of accidental spills. Although inhalation is the major route of entry, it can also be absorbed through the skin. Exposure can produce severe respiratory irritation, digestive disturbances, and marked renal damage. Mercury has also been reported as a cause of sensitization dermatitis.
   - OSHA has set a transitional limit of 0.10 mg/m^3 as a ceiling value under 29 CFR 1910.1000, Table Z-2.
   - The NIOSH REL is 0.05 mg/m^3 as an 8-hour TWA.
   - The ACGIH is currently considering lowering the TWA to 0.025 with no allowable STEL.

2. *Protection*—Gloves should be used when handling methyl methacrylate, and the material should be kept away from the skin. Disposable protective equipment, such as shoe covers, gloves, and special mercury vapor respirators, should also be used.

- Gowns and hoods should be used when cleaning up mercury spills.
- Spills should be cleaned up promptly with special mercury vacuum cleaners, disposable protective equipment, and a water-soluble mercury decontaminant.
- Exposure to air concentration of 0.5 mg/m^3 may require the use of a hazard-specific air-purifying respirator. Exposure to higher concentration requires the use of a self-contained breathing apparatus.

3. **Disposal**—Mercury wastes must be disposed of according to EPA regulations (40 CFR 261.24).

Lead (29 CFR 1910.1025)—OSHA Standard 29 CFR 1910.1025 requires employers to provide initial and annual training to all employees exposed to an airborne concentration of lead of 30 μg/m^3 averaged over an 8-hour period. Healthcare facilities use lead in a number of locations, especially for shielding in surgical areas and radiology departments. In some very old facilities, exposure can come from lead-based paint or lead-soldered pipe connections. Construction and renovation projects can release lead particles into the air. Some facilities also use lead in casting procedures in oncology or radiology departments.

1. **Exposure Information**—Lead includes metallic lead, all inorganic compounds, and organic lead soaps. Healthcare facilities should evaluate potential lead hazard areas. All areas where workers could be exposed should be tested periodically. The PEL is 50 μg/m^3 of air averaged over an 8-hour period.

2. **Protection**—Respiratory protection and protective clothing should be provided. Organizations must also confine lead to a specific area. Decontamination and shower facilities should be provided to keep lead from being tracked to other areas.

3. **Training Requirements**—The OSHA standard requires employee training in the following areas:
   - Specific workplace hazards
   - Protective measures and equipment
   - Dangers associated with lead exposures
   - Employee rights under the OSHA standard
   - Employee access to the standard and it appendices

4. **Surveillance**—All personnel exposed to an airborne concentration of lead of 30 μg/m^3 of air for an 8-hour TWA for more than 30 days per year must be placed in the medical surveillance program. Employers must ensure that all medical tests, examinations, and procedures are provided by a company physician at no cost to the employee.

5. **Biological Monitoring**—For each employee exposed to the action level described above, blood lead and zinc protoporphyrin sampling must be done at six-month intervals.
   - An employee whose sampling analysis indicates a blood lead level of 40 μg/100 g whole blood must be tested every two months until two consecutive analyses indicate readings below 40 μg.

- Monthly tests are required for an employee removed from exposure to lead due to an elevated lead level at or above 50 μg/100 g whole blood based on an average of the last three blood sampling tests. The employee need not be removed if the last test was at or below 40 μg/100 g whole blood.

6. *Employee Notification*—Within five working days after receiving test results, the employer must notify in writing any employee whose blood lead level exceeds 40 μg/100 g.

7. *Medical Examinations/Requirements*—All employees with blood lead levels at or above 40 μg/100 g at any time during the preceding 12 months must receive an annual medical examination or consultation. Pre-assignment examinations must be given to all employees exposed above the action level. An employee may request an examination/consultation at any time to address signs or symptoms of lead intoxication, receive information on current/past exposures or his or her ability to have healthy children, or when breathing difficulty is demonstrated during a respirator fit.
   - Detailed work and medical histories should focus on past exposures, including non-occupational exposures, and personal habits (smoking/hygiene), with emphasis on history of gastrointestinal, hematologic, renal, cardiovascular, neurological, and reproductive problems.
   - A thorough physical exam should include pulmonary function, blood pressure, routine urinalysis with microscopic examination, and a blood sample analysis that includes the following:
     1. Blood lead level
     2. Hemoglobin
     3. Hematocrit
     4. Zinc protoporphyrin
     5. Blood urea nitrogen
     6. Serum creatinine
   - Other tests should be conducted as deemed necessary by sound medical practice. If requested by an employee, tests shall include pregnancy testing or laboratory evaluation of male fertility.
   - If the employer selects the initial physician, the employee may select a second physician to review findings or provide a second opinion.

8. *Written Medical Opinions*—The facility must provide information to the employee regarding:
   - Opinion as to whether the employee has a medical condition related to lead exposure
   - Limitations to be placed on the employee
   - Recommended special protective measures
   - Recommended limitations on respirator use
   - Blood lead determinations

9. *Medical Information Disclosure*—The physician should not reveal to the employer any medical information not related to the employee's occupational exposure to lead. The physician should advise the employee of any medical condition that requires further examination or treatment.

K. OXIDES

Ethylene Oxide (29 CFR 1910.1047)—This oxide gas is colorless with a distinctive sweet, ether-like odor. It is used to sterilize medical instruments, especially those made of heat-labile materials. It is regulated by OSHA as a carcinogen under 29 CFR 1910.1047. For years, ethylene oxide has been typically supplied to U.S. hospitals in compressed gas cylinders that contain 88% freon and 12% ethylene oxide. It is also supplied in single-dose cartridges of 100% ethylene oxide.

1. *Key Provisions of the OSHA Standard*—
   - The limit on workplace exposure is 1 ppm averaged over an 8-hour TWA. The action level is 0.5 ppm.
   - The STEL is 5.0 ppm averaged over a sampling period of 15 minutes. Employee rotation is prohibited as a means of compliance with the excursion limit.

2. *Exposure Information*—Workers in central supply, dental operations, and surgical suites who use ethylene oxide are at risk of potential exposure. The typical source of ethylene oxide exposure in the hospital environment comes from sterilizing equipment. Even when engineering controls and good work practices are used, workers may encounter relatively high concentrations of ethylene oxide over relatively brief periods.
   - Exposure to ethylene oxide occurs primarily through inhalation. Ethylene oxide has an odor threshold of about 700 ppm, but exposure at 200 ppm may cause irritation of the eyes and upper respiratory system.
   - The OSHA PEL for ethylene oxide is an 8-hour TWA of 1 ppm with an excursion limit of 5 ppm for any 15-minute period. The OSHA action level of 0.5 ppm requires employers to implement exposure monitoring and medical surveillance programs.
   - The NIOSH recommendation is a ceiling of 5 ppm for no more than 10 minutes in any working day and an 8-hour TWA less than 0.1 ppm.

3. *Hazards and Protection*—High concentrations can cause severe skin burns, rashes, sores, headache, nausea, and hemolysis. Very high exposures may cause vomiting, shortness of breath, weakness, drowsiness, lack of coordination, cyanosis, bluish skin color resulting from oxygen insufficiency, and pulmonary edema.
   - Contact with ethylene-oxide-sterilized equipment or wrappings that have not been adequately aerated to remove residual ethylene oxide may cause severe skin burns with large blisters and peeling skin.
   - Workers should use protective gloves and splash-proof goggles and/or a face shield when changing ethylene oxide supply cylinders. If good engineering controls are used and if the cylinder is located in a ventilated hood, a respirator should not be necessary.
   - A chemical cartridge respirator (NIOSH approved) with an end-of-service-life indicator should be used. The end-of-service-life indicator is necessary because the odor threshold for ethylene oxide is about 700 ppm and failure of the absorbent material will not be detected by the user.

- Protective gloves and long-sleeved garments should be worn when removing items from a sterilizer or transferring them to an aerator.

4. *Cleaning Spills*—When cleaning up liquid spills, workers should wear protective outer clothing. A positive-pressure self-contained breathing apparatus should be available for emergency situations and should be stored in an area away from the sterilizer and the ethylene oxide supply. The following procedures should be followed in the event of a major gas leak, suspected leak, or gas spill, such as but not limited to equipment failure, defective cylinders, canisters, or ventilation alarm malfunctions:
   - Alert all personnel in the immediate area of the ethylene oxide release and evacuate the area immediately. DO NOT RE-ENTER THIS AREA UNTIL THE HOSPITAL SAFETY OFFICER PERMITS RE-ENTRY.
   - Close doors to the affected area.
   - Implement the appropriate emergency response program.
   - Wait for emergency response personnel to arrive and identify the location of the leak.

5. *Operating Sterilizers*—Sterilizers should be operated only by personnel trained in sterilization procedures and safety hazards of ethylene oxide. To clean the sterilizer, a worker must often reach inside the chamber with the whole upper body. Ethylene oxide exposure during this cleaning can be controlled by:
   - Scheduling the cleaning activity as long as possible after processing a load
   - Leaving the sterilizer door fully open for at least 30 minutes before cleaning
   - Wearing a respirator

6. *Employee Monitoring*—Employers should obtain pre-employment baseline data on workers who will be handling ethylene oxide. This information should include data on the eyes, skin, blood, and respiratory tract.

7. *Exposure Monitoring*—Annual monitoring is required for each job classification in a work area during each shift. Representative sampling is permitted under certain circumstances. Subsequent monitoring depends on the initial sampling results.

8. *Medical Surveillance*—The OSHA standard requires that a comprehensive program be conducted under the supervision of a licensed physician. Workers must be given an examination before entering an area where exposure is at or above the action level. An annual examination must be given to all workers exposed for 30 days or more during the year. Workers must be given an examination upon request if they develop overexposure symptoms or want medical advice concerning the reproductive hazards of ethylene oxide.

9. *Communication of Hazards to Workers*—Employers must provide information through signs and labels that clearly indicate the carcinogenic and reproductive hazards of ethylene oxide. Initial and annual training should be given to workers who may be exposed at the action level.

10. *Warnings*—Employers must post and maintain signs that show regulated areas and entrances to regulated areas. The warning signs must read as follows:

DANGER—ETHYLENE OXIDE—CANCER AND REPRODUCTIVE HAZ-
ARD—AUTHORIZED PERSONNEL ONLY—RESPIRATORS AND PROTEC-
TIVE CLOTHING MAY BE REQUIRED TO BE WORN IN THIS AREA.

11. *Container Labels*—Container labels must comply with the requirements of 29
    CFR 1910.1200 and include the following: DANGER—CONTAINS ETHYL-
    ENE OXIDE—CANCER AND REPRODUCTIVE HAZARD.

Nitrous Oxide—When mixed with oxygen, nitrous oxide, or laughing gas, acts as a
sedative and painkiller. NIOSH estimates that over 500,000 dental workers are exposed
to potentially harmful amounts. NIOSH recommends that exposure levels be kept under
25 ppm through the use of engineering controls, proper ventilation, and work practices.
Long-term exposure to nitrous oxide is suspected of causing headaches, infertility,
cervical cancer, and neurological disease.

1. *Exposure Information*—NIOSH recommends a ceiling limit of 25 ppm in a
   surgical room environment. Halogenated agents have a NIOSH exposure level of
   2 ppm.

2. *Safety Precautions*—Several precautions can help control worker exposure:
   • Visually check all equipment and replace defective or worn parts.
   • Check the nitrous oxide tank for leaks.
   • Select a scavenging system that operates at a flow rate of 45 liters per minute.
   • Ensure that various sizes of nasal masks are available to fit patients.
   • Connect the mask to the hose and start the vacuum pump before turning on
     the gas. The pump should maintain a rate of 45 liters per minute at each work
     station.
   • Ensure that exhaust fans are not near air supply vents.
   • Monitor worker exposure by using a diffusive sampler or infrared spectro-
     photometer.

3. *Controls*—Most anesthetic agents are controlled by a central evacuation system
   that scrubs the gases and vents to the exterior environment.

L. ASBESTOS

Asbestos includes six naturally occurring fibrous minerals found in certain rock forma-
tions. When mined and processed, it is typically separated into very thin fibers normally
invisible to the naked eye. Asbestos was commonly used in older buildings as an
insulating material for steam pipes. When the insulation is torn off, asbestos fibers may
be released into the air. The fibers can remain in the air for hours and can easily be
inhaled.

Asbestos can be divided into two major groups: serpentine and amphibole. Serpentine minerals have a layered structure, whereas amphiboles have a chain-like structure.
More than 95% of the asbestos found in buildings in the United States is referred to as
white asbestos or chrysotile, which is of the serpentine group. The EPA requires materials suspected to contain asbestos to be analyzed using polarized light microscopy

(PLM) or an equivalent technique. PLM can determine both the percent and type of asbestos in the material.

OSHA Exposure Limits—29 CFR 1910.1001 lists the PEL as 0.1 fibers per cubic centimeter of air (f/cc) averaged over an 8-hour day. The excursion limit (EL) is 1.0 f/cc averaged over 30 minutes. OSHA recognizes the need to group asbestos jobs into categories based on objective criteria. A risk-based classification system includes factors such as type of activity and the amount of asbestos-containing material to be distributed.

Hazard Classes—Asbestos exposures can be better managed by classifying risks.

Class I—This class poses the greatest hazard to workers and involves removal of surface materials or thermal system installation. Surfacing materials include decorative plaster on ceilings, acoustical materials on decking, and fireproofing materials on structural surfaces. Basic safety precautions include:
- Establishing a regulated area with signs
- Presuming exposure potential in excess of the PEL
- Daily monitoring
- Providing abatement worker training (four days)
- Supervision of all activities by a competent person
- Medical surveillance

Class II—This class involves the removal of any material that is not thermal system insulation or surfacing material. Examples include removal of floor/ceiling tiles, siding, roofing, and transite panels. Safety precautions are the same as for Class I except:
- Worker abatement training can be reduced when working with only one generic category of building material.

Class III—This class is defined as repair and maintenance activities that involve intentional disturbance of asbestos or presumed asbestos-containing materials. Class III requires incidental cutting away of small amounts of material. Safety precautions are similar to Classes I and II except:
- Periodic monitoring when the PEL is expected to be exceeded
- Operations and maintenance training course (16 hours)

Class IV—This class is defined as maintenance and custodial activities during which workers come into contact with materials during cleanup operations. Safety precautions are similar to Class III except that worker training should be at the awareness level for maintenance/custodial workers (two hours).

Facility Management—Each healthcare facility must reduce worker exposure to within the PEL. Implementing one or any combination of the following controls should be considered:
- Local exhaust ventilation equipped with a high-efficiency particulate air (HEPA) filter dust collection system
- Dedicated vacuum cleaners with HEPA filters

- Enclosure or isolation of processes that produce asbestos dust
- Use of wet methods, wetting agents, or removal of encapsulants during asbestos operations
- Prompt disposal of asbestos wastes in leak-tight containers
- Use of respiratory protection where other controls cannot reduce exposure below the PEL

Worker Notification—A facility must notify employees of the presence of asbestos materials by providing the following information:

- The exact locations of asbestos-containing materials
- How individuals can avoid disturbing asbestos
- How to recognize and report damage
- What will be done to protect the safety of all occupants
- Name and telephone number of the person coordinating asbestos-related activities for the facility

Medical Surveillance—Medical examinations must be provided annually to all employees who are required to wear respirators or will be exposed to airborne concentrations of asbestos at or above the PEL for 30 or more days per year.

Maintenance/Renovation Permit System—Facilities should use a permit system when work involves exposure or possible disturbance of asbestos fibers. All requests for maintenance/renovation activities should be given to the asbestos program manager prior to issuance of a work order. The building records are then checked for information about the presence of asbestos. If none is found, a work order can be issued, but if asbestos is present, abatement operations must be conducted prior to the maintenance/renovation activities.

Types of Asbestos—

- Actinolite
- Amosite
- Anthophyllite
- Chrysotile
- Crocidolite
- Tremolite

Diseases Caused by Asbestos—Asbestos causes asbestosis (a fibrosis or scarring of the lung tissue) and cancer. These diseases may develop 15 to 30 years after the first exposure.

- Asbestosis belongs to the group of pulmonary diseases called pneumoconioses; these include pneumoconiosis (often called black lung disease) among coal workers and silicosis among workers with prolonged exposure to sand-blasting or other operations in which silica-containing rock is crushed, drilled, or used.
- Pneumoconiosis is characterized by restriction of lung function, which eventually increases the load on the circulatory system so that the fully developed disease usually involves heart failure as well.
- The only hospital workers most likely to encounter enough asbestos to produce asbestosis are engineers who work in furnace rooms where boilers are lined with asbestos and maintenance workers who frequently repair old piping or do minor

renovation. These workers must take special care to protect themselves and to ensure that asbestos is not spread throughout the facility when they perform tasks which involve this substance.

Protection—To reduce asbestos exposure, workers should wear a NIOSH-approved positive-pressure air-supplied respirator, and the insulation material should be dampened before it is cut or torn apart.

- Areas containing asbestos should be vacuumed rather than swept, and waste material should be discarded in sealed plastic bags.
- State agencies or other responsible jurisdictions should be contacted before asbestos removal operations begin.
- Many states certify companies engaged in asbestos removal.
- Because asbestos is an extremely hazardous material and compliance with all relevant aspects of the OSHA asbestos regulations must be assured, hospitals should develop a policy for working with asbestos.
- All workers who may have reason to work with this substance should receive training.

Hazard Control—Whenever asbestos fibers are exposed, they present a hazard that can be eliminated by removing or encapsulating (covering) them so that they will not be released.

- Asbestos must be removed only by fully trained personnel using methods and protective equipment mandated in OSHA 29 CFR 1910.1001.
- Complete physical covering and a NIOSH-certified positive-pressure air-supplied respirator are required for any worker exposed to asbestos.
- The OSHA asbestos standard should be consulted, along with the NIOSH/EPA document entitled "A Guide to Respiratory Protection for the Asbestos Abatement Industry" (NIOSH/EPA 1986).
- Only workers fully trained in asbestos handling should be allowed in areas where asbestos is exposed. The work practices appropriate for handling asbestos are set out in detail in OSHA 29 CFR 1910.1001.

Exposure Sampling—Sampling should be conducted in a manner and on a schedule that will provide an accurate depiction of job-specific asbestos exposures. All analyses should be done by laboratories accredited by the American Industrial Hygiene Association. The minimum schedule for monitoring is established by OSHA in 29 CFR 1910.1001. The PEL for asbestos was recently lowered to 0.1 f/cc effective October 11, 1994.

- Entrances to regulated areas must be posted with signs that read as follows: DANGER—ASBESTOS—CANCER AND LUNG DISEASE HAZARD—AUTHORIZED PERSONNEL ONLY—RESPIRATORS AND PROTECTIVE CLOTHING ARE REQUIRED IN THIS AREA.

- Labels on scrap, materials, and products containing asbestos must contain the following information: DANGER—CONTAINS ASBESTOS FIBERS—AVOID CREATING DUST—CANCER AND LUNG DISEASE HAZARD.

Asbestos Management Policy—A hospital asbestos policy must outline specific OSHA requirements (29 CFR 1910.1001) for the following:

- Reports of each asbestos use or exposure, along with all jobs in which personnel are exposed
- Work practices for handling asbestos, such as wet handling, development of cleanup protocols, use of plastic sheeting to seal off work areas, and bagging to remove insulation during routine operations, maintenance, and repair
- Asbestos waste collection, labeling, and disposal
- Respiratory protective equipment (types of respirators), maintenance, training programs, use, and recordkeeping
- Dressing rooms and special clothing
- Air monitoring
- Recordkeeping and maintenance of records (30 years)
- Medical surveillance (requirements are set by OSHA according to the level of asbestos exposure)
- Training

Summary of the EPA Policy—The Advisory to the Public on Asbestos in Buildings (1991) covers five areas:

- The risk of asbestos-related disease depends upon exposure to airborne asbestos fibers.
- According to surveys the average airborne level in most buildings seems to be very low.
- Removal is often not the best course of action and can create a dangerous situation.
- The EPA only requires asbestos removal to prevent significant public exposure.
- The EPA does not recommend a proactive, in-place management program when asbestos is discovered.

Position of NIOSH on Asbestos Exposure—

- NIOSH recommends the goal of eliminating exposure in the workplace if possible.
- NIOSH contends that there is no safe airborne fiber concentration.
- NIOSH contends that there is no scientific basis to support differentiating health risks between types of asbestos for regulatory purposes.

M. MISCELLANEOUS MATERIALS

Drain-Cleaning Chemicals, Paints, and Adhesives—Drain-cleaning chemicals can burn the skin and damage the eyes. Workers should wear rubber gloves and goggles or face shields when they use drain cleaners because splashing is possible. Product information sheets or MSDSs contain additional information. Paints and adhesives contain a wide variety of solvents and should be used only in areas with adequate ventilation. If ventilation is inadequate, workers should wear respirators approved for use with organic vapors. Skin contact with epoxy paints and adhesives can be prevented by using gloves and other personal protective clothing.

Gasoline—Gasoline should never be used for cleaning floors, tools, clothing, or hands. It should be stored in an approved, labeled, closed container in an outbuilding or flammable storage area. Gasoline spills should be cleaned up immediately.

Pesticides—Insecticides, herbicides, fungicides, disinfectants, rodenticides, and animal repellents are all classified as pesticides. They are considered hazardous substances under the OSHA hazard communication standard. Pesticides are regulated by the EPA under regulations in the Federal Insecticide, Fungicide, and Rodenticide Act. Responsibility for the safe use of these toxic materials begins with purchase and continues until the empty container is properly discarded. All pesticides sold in the United States carry an EPA regulation number which shows they have been deemed safe and effective when used according to directions. Pesticides labeled DANGER—POISON are highly toxic. If inhaled, ingested, or left on the skin, they can be lethal.

- *Labels*—Poisons marked WARNING are moderately toxic and can be very hazardous. A CAUTION label denotes low toxicity, but the chemical can be harmful if ingested or grossly misused.
- *Precautions*—Pesticides should be used only for purposes listed on the label and should be kept in the original container. They should be mixed carefully, outdoors if possible. Protective clothing and equipment should be used as required, and traffic in pesticide applications areas should be controlled. In case of spills, hands should be washed at once with soap and water. Workers should never smoke or eat while spraying or dusting, stay out of spray drift, and avoid using pesticides on windy days. After using pesticides, workers should not smoke, eat, or drink until they have washed their hands.
- *Storage and Disposal*—Pesticides should be stored in a well-ventilated and locked area. Containers should be tightly closed and labeled. Clothing, equipment, or food should not be stored near pesticides. Plenty of soap and water should be available. Pesticides should never be poured into a sink or toilet. Pesticide boxes should be burned outdoors or in incinerators. Empty containers should be wrapped in newspaper before disposal. Pressurized containers should not be punctured or incinerated. All personnel should be instructed as to emergency actions. The number for the Poison Control Center should be posted by the phone.

Compressed Gases—Because some compressed gases are flammable and are under pressure, they must be handled with extreme care. An exploding cylinder can have the same destructive effect as a bomb. Compressed gases used in hospitals include acetylene, ammonia, anesthetic gases, argon, chlorine, ethylene oxide, helium, hydrogen, methyl chloride, nitrogen, and sulfur dioxide. Acetylene, ethylene oxide, methyl chloride, and hydrogen are flammable, as are the anesthetic agents cyclopropane, dimethyl ether, ethyl chloride, and ethylene. Although oxygen and nitrous oxide are labeled as non-flammable, they are oxidizing gases that will aid combustion. Compressed gas safety is covered more fully in Chapter 5.

N. JOINT COMMISSION STANDARDS

A summary of the Joint Commission's hazardous materials and waste management standard is provided below.

- *Orientation and Education*—The Joint Commission requires facilities to conduct an orientation and education program to address handling, storage, usage, and disposal requirements for all hazardous materials. Facilities must also establish emergency procedures to be followed during a spill or exposure situation. Workers must be provided with information on health hazards and guidance on how to report hazardous material spills and exposures.
- *Hazardous Materials*—The Joint Commission requires organizations to provide workers with information on their role, participation, and responsibilities in dealing with hazardous materials. Facilities must establishing monitoring procedures, an inspection program, emergency protocols, and reporting requirements. Facilities must also develop inspection, preventive maintenance, and testing procedures for all applicable equipment.
- *Reporting and Evaluation Requirements*—All hazardous material incidents must be reported and investigated. Facilities must ensure that an annual evaluation is conducted to assess objectives, scope, performance, and effectiveness of the program. Facilities may address the management plan in a number of ways, including written plans, policies, procedures, performance standards, criteria, goals, and objectives.
- *Other Requirements*—Organizations must establish a written program to addresses criteria, procedures, permits, licenses, manifests, and applicable regulations relative to hazardous materials management. Facilities must also monitor/dispose of hazardous gases and vapors. Emphasis must be on the management of chemical, chemotherapeutic, radioactive, and medical waste materials. Hazardous materials managers must ensure that all hazardous materials are properly labeled and adequately stored. Hazardous materials storage and processing areas must be separate from other facility areas.

O. WASTE MANAGEMENT

Virtually all healthcare facilities generate solid and hazardous wastes as defined by the Resource Conservation and Recovery Act (RCRA). The CDC estimates that a typical hospital generates more than ten pounds of solid and hazardous waste per patient each day. These wastes may contain etiologic and chemical agents that could cause injury if improperly handled.

Medical wastes found in healthcare facilities include excreta, blood, exudates, secretions, and solid wastes such as linens and paper/plastic materials.

Because hazardous chemicals are used throughout healthcare facilities, administrators, hazard control managers, and safety specialists must control all such exposure potentials. Management personnel should develop hazardous waste management procedures to address the generation, use, storage, transportation, and proper disposal of all hazardous or potentially hazardous materials.

1. **Cradle-to-Grave Approach**—This approach begins tracking all materials when they enter the facility to ensure compliance with all regulations until the materials have been properly disposed of as waste.

2. **Joint Commission Requirements**—The Joint Commission requires facilities to have "a hazardous materials and wastes program, designed and operated in accordance with applicable law and regulation, to identify and control hazardous materials and wastes." The program must include the following components:
   • Policies and procedures for identifying, handling, storing, using, and disposing of hazardous wastes from generation to final disposal
   • Training for and, as appropriate, monitoring of personnel who manage and/or regularly come into contact with hazardous materials and/or wastes
   • Monitoring compliance with program requirements
   • Evaluating the effectiveness of the program, with reports to the safety committee and those responsible for monitoring activities

Resource and Conservation Recovery Act—The RCRA was passed in 1976 as an amendment to the Solid Waste Disposal Act.

1. **RCRA Goals—**
   • To protect human health and the environment
   • To reduce waste and conserve natural resources and energy
   • To reduce or eliminate hazardous waste as expeditiously as possible

2. **Solid Waste**—The term solid waste is very broad. It includes not only the traditional non-hazardous solid wastes, such as municipal garbage, but hazardous solid wastes as well. Subtitle C of the RCRA regulates hazardous wastes and Subtitle D covers all other solid wastes. Subtitle D covers certain hazardous wastes excluded from regulation under Subtitle C such as household hazardous wastes and those generated by small-quantity generators. Section 1004(27) of the RCRA defines solid waste as:
   • Garbage such as milk cartons and coffee grounds
   • Refuse including scrap metal, wallboard, and empty containers
   • Sludge from waste treatment and water supply plants
   • Other materials including solid, semi-solid, liquid, or contained gaseous material resulting from industrial, commercial, mining, agricultural and community activities

3. **RCRA and Solid Waste**—To understand the RCRA definition of solid waste, it is important to keep in mind that not all solid waste is solid. As noted above, many solid wastes are liquid, while others are semi-solid or gaseous. If the definition were limited to that given above, just about every type of waste produced by man would qualify as solid waste. This, however, is not the case, and Section 1004(27) contains a number of exceptions to the definition. Specifically, the following materials are not considered solid waste under the RCRA:
   • Domestic sewage (defined as untreated sanitary wastes that pass through a sewer system)
   • Industrial wastewater discharges regulated under the Clean Water Act

- Irrigation return flows
- Nuclear materials, or by-products, as defined by the Atomic Energy Act of 1954

P. HAZARDOUS WASTE REGULATION

Subtitle C of the RCRA focuses on the management of wastes with hazardous properties. It directs the EPA and the states to protect human health and the environment from mismanagement of hazardous wastes. Generators normally cannot store hazardous waste for more than 90 days. Under the authority vested by Subtitle C, the EPA has established four hazard characteristics: corrosiveness, ignitability, reactivity, and toxicity. Criteria or test methods have also been established for determining if a waste possesses any of these hazardous characteristics. Specific wastes and chemical constituents that are known to bear one or more of these hazardous characteristics have been identified and listed as hazardous.

Hazardous Waste Definition—Congress defined the term hazardous waste in Section 1004(5) of the RCRA as a "solid waste, or combination of solid wastes, which because of its quantity, concentration, or physical, chemical or infectious characteristics may:

- Cause or significantly contribute to an increase in mortality or an increase in serious irreversible, or incapacitating reversible, illness
- Pose a substantial present or potential hazard to human health or the environment when improperly treated, stored, transported, or disposed of, or otherwise managed"

Solid Hazardous Waste—40 CFR Part 261 specifies that a solid waste is hazardous if it is not excluded from regulation as a hazardous waste and meets any of the following conditions:

- Exhibits any of the characteristics of a hazardous waste
- Has been named as a hazardous waste and listed as such in the regulations
- Is a mixture containing a listed hazardous waste and a non-hazardous solid waste
- Is a waste derived from the treatment, storage, or disposal of a listed hazardous waste

1. *RCRA Authority*—The RCRA applies to any facility that generates, stores, transports, treats, or disposes of:
   - A listed hazardous waste
   - A waste containing a known hazardous constituent
   - A waste that exhibits a hazardous characteristic (the EPA must officially be notified of such activity)

2. *Compliance with RCRA*—Violation of RCRA regulations is strictly prohibited. The EPA and state environmental agencies have the right to inspect a facility and its records at any reasonable time (i.e., during normal operating hours). If an inspector finds that a facility is in violation of the RCRA or its permit, enforce-

ment action in the form of a compliance order (including an administratively imposed injunction) or action in court may follow.

3. *Regulatory Requirements*—There are basically five ways that a facility can fall under the umbrella of federal or state hazardous waste regulation:
   * As a generator of hazardous waste (cannot be stored over 90 days)
   * As a transporter of hazardous waste
   * As a facility that stores hazardous waste
   * As a facility that treats hazardous waste
   * As a facility that disposes of hazardous waste

Hazardous Waste Categories—

1. *Ignitable Solid Waste*—A solid waste that exhibits any of the following properties is considered a hazardous waste due to its ignitability:
   * A liquid, except an aqueous solution containing less than 24% alcohol that has a flash point less than 60°C (140°F)
   * A non-liquid capable, under normal conditions, of spontaneous and sustained combustion
   * An ignitable compressed gas per DOT regulations
   * An oxidizer per DOT regulations

2. *Mixed Waste*—One of the more difficult issues facing the EPA is the regulation of mixtures of radioactive and hazardous wastes, called mixed waste.
   * Mixed waste is considered hazardous under the RCRA and radioactive under the Atomic Energy Act. Both the Nuclear Regulatory Commission (NRC) and the EPA work together to address the management of these wastes.
   * Mixed waste is most often produced by laboratories and by nuclear energy production sources.
   * Laboratories may produce scintillation solvents that contain organic reagents and low-level radioactive wastes.
   * In nuclear energy production, discarded lead shielding and cooling materials may contain heavy metals and radioactive wastes.
   * Generators of mixed waste must comply not only with the minimum technical requirements of the RCRA, but with NRC regulations as well.
   * Because hazardous and radioactive waste components cannot be readily separated, design of facilities, drafting of operating requirements for RCRA permits or NRC licenses, and development of cleanup solutions must be done in a manner that adequately addresses the hazards of both components.

3. *Waste Gases*—Waste gases are those that are released through expiration, escape, leaching, decomposition, or from an accidental spill. The waste management coordinator and fire safety officer should have the authority to implement the emergency plan in the event of a major waste incident. Each department that generates gas should have specific rules and policies concerning the safe handling and disposal of these wastes. Fire-fighting equipment must be available in all areas where waste may be generated. Gases should be classified as flammable, non-flammable, compressed, or toxic.

- **Storage, Handling, and Disposal Requirements—**
  1. Only disposable plastic items should be used with inhalation agents.
  2. Controls should be established in all areas where gases are generated to reduce the risk of fire and explosion.
  3. All gas scavenging systems should be vented to outside air.
  4. Ethylene-oxide-sterilized items should be aerated according to specific time requirements to eliminate toxic residue hazards.
  5. Plastic items containing gas residue should be disposed of with other regular solid waste.
  6. Empty tanks and cylinders should be handled and stored carefully until returned to the distributor.

4. *Antineoplastic Wastes*—These wastes are defined as those chemicals that remain in containers, tubes, and vials or are due to accidents/spillage.
   - Persons who handle these agents must be given detailed orientation, appropriate equipment, and on-the-job training.
   - These agents should be classified as hazardous by the pharmacy and the safety committee.
   - All mixing of drugs should be performed in a biological safety cabinet.
   - Disposable gloves should be worn for all procedures. OSHA-approved chemotherapy gloves are available. Hands should be thoroughly washed with a disinfectant before and after gloving.
   - Contaminated clothing must not be worn outside the work area.
   - All antineoplastic agents, wastes, or materials should be disposed of by incineration.

Q. MEDICAL WASTES

Medical waste has been defined in 40 CFR 259.10 and 40 CFR 22 as any solid waste that is generated in the diagnosis, treatment, or immunization of human beings or animals, in related research, biological production, or testing. Currently, medical waste is regulated by most states under the EPA's Infectious Waste Guidelines of 1982. Some states regulate medical or infectious waste under their solid waste regulations.

Regulated Medical Waste—Regulated medical waste is a subset of all medical wastes and includes seven distinct categories:

- *Class 1*—Cultures and stocks of infectious agents
- *Class 2*—Human pathological wastes (e.g., tissues, body parts)
- *Class 3*—Human blood and blood products
- *Class 4*—Sharps (e.g., hypodermic needles and syringes used in animal or human patient care)
- *Class 5*—Certain animal wastes
- *Class 6*—Certain isolation wastes (e.g., wastes from patients with highly communicable diseases)
- *Class 7*—Unused sharps (e.g., suture needles, scalpel blades, hypodermic needles)

Medical Waste Generators—A medical waste generator is a medical facility or person who produces medical wastes. The term includes but is not limited to:

- Hospitals
- Nursing and convalescent facilities
- Intermediate care facilities
- Dialysis clinics
- Blood banks
- Home health agencies
- Physician offices
- Laboratories
- Funeral homes
- Veterinary clinics
- Emergency medical services

Animal Wastes—Animal wastes are the carcasses and body parts of animals exposed to human infectious agents as a result of being used for the production and/or testing of biological substances and pharmaceuticals or in research.

- Bulk blood, blood components, and potentially infectious body fluids from these animals must be handled as specified for human blood and body fluids.
- All materials discarded from surgical procedures involving these animals that are grossly contaminated with bulk blood, blood components, or body fluids must be treated as specified surgical waste.

Blood and Body Fluids—These wastes are human bulk blood and bulk blood components such as serum and plasma.

- Bulk laboratory specimens of blood tissue, semen, vaginal secretions, cerebrospinal fluid, synovial fluid, pleural fluid, peritoneal fluid, pericardial fluid, and amniotic fluid are included.
- Free-flowing materials or items saturated to the point of dripping liquids that contain visible blood or blood components would be treated/handled as bulk blood and bulk blood components.
- Precautions do not apply to feces, nasal secretions, sputum, sweat, tears, urine, or vomitus unless they contain visible blood.

Microbiological Wastes—These are discarded cultures and stocks of human infectious agents and associated microbiological substances, including:

- Human and animal cell cultures from medical and pathological laboratories
- Cultures and stocks of infectious agents from research and industrial laboratories
- Discarded live and attenuated vaccines
- Dishes used to transfer, inoculate, or mix cultures
- Animal vaccines that are potentially infectious to humans
- Waste from the production of biologicals

Pathological Wastes—All discarded waste from renal dialysis contaminated with peritoneal fluid or blood visible to the human eye is considered medical waste. Solid renal dialysis waste is considered medical waste if it is saturated and has the potential to drip or splash blood or regulated body fluids.

Waste Sharps—Any used or unused discarded article that may cause punctures or cuts and with intention for use in human and animal medical care is included here.

- Articles such as needles, intravenous tubing with needles attached, scalpel blades, and syringes with or without attached needles
- Any item described above that has been removed from its original sterile container
- Glassware, blood vials, pipettes, and similar items that have been contaminated with blood or body fluids

Surgical Wastes—All materials discarded from surgical procedures which are contaminated with blood or body fluids.

- Includes but not limited to disposable gowns, dresses, sponges, lavage tubes, drainage sets, underpads, and surgical gloves

R. WASTE DISPOSAL

Hazardous chemicals should never be discarded down a drain, in a toilet, on the ground outside, or in trash cans or garbage containers destined for landfills, nor should they be burned as a means of disposal. It is important to read the label, check the MSDS, and follow established facility procedures.

1. ***Empty Container Disposal***—Even a small amount of some chemicals when left in a container can be dangerous. They can produce toxic fumes or be ignitable, reactive, or explosive.
   - All containers should be disposed of according to required procedures.
   - Many containers must be cleaned first.

2. ***Separating Waste Materials***—Waste materials can react with one another and burn, release toxic vapors, or explode. Instructions on disposal containers should be followed.

3. ***Hazardous Waste Labels***—During handling or spills, workers need to know what is in the container. Samples of waste labels used in the facility should be displayed and employees should be instructed how to read the labels. Hazard warning signs that apply to chemical waste storage and disposal should be displayed and the importance of obeying these signs should be discussed.

4. ***Unlabeled Containers***—An unlabeled container should never be opened because the fumes could be poisonous or the material could be toxic.

5. ***Report Unsafe Conditions***—Leaking disposal containers, missing labels, chemical containers that have been improperly disposed of, fire hazards, or anything else that has the potential to cause an accident, injury, or damage to plant and equipment should be reported.

General Medical Waste Handling Procedures—

1. *Packaging Requirements*—Untreated waste intended for transport to off-site locations should be packaged in the following manner:
   - Outermost layer of packaging for medical containers excluding sharps should be labeled with a red background color or red lettering with a contrasting background color. The container must be clearly labeled as follows: INFECTIOUS WASTE, MEDICAL WASTE, or BIOHAZARDOUS. The label should also contain the universal biohazard symbol.
   - Wording should be printed on the container or securely attached to the label on two or more sides. The symbol should be at least six inches in diameter and in a color contrasting with the background.
   - Containers should be impermeable to moisture and strong enough to resist ripping, tearing, or bursting under normal conditions of use.
   - Sharps should be placed in rigid, leak-proof, puncture-resistant, sealed containers to prevent loss during handling. The container should be clearly labeled as described above.
   - Small containers used to collect untreated medical waste should be placed inside larger containers during storage, transportation, and disposal.

2. *Outer Packaging Requirements*—The outermost layer of packaging should be properly identified with the following information:
   - Generator's name and address.
   - Date packed in the outermost container.
   - Biohazard symbol and warning: MEDICAL WASTE, INFECTIOUS WASTE, or BIOHAZARDOUS.

TABLE 7.3 Infectious Waste Recommended Treatment Methods

Type of Infectious Waste	Autoclaving	Incineration	Chemical Disinfection	Flush into Sewer System
Waste blood/components			X	
Cultures/stocks of etiologic agents	X			
Contaminated laboratory waste	X	X	X	
Dialysis solid waste		X		
Dialysis coils			X	
Dialysis liquids				X
Isolation solid waste		X		
Isolation liquid waste				X
Needles and sharps		X		
Surgical specimens		X		
Pathological specimens		X		
Suctioned fluids				X
Specimen fluids			X	
Disposable soiled washcloths		X		
Dressings		X		
Linen savers contaminated with feces		X		

- A disposable single-use container used for storage or transportation of untreated waste must be rigid, leak resistant, and tear resistant under normal handling conditions.
- Reusable containers must meet the above requirements and must also be easily cleanable and resistant to corrosion.
- Fiberboard containers must meet the requirements in 49 CFR 178.210 and be classified for a strength test of at least 200 pounds. They must also be marked DOT-12A65. All containers must be sealed prior to shipment.
- The maximum gross weight of fiberboard containers is 65 pounds.

3. *Waste Containers*—Containers used for collection, storage, and transportation must be constructed of materials compatible with the treatment method utilized:
   - Single-use containers destined for incinerators should be burnable.
   - Containers destined for steam sterilizers should allow waste to be properly treated.
   - Reusable containers should be decontaminated after each use with approved methods.
   - Reusable containers should not be used for other purposes unless decontaminated and all medical waste labeling is removed.

4. *Storage of Untreated Waste*—Persons and facilities engaged in storing untreated waste must comply with the following:
   - No one should operate a storage facility without a proper permit.
   - Storage facilities must be fully enclosed.
   - Designated storage facilities should not be used for other purposes.
   - Storage facilities must be identified with signs that contain the universal biohazard symbol or wording as follows: MEDICAL WASTE, INFECTIOUS WASTE, or BIOHAZARDOUS.
   - Storage facility surfaces should be constructed of easily cleaned materials that are impervious to liquids.
   - Facilities should be adequately secured to prevent entry by unauthorized persons.
   - Storage facilities must ensure that actions are taken to minimize or prevent odors from migrating off-site.
   - Handlers of untreated waste should wear gloves and other protective clothing as required.
   - Containerized treated medical waste may be mixed with other solid waste for storage prior to transportation to an approved facility.

5. *Transportation of Untreated Waste*—Transporters should accept untreated waste which contains hazardous or radioactive waste for transportation to a medical treatment facility in accordance with the following guidelines:
   - No transporter may transport untreated medical waste in the same transport vehicle with other solid waste unless all waste is managed as untreated medical waste.
   - No transporter should accept any container which shows visible signs of leakage or is not properly labeled and sealed.
   - No transporter should compact untreated medical waste in a transport vehicle.

- Transporters should not allow untreated medical waste to escape from the vehicle during transport. All vehicles must be enclosed.
- Transporters should not deliver medical waste to an unlicensed storage, treatment, or disposal facility.
- Transport vehicles that become contaminated must be cleaned and decontaminated immediately.
- All transport vehicles must be properly labeled and identified with name of the business, biohazard warnings, and permit number.

6. **Separating Waste**—Infectious and non-infectious wastes should be separated at the point of generation. If the infectious waste contains other non-infectious hazards, it should be properly labeled and treated as required. Infectious waste should be disposed of in identifiable containers or plastic bags that are leak-proof and puncture-resistant.
   - OSHA indicates that red or red-orange containers should be used for bloodborne pathogen waste materials. Such containers must also be marked with the universal biohazard symbol.
   - Packaging should be appropriate for the type of waste and should be durable in handling, storage, transportation, and treatment.
   - Liquid waste should be placed in appropriately marked and sealed bottles, barrels, or drums.

7. **Solid and Semi-Solid Wastes**—Solid and semi-solid waste materials may be placed in plastic bags, but the following guidelines apply:
   - Use tear-resistant bags selected for their durability or thickness.
   - Place sharps in specially designed sharps containers.
   - Never fill a bag beyond its weight or volume capacity.
   - Keeps bags away from sharp external objects.
   - Double bag as required to ensure the contents are safe.

Medical Waste Disposal Methods—

1. **Autoclaving**—Steam sterilization uses saturated steam within a pressure vessel at temperatures high enough to kill infectious agents. Steam sterilization is most effective with low-density materials such as plastics. Containers that can be effectively treated include plastic bags, metal pans, bottles, and flasks. High-density polyethylene and polypropylene should not be used in the process because they do not allow steam penetration. Autoclaving can be used provided the following guidelines are followed:
   - The waste cannot contain hazardous chemicals or radioactive materials.
   - Medical waste such as sharps, recognizable human tissue, and infected animals must be further processed.
   - Steam sterilizers should be equipped to continuously monitor and record temperature and pressure during the process.
   - Some states require that a written log or other documentation be maintained for each sterilization unit.
   - A sterilizer used for waste treatment cannot be used to sterilize equipment, food, or other items.

2. **Autoclaving Precautions**—
   - Plastics bags should be placed in a rigid container to prevent spillage and drain clogging.
   - Infectious waste should be separated from other hazardous waste.
   - Infectious waste that contains non-infectious hazards should not be steam sterilized if it could expose personnel to hazardous materials.
   - Waste that contains antineoplastic substances, toxic chemicals, or chemicals volatilized by steam should not be steam sterilized.
   - Persons operating autoclaving equipment should be trained in handling and operating procedures.
   - Autoclave temperatures should be checked with a recording device to ensure that proper temperatures are being maintained during the process.
   - Steam sterilizers should be routinely inspected and serviced.

3. **Incineration**—The primary method of disposal converts combustible materials into residue or ash. Gases are ventilated through the incinerator stacks and the ash is disposed of as solid waste in a sanitary landfill. Incinerators that are properly designed and maintained are effective in killing infectious organisms.
   - The EPA recently published a proposed emission control standard in 40 CFR Part 60, Standards of Performance for New Stationary Sources and Emission Guidelines for Emission Sources—Medical Waste Incinerators. The proposed rule would affect almost 3700 medical waste incinerators in the United States. Some estimate that as many as 90% could be shut down because the costs to upgrade the facilities could be prohibitive.
   - Incineration has been an effective waste disposal method because it is useful in disposing of pathological waste such as tissue and body parts. Incineration also renders contaminated sharps unusable.
   - The principal factors to consider are variations in waste composition, feed rate, and combustion temperature.
   - Incinerators are operated under permits granted by state environmental agencies and must meet all regulations for air pollution control.
   - Combustible medical waste must be rendered non-recognizable prior to final disposal.

4. **Thermal Inactivation**—This process involves the treatment of waste with high temperatures to destroy the infectious agents. It is normally used for large volumes of waste.
   - Liquid wastes are collected in a vessel and warmed by heat exchangers or a steam jacket surrounding the vessel.
   - The types of pathogens treated determine the temperature and duration of the process.
   - After successful treatment, the contents normally can be discharged into a sewer system in a manner that complies with state and local regulations.
   - Solid waste is treated with dry heat in an oven. This methods requires longer treatment times at much higher temperatures.

5. **Gas/Vapor Sterilization**—This method uses gaseous or vaporized chemicals. Ethylene oxide is the most common agent used. It can be adsorbed on surfaces

of treated materials and workers may be exposed when handling sterilized materials or instruments.

6. ***Chemical Disinfection***—Chemical disinfection is a preferred method for treating liquid infectious wastes, but it can also be used to treat some solid infectious wastes. Some experimental methods that use a combination of shredding and chlorine dioxide are being tested. Factors to consider include:
   - Type of pathogen or microorganism
   - Degree of contamination
   - Amount of proteinaceous material present
   - Type of disinfectant used
   - Contact time
   - Temperature, pH, and mixing requirements
   - Biology of the organism
   - Final disposal of treated wastes must follow environmental laws

7. ***Irradiation (Microwave)***—This method is gaining popularity; dozens of healthcare facilities and some private waste contractors now have microwave disinfecting systems. The following factors should be considered:
   - Minimal electricity requirements and steam not required
   - No heat or chemicals remain in the treated wastes
   - High costs, and procedures require extensive training of personnel
   - Potential for worker exposure
   - Large space requirements
   - Disposal of radiation waste may pose problems

8. ***On-Site Disposal***—
   - Discharge materials such as suctioned fluids, waste in liquid apparatus, bodily discharges, and dialysate liquids.
   - Sterilize substances such as cultures, etiologic agents, and other laboratory wastes.
   - Chemically disinfect wastes from dialysis equipment and specimen spills prior to cleaning.
   - Compactors or grinders should not be used to process infectious wastes.

9. ***Off-Site Disposal***—
   - Infectious waste to be disposed of off-site is to be collected, transported, and stored in a manner prescribed by the contractor.
   - The contractor picks up the waste in leak-proof and fully enclosed containers and transports it to a site approved for handling and disposal of hazardous medical wastes.
   - The contractor is responsible for maintaining all required permits relevant to medical waste disposal.
   - The contractor must maintain all regulatory documentation and records to ensure complete disposal, including incineration waste disposal.

10. ***Final Disposal***—The EPA recommends that each facility contact state and local environmental agencies to identify approved disposal options. The EPA also recommends:

- Discharge of treated liquids and ground solids into sewer systems
- Landfill disposal of treated solids and incinerator ash
- Placing only treated infectious waste in landfills
- Use reputable and permitted hazardous waste handlers

Medical Waste Program Management—Each healthcare facility must operate an effective waste management program. The program should be developed and updated quarterly or as required by the safety committee.

1. *Program Components*—
   - Inventory and categorization of infectious substances by department
   - Identification of hazards during daily activities
   - Purchase of materials and labeling if transferred
   - Storage requirements
   - Distribution and special handling
   - Special cleanup procedures
   - Disposal procedures, identification, transport, pickup, and end point
   - Accident/exposure procedures as part of an exposure control plan
   - Training to be coordinated with the safety committee

2. *Evaluating Medical Waste Management Programs*—Each segment of the infectious waste management program should be reviewed on an annual basis. The evaluation should emphasize the following:
   - Problem-solving and -identification processes
   - Efficiency and productivity of the program
   - Waste management reporting procedures
   - Assessment of responsibilities and authority
   - Review of major incidents or accidental spills
   - Evaluation of contractor performance and compliance
   - Identification of problems that need immediate attention
   - Review of transport records and permits

3. *Contingency Planning*—OSHA and EPA standards require facilities to have an emergency response plan that provides guidance in case of a hazardous materials release or spill:
   - Spill and containment actions should be reported to the appropriate action department.
   - The emergency action plan should address each hazardous substance including biohazards.
   - The facility should determine the level of response necessary and train responders according to the requirements in 29 CFR 1910.120.
   - All personnel working in areas where spills could occur should receive exposure control training.
   - The contingency plan should outline basic containment actions.

4. *Recordkeeping Summary*—Several local, state, and federal agencies require facilities to maintain a variety of hazardous material records. These records can include the following:

- Compliance management plans, manuals, and procedures
- Copies of hazardous waste disposal and release permits
- Inspection and evaluation reports
- Personnel and training information
- Hazardous substance monitoring documentation
- Reports on spills, releases, and discharges
- Material Safety Data Sheets
- SARA Title III reports if applicable
- Emergency responder notification procedures
- Asbestos management plan
- Hazardous waste contingency plan
- Waste handling management plans and documentation

S. HAZARDOUS WASTE TRAINING

Worker Training—Training employees on the specifics of hazardous waste management is vital to the performance of a facility. Personnel should be thoroughly trained to safely and effectively manage hazardous wastes. As new personnel are hired or transferred to work with hazardous waste, training takes place on the job. Trainees should be supervised at all times until they complete the training program.

1. *General Training Topics*—Training for normal or routine operating conditions includes the following topics:
   - Proper handling and safety procedures
   - Waste analysis plan
   - Monitoring and recordkeeping requirements

2. *Hazardous Materials Training*—All workers who handle hazardous waste should receive training. This training should include a session on hazard communication standards. The training should also include the following:
   - Identification of various types of waste
   - How to segregate and prepare for transport
   - How to implement emergency contingency plans
   - Transport and disposal procedures and methods
   - Explanation of hazardous waste management plan and procedures
   - Assignment of roles and responsibilities under the plan
   - Methods for employees to protect themselves from exposure to hazardous materials
   - Personal protective equipment selection and use

3. *Scope of Training*—Training must familiarize employees with:
   - The purpose of hazardous waste laws and regulations
   - The nature of the hazardous wastes generated at the facility
   - Proper handling procedures for wastes
   - Emergency response procedures and contingency planning

4. *Other Training Topics*—To ensure that facility personnel are able to respond effectively to emergencies, the following should be part of the training program:

- Use, repair, and replacement of emergency equipment
- Key parameters for automatic waste feed cutoff systems
- Communications and alarm systems
- Response to fires and explosions
- Response to contamination incidents

5. *Training Frequency*—Training should teach facility personnel about waste disposal procedures relevant to the positions in which they are employed. All personnel involved in hazardous waste management should be adequately trained as outlined above within six months of employment and should undergo an annual review of the initial training as required by the regulations. Training courses should be held annually or as required.

6. *Documentation*—The following documents and records should be maintained at the facility:
   - The job title for each position at the facility related to hazardous waste management, and the name of the employee filling each job
   - A written job description for each position, including the requisite skills, education, or other qualifications and duties of employees assigned to each position
   - A written description of the type and amount of both introductory and continuing training that will be given to each person filling a position
   - Training information to be kept in the personnel office for a period of three years

7. *Infectious Waste Disposal Training*—All workers who handle infectious waste should receive infectious waste management training. This training should include information on OSHA's bloodborne pathogens (29 CFR 1910.1030) and emergency response (29 CFR 1910.120) standards. The training should also include the following:
   - Explanation of infectious waste management plan and procedures
   - Assignment of roles and responsibilities under the plan
   - Procedures to follow to protect workers from exposure
   - Personal protective equipment selection and use
   - Handwashing and precautionary measures
   - Review of standards (universal precautions)
   - Exposure reporting procedures
   - Laws and regulations, and enforcement provisions
   - Modes of transmission and prevention of hepatitis B virus and HIV infections
   - Collection and segregation procedures
   - Sanitation and disinfection procedures
   - Overview of work practices, packaging, and holding requirements

DOT Hazardous Materials Markings—Employers that receive packages or other containers of hazardous materials are required to retain the DOT markings on the containers. Effective October 17, 1994, employers that receive packages, transport vehicles, freight containers, motor vehicles, or rail freight cars which contain hazardous materials marked, placarded, or labeled according to DOT's hazardous materials regulations are required to retain such markings until all hazardous materials have been removed. To be considered a hazardous material under this OSHA regulation, the material must meet the criteria defined in Hazardous Materials Regulations, 49 CFR Parts 171 through 180. This final rule applies to the employer that receives the containers of hazardous materials and not the person responsible for transporting such material unless the material is still under the control of the transporter at its final destination.

DOT provisions require:

- The receiver of a container of hazardous materials marked, labeled, or placarded according to DOT hazardous materials regulations not to remove markings or labels until all hazardous materials are removed, cleaned of residue, or purged of vapors
- The receiver of a freight container, rail freight car, motor vehicle, or transport vehicle which is marked or placarded according to DOT hazardous materials regulations not to remove markings until the hazardous materials are removed
- Markings, placards, and labels to be located in a place that ensures that they are readily visible
- Non-bulk packages not being reshipped need to retain labels or markings according to the requirements of the OSHA HAZCOM standard

SUMMARY

This chapter addressed many of the hazardous materials found in healthcare facilities, including disinfectants, solvents, antineoplastic drugs, formaldehyde, ethylene oxide, and asbestos. Hazardous materials can exert acute or chronic effects that can seriously affect human health. Most hazardous materials found in healthcare settings pose more serious occupational than environmental risks. OSHA and the Joint Commission require that healthcare workers receive training in recognizing and protecting themselves from exposure. Training should include description of hazards, symptoms of overexposure, selection and use of protective equipment, emergency procedures, and first-aid responses. This chapter also covered waste handling regulations and requirements, including sections on managing hazardous and medical wastes. In summary, healthcare administrators and safety management personnel must ensure that all workers understand potential exposure risks. Workers must know the procedures for handling and disposing of hazardous or infectious materials.

REVIEW QUESTIONS

1 Which of the following hazardous materials is not regulated by an OSHA standard?
   a. benzene
   b. ethylene oxide
   c. ribavirin
   d. formalin

2 Substances that easily release oxygen and should not be stored near flammable substances are:
   a. compressed gases
   b. oxidizers
   c. reactives
   d. corrosives

3 The process that equalizes the electrical charge between a storage drum and a transfer container is known as:
   a. grounding
   b. bonding
   c. vapor equalization
   d. ignition temperature stabilization

4 The term "fumes" refers to:
   a. gases released into the atmosphere
   b. carbon or soot particles
   c. the result of volatile solids condensing in cool air
   d. odors caused by toxic vapors

5 Airborne concentrations of hydrocarbons are usually expressed as:
   a. mg/m^3
   b. PEL
   c. (f/m^3)
   d. STEL

6 The OSHA hazard communication standard requires that workers be trained:
   a. upon initial assignment or transfer to another department
   b. when a new hazardous substance is introduced into the workplace
   c. at least annually
   d. a and b

7 The organizational HAZCOM plan must:
   a. be available for worker review during normal business hours
   b. be posted on the employee bulletin or information board
   c. provide information on the location of the MSDS file
   d. all of the above

8 The color *blue* in NFPA 704, Hazard Identification System indicates:
   a. a flammability hazard
   b. a reactivity hazard
   c. the presence of a unique or specific hazard
   d. a health hazard

9 What is the OSHA exposure limit for formaldehyde based on an 8-hour TWA?
   a. 0.75 ppm
   b. 0.95 ppm
   c. 1.0 ppm
   d. 2.0 ppm

10   Which of the following is used in many hospitals to remove adhesives from topical skin areas of patients?
   a.   benzene
   b.   xylene
   c.   perchloroethylene
   d.   acetone

11   At what level must employers implement ethylene oxide exposure monitoring and medical surveillance programs?
   a.   0.1 ppm
   b.   1.0 ppm
   c.   0.5 ppm
   d.   5.0 ppm

Answers: 1–c   2–b   3–b   4–c   5–a   6–d   7–c   8–d   9–a   10–d   11–c

CHAPTER *8*

BIOLOGICAL
HAZARD CONTROL

A. INFECTION CONTROL GUIDELINES

Centers for Disease Control and Prevention (CDC) Guidelines—"Infection Control in Hospital Personnel," which is part of the "Guidelines for Prevention and Control of Nosocomial Infections," has been used as a guide in healthcare settings for a number of years. This CDC publication provides an excellent reference manual for personnel involved in infection control. The following should be considered when referring to CDC guidelines:

- CDC guidelines represent the advice of the center on commonly asked questions and do not have the force of law or regulation.
- CDC recommendations are based on scientific studies.
- Some published CDC recommended practices that have not been adequately evaluated by controlled scientific trials are based on inherent logic and the broad experience of experts.
- Infection control practices that are applicable in one setting may not necessarily be applicable or practical in another.
- The guidelines provide useful information for hospital officials who must decide on the recommendations that best suit their facility's needs and resources.
- Reduction of the risk of nosocomial infection depends largely on the performance of correct patient care and handling practices.
- Personnel will be more likely to follow effective practices if given proper training followed by periodic in-service education sessions.
- Continuous evaluation of care practices under the supervision of the infection control staff will assure continued adherence to correct practices.

New CDC Infection Control Guideline—The newly adopted guideline was first published in the Federal Register, Vol. 59, No. 214, November 7, 1994. The updated guideline replaces the CDC's "Guideline for Isolation Precautions in Hospitals," published in 1983. The new recommendations generally apply to acute care hospitals,

265

subacute facilities, and most long-term nursing facilities. The new guideline was published as a special report issued by the CDC in January 1996.

1. ***Design of the New Guideline***—The new guideline has been designed to:
   - Be epidemiologically sound
   - Recognize the importance of all body fluids, secretions, and excretions in the transmission of nosocomial pathogens among workers and patients
   - Contain adequate precautions for infections transmitted by the airborne, droplet, and contact routes of transmission
   - Be as simple as possible to avoid confusion with existing infection control programs and isolation systems

2. ***Major Changes from Previous CDC Recommendations***—
   - Synthesizes major features of universal precautions and body substance isolation espoused by some hospitals into standard precautions
   - Combines the old categories of isolation precautions (strict, respiratory, tuberculosis, enteric, and drainage/secretions) and the old disease-specific precautions into three sets of precautions based on route of transmission
   - Lists specific syndromes in adult and pediatric patients that are highly suspicious for infection and identifies transmission-based precautions to use on an empirical basis until diagnosis can be determined

3. ***Standard Precautions***—
   - The first tier of precautions under the new guideline is referred to as standard precautions as opposed to universal precautions (the previous term).
   - Standard precautions are designed for the care of all patients regardless of diagnosis or presumed infectious status.

4. ***Transmission Precautions***—
   - Table 1 of the CDC guideline provides a synopsis of the various types of precautions.
   - Table 2 of the guideline provides information on clinical syndromes and empiric precautions.
   - Transmission precautions should be used when caring for patients known or suspected to be infected or colonized with pathogens that can be transmitted by airborne transmission, droplet transmission, or by dry skin contact with contaminated surfaces.
   - The transmission-based precautions are used in addition to standard precautions and include airborne, droplet, and contact precautions.

5. ***Recommendations***—Each tier has guidelines that contain recommendations for:
   - Handwashing and glove use
   - Patient placement procedures
   - Transporting infected patients
   - Masks and other personal protective equipment
   - Gowns and other protective apparel
   - Patient care equipment and articles
   - Linen and laundry procedures
   - Dishes, glasses, cups, and eating utensils

B. INFECTION CONTROL

Healthcare Disinfecting—Healthcare facilities use disinfectants to control organisms that cause disease and infection. Cleaning alone reduces bacteria populations to an acceptable level, but bacteria immediately start growing again.

Bacteria cannot live in the presence of certain chemical compounds. Disinfectants used at the proper dilution will destroy bacteria. Extreme heat also kills bacteria.

1. *Disinfectants*—
   - Synthetic phenols (carbolic) have been developed for safe use by humans. These solutions have a high disinfectant power, even when diluted with water. Phenols can dull surfaces and finishes and also cause skin irritations.
   - Iodine has the reputation of being one of the most powerful disinfectants. Iodine exposure can be hazardous to humans.
   - Iodophors have been developed in recent years and can be used without the strong odor of iodine. They are relatively non-staining and do not corrode metal surfaces.
   - Household bleach is an effective disinfectant but can be toxic in certain circumstances.

2. *Quaternary Ammonia Compounds*—Quaternary ammonia compounds are now the safest and most widely used disinfectants. Their advantages include:
   - No odor
   - More effective than phenol
   - Non-corrosive on most surfaces, as well as non-staining
   - Equal killing power against all types of bacteria
   - Can be safely stored for months

3. *Disinfection Terms*—
   - **Germicide**—Kills germs, especially those that cause disease
   - **Antiseptic**—Destroys germs in living tissue
   - **Bactericide**—Kills bacteria
   - **Bacteriostat**—Inhibits the growth of bacteria
   - **Disinfectant**—Kills bacteria that cause infections
   - **Sterilization**—Renders a surface completely free of all bacteria
   - **Sanitizing**—Reduces bacteria on a surface to an acceptable level

4. *Bacteria*—These tiny invisible plants are members of the fungus family. They do not make their own food and therefore live in the presence of other plant and animal life. A million bacteria could be placed on the head of a pin. Bacteria grow when they find food and favorable conditions in certain places. The human body is a welcome host. Each time a person coughs or sneezes, millions of bacteria are released from the body. Bacteria can be identified by their shape and size.
   - *Staphylococcus aureus*—Commonly called cocci. These round organisms sometimes cluster together. They are found in boils and food poisoning.
   - **Bacilli**—Rod-shaped organisms. They cause typhoid fever and tuberculosis (TB). Bacilli sometimes form hard, resistant bodies known as spores. A disinfectant rated for killing spores is sometimes used to kill bacilli.

- **Sprillae**—Spiral-shaped bacteria that have a corkscrew appearance. Syphilis and cholera are caused by sprillae bacteria.
- **Gram-negative bacteria**—Refers to a chemical difference in certain bacteria which shows up by the Gram-staining procedure. Gram-negative bacteria stain red whereas Gram-positive bacteria stain blue. Bacteria grow when they find food and favorable conditions in certain places. The human body is a welcome host. Each time a person coughs or sneezes, millions of bacteria are released from the body.
- **Virus**—A chemical compound of the protein variety. Viruses cannot grow, but under certain conditions they appear to have the ability to reproduce. They are easily destroyed by disinfectants.

Infection Control Procedures—Most healthcare organizations have some type of written infection control program. It is recommended that procedures be published for each patient care department, including the pharmacy. At many acute care facilities, the infection control committee participates in employee safety and health program activities.

1. *Infection Control Programs*—
   a. An effective infection control program should stress sound personal hygiene, individual responsibility, monitoring, and investigating infectious diseases with potentially harmful infectious exposures.
   b. The program should also stress providing care for work-related illnesses, identifying occupational infection risks, instituting preventive measures, eliminating unnecessary procedures, and preventing infectious diseases.
   c. Organizations should consider the following to be important in their infection control program:
      - Placement evaluations
      - Personnel health and safety education
      - Immunization programs
      - Protocols for surveillance
      - Management of job-related illnesses
      - Exposures to infectious diseases
      - Counseling regarding infection risks
      - Guidelines for work restrictions due to infections
      - Maintenance of health records

2. *Infection Control in Long-Term Care Facilities*—
   - Infections can occur more often in people who have impaired immunity due to the aging process as well other medical complications, from diabetes to malignancy.
   - The range and number of medications residents in such facilities take may also predispose them to nosocomial infection.

3. *Environmental Controls*—Recognizing the need for environmental control is an essential part of protecting workers and patients. Frequent causes of infection include:
   - Pathogenic agents may become airborne through spills or broken containers. TB is a serious disease caused by airborne bacteria.

- Workers and patients can ingest agents as they eat. A major cause is eating with unwashed hands after handling infected materials.
- Employees should never eat in work areas or store lunches with non-food substances such as medications, solvents, specimens, or other chemicals.
- Sudden movement of a patient may result in a needlestick injury to nursing personnel.
- Infecting agents can enter the body through the skin and the eyes. Small scratches and cuts on the hands are common points of entry for infectious organisms.

4. *Infection Control and Safety*—Historically, infection control has not been considered a part of the safety management program.
   - The safety program must work closely with the infection control officer.
   - At facilities with no infection control coordinator, the safety committee must see that someone is designated to perform this function.
   - Safe practices and proper infection control measures intersect with nearly every area of nursing home operation, from catheterizing a resident to cleaning residents' rooms and bathrooms.

5. *Infection Control Committee*—
   a. The infection control committee establishes policies and procedures for investigating, controlling, and preventing infections in healthcare organizations.
   b. Membership should include representation from the medical staff, nursing services, pharmacy, quality improvement, safety, and administration. The committee should:
      - Monitor staff performance to ensure proper execution of policies and procedures
      - Meet as required but at least quarterly and document content of meetings in written form, with copies maintained in facility files.
      - Review and update policies and procedures as required but as least annually
      - Establish procedures for aseptic and isolation techniques as established by CDC guidelines
      - Review patient infections to determine if nosocomial in nature
      - Ensure that all personnel are trained and educated in infection control policies and control techniques
      - Review clusters of infections, infectious outbreaks, and epidemics that exceed usual levels

Joint Commission Infection Control Guidelines—Healthcare organizations run the risk of exposing patients, visitors, and staff members to nosocomial and other infections. Infections can be endemic or epidemic in nature. An effective infection control program must strive to identify and reduce the risks of acquiring and transmitting infections among hospital employees, staff members, and visitors. An effective program must stress surveillance, prevention, and control activities. Joint Commission standards require an organization to use a coordinated prevention process that is managed by one or more qualified individuals. The program should take into consideration sound epidemiologic principles and data on nosocomial infection. Some organizations require infection control coordinators to be certified by organizations such as the Certification Board for Infection Control. The scope of a particular program will vary but should consider the following factors:

- Geographic location of the facility
- Type of population served
- Size and type of the facility
- Number of employees

Epidemiological considerations:

- Infections that are device or surgery related
- Infections that occur in special care units
- Infections that result from antibiotic-resistant organisms
- Infections such as TB or other communicable diseases
- Infections that occur in neonatal units

The infection control program should be integrated into the hospital's efforts to improve organizational performance. The organization must track data and information relating to risks, rates, and trends in nosocomial infections. The infection control process should have the necessary management systems and staffing to achieve the objectives of the program.

1. ***Basic Requirements—***
   - The organization must have a functioning, coordinated process in place to reduce the risks of endemic and epidemic nosocomial infections in patients and healthcare workers.
   - This infection control function must be managed by a qualified individual.
   - A facility must have mechanisms for patient care and employee health activities designed to reduce risks of endemic and epidemic nosocomial infections.

2. ***Management Requirements—***
   - Case findings and identification of nosocomial infections must be tracked according to demographics, so as to provide data for the organization.
   - Organizations must report information about infections to public health agencies.
   - Healthcare facilities must implement strategies to reduce risks and/or prevent nosocomial infections in patients, employees, and visitors.
   - Organizations must implement strategies to control outbreaks of nosocomial infections when identified.
   - Facilities must measure their infection risks, trends, and rates of epidemiologically significant infections, where appropriate.
   - Appropriate management systems must be established to support nosocomial infection risk reduction processes to ensure adequate analysis and interpretation of data and presentation of findings.

Isolation and Infection Control Principles—

1. *Handwashing*—The CDC emphasizes that handwashing before and after contact with patients or potentially infectious materials is the most important means of preventing the spread of infection.

2. *Transmission Routes*—Federal, state, and local health agencies have published rules and guidelines that define isolation procedures. Healthcare organizations should follow these guidelines because infectious agents can be transmitted by several routes:
   - **Contact Transmission**—Actual contact with a person with a contagious disease by a susceptible person.
   - **Indirect Contact Transmission**—Contact with a contaminated object used by an infected person.
   - **Droplet Transmission**—The spray of mist ejected from the nose or mouth when coughing, sneezing, or talking. Droplets usually do not travel more than three feet.
   - **Common Vehicle Transmission**—Disease spread through contaminated food, water, drugs, devices, and equipment.
   - **Airborne Transmission**—Droplets of an infectious agent lodge in dust and are inhaled or digested by a susceptible host.
   - **Vectorborne Transmission**—Disease organisms carried by an animal or insect, such as a tick or mosquito, infect a person.

3. *Immunizations*—Hospitals should develop a comprehensive written policy on immunizing their personnel. Healthcare personnel are not at a substantially higher risk than the general population for many diseases, including diphtheria, mumps, tetanus, and pneumococcal diseases. However, healthcare organizations should be aware of the following recommendations concerning immunizations:
   - **Rubella**—All personnel considered to be at risk or who have direct contact with pregnant patients should be immune to rubella. Serologic screening is not recommended unless determined to be cost effective by the hospital. Persons need to be immunized unless they can provide laboratory evidence of immunity or documented immunization with live virus on or after their first birthday.
   - **Hepatitis B**—All personnel exposed to bloodborne pathogens should be offered the vaccine within ten days of job assignment. Refer to Section E in this chapter for additional information.
   - **Measles**—All persons susceptible by history or serology who are considered to be at risk should be immunized. Most persons born before 1957 have probably been naturally infected and need not be considered susceptible. Others should be immunized unless they have documentation of physician-diagnosed measles, laboratory evidence of immunity, or adequate immunization after their first birthday.
   - **Influenza**—Healthcare personnel should consider taking flu immunization to help prevent the spread of influenza from personnel to patients. Hospitals should promote such a program and provide vaccine during the fall of each year.

Types of Isolation—Isolation is a means of preventing the spread of a communicable disease in the facility. There are several types of isolation:

1. *Contact Isolation*—Can protect against severe respiratory infections and scabies. Masks are normally used for close contact. Gowns and gloves should be used if contact is likely. Contaminated articles must be discarded or bagged for decontamination.
2. *Secretion Precautions*—Protection against "staph and strep" infections of the skin. Masks are not required, but gloves and gowns must be used if contact is likely. Contaminated articles should be discarded or bagged for decontamination.
3. *Strict Isolation*—Protects against serious infections. Masks, gloves, and gowns are required. Entry into patient's room is restricted. Contaminated articles should be discarded or bagged for decontamination.
4. *Respiratory Isolation*—Protects against airborne infections such as pneumonia. Masks or appropriate respirators are required. Gloves must be used, and universal precautions must be followed for blood or body fluid contact. Contaminated articles should be discarded or bagged for decontamination.
5. *Acid-Fast Bacillus Isolation (Tuberculosis)*—Approved respirators must be worn. Gloves are indicated for contact with blood or body fluids. Articles must be discarded or bagged for decontamination.
6. *Enteric Isolation*—Prevents the spread of pathogens through contact with feces. Gloves must be worn and hands washed upon entering and leaving the room.

Methicillin-Resistant *Staphylococcus aureus* (MRSA)—This disease is resistant to many antibiotics. The disease occurs in warm, moist areas (i.e., the nose or mouth). It primarily affects persons with an impaired immune system and occurs frequently in long-term nursing facilities. Symptoms include skin redness, drainage, and fever. MRSA pathogens are highly contagious and can be spread on hands of healthcare workers. Preventive practices include proper handwashing and using aseptic techniques.

Healthcare Immunizations—The Advisory Committee for Immunization Practice of the U.S. Department of Health and Human Services suggests the following:

- Healthcare workers should meet guidelines for immunization against mumps, rubella, diphtheria, and measles.
- Workers should get annual influenza immunization.
- Exposed workers should receive hepatitis B immunization (required by OSHA under its bloodborne pathogens standard).

C. WORKING SAFELY WITH SHARPS AND NEEDLES

Hazards of Needles and Sharps—Cuts, lacerations, and punctures are common among healthcare workers. Many of these injuries go unreported.

1. *Needlesticks*—Needlesticks continue to be a major concern in healthcare facilities.

- According to the CDC, over 2500 healthcare workers reported HIV exposure from needlesticks between 1983 and 1993.
- There are options to the standard hollow-bore needle and syringe, such as safety needles with sliding sleeves and self-capping devices.
- Medical suppliers are providing more advanced systems, such as the bloodless system which uses a stop device when the needle is withdrawn.
- Another effective system uses a plastic cannula to penetrate intravenous lines.

2. **Sharps Safety**—Sharp instruments should be discarded in designated puncture-resistant containers and not in trash cans or plastic bags.
   - Hospitals should establish and enforce policies to prevent the recapping of needles.
   - Rules for safe disposal and collection of sharp instruments or other hazardous materials should be reviewed regularly.
   - Workers should examine and handle soiled linens and similar items as if they were hazardous.

3. **Sharps or Needlestick Exposure Incidents**—The following guidelines apply when the skin of a worker is punctured or lacerated with a contaminated sharp or needle:
   - The wound should be allowed to bleed and then be thoroughly cleaned.
   - A suitable topical disinfectant or antiseptic and proper bandaging should be applied.
   - Medical information should be obtained from the person on whom the medical device was used.
   - Special attention should focus on determining the possible risk of acquiring hepatitis, AIDS, or other infectious diseases.
   - An incident report should be completed to document as much information about the exposure as possible.
   - When it is determined that a person is at risk of becoming infected, a medical evaluation should be undertaken immediately and should include prophylactic treatment or immunization if applicable.
   - A blood sample should be taken from the injured person to provide for a baseline sample for serological evaluation.
   - Exposures should be monitored to evaluate health status as it relates to the possible infection as well as to document for a future workers' compensation claim.

4. **Sharps Handling**—The following safety rules apply when handling sharps:
   - Never remove the protective sheath from a needle by placing the sheath in the mouth or teeth.
   - Under no circumstances should a used needle be re-inserted into its original protective sheath.
   - Directly following use, needles should be destroyed in a needle-destroying device or placed directly into a puncture-proof container designed specifically for this purpose.
   - Simple protective latex gloves should be worn. They will protect against all contamination unless the gloves are torn by a sharp object and injury occurs.

- Any existing injury or open cut should be protected when handling potentially contaminated materials.
- Hands should be washed immediately following completion of a procedure or upon removal of protective gloves.
- Used needles, syringes, scalpels, or other sharps should be discarded in a puncture-proof container. When the container becomes filled, it should be secured shut and disposed of according to regulations.
- No sharps should ever be disposed of by placing them in standard trash receptacles where custodial or other housekeeping personnel could sustain injuries.
- Individuals who are at a high risk of exposure to potentially infected persons, human blood, or other bodily fluids should obtain adequate immunization.
- Healthcare workers should receive a course of hepatitis B vaccine as a pre-exposure measure.

D. BLOODBORNE PATHOGENS

The Occupational Safety and Health Administration (OSHA) estimates more than 5.6 million workers in healthcare and related occupations are at risk of exposure to bloodborne pathogens.

- These exposures can put workers at risk for HIV, hepatitis B and C viruses, and other infectious diseases.
- OSHA recognizes the need for a regulation that prescribes safeguards to protect these workers against health hazards due to exposure to blood and certain body fluids, including bloodborne pathogens.

Public Law 102-141—This federal law became effective on October 21, 1991 and requires states to adopt the CDC's Recommendations for Preventing Transmission of HIV and HBV to Patients During Exposure-Prone Invasive Procedures.

- Compliance is the responsibility of each state's public health department. Each state must develop a set of rules to regulate the implementation of the law.
- The law requires that any healthcare worker infected with HIV/hepatitis B virus who performs invasive procedures or any physician providing the infected worker care must notify the state health officer of the infection.
- Each infection situation will be thoroughly reviewed by an expert panel. Based on the panel's finding, the state public health officer could place restrictions on the practice of the infected worker.

OSHA Bloodborne Pathogens Standard—OSHA Standard 29 CFR 1910.1030 protects employees who may be occupationally exposed to blood and other potentially infectious materials.

- Blood means human blood, blood products, or blood components.
- Other potentially infectious materials include human body fluids such as saliva,

semen, vaginal secretions, and cerebrospinal, synovial, pleural, pericardial, and amniotic fluids.

- The standard covers body fluids visibly contaminated with blood, unfixed human tissue and organs, HIV-containing cells or tissue cultures, and HIV/hepatitis B virus culture media or other solutions.
- Occupational exposure means "reasonably anticipated skin, eye, mucous membrane, or parenteral contact with blood or other potentially infectious materials that may result from the performance of the employee's duties."
- OSHA authority extends to all private sector employers with one or more employees, as well as federal civilian employees.
- The standard specifies labeling requirements as listed in Table 8.1.

The Written Exposure Control Plan—A written exposure control plan is necessary for the safety and health of workers. The plan must include the following:

- Job classifications where there is exposure to blood or other potentially infectious materials
- Protective measures currently in effect in an acute care facility and methods of compliance
- Hepatitis B vaccination and post-exposure follow-up procedures
- How hazards are communicated to employees
- Personal protective equipment, housekeeping procedures, and recordkeeping policies
- Procedures for evaluating the circumstances of an exposure incident

Occupational Exposure—Exposure determinations must be based on the definition of occupational exposure *without regard to personal protective clothing and equipment.*

1. Exposure determination begins by reviewing job classifications of employees within the work environment.
2. Healthcare facilities should evaluate all job classifications and divide them into three categories:
   - *Category I*—Includes nurses, nursing assistants, and home health caregivers. The list must contain all job classifications in which *all* of the employees have occupational exposure.
   - *Category II*—Job classifications in which *some* of the employees have potential occupational exposure. These jobs can include physical therapists, social workers, occupational therapists, environmental services personnel, housekeepers, and speech therapists.
   - *Category III*—Includes healthcare workers in clerical or administrative positions who have no exposure to blood or body fluids in the normal performance of their jobs.

Training Requirements—When employees with occupational exposure have been identified, the next step is to communicate the hazards of exposure to them.

1. Training must be comprehensive and must include information on bloodborne pathogens, the OSHA standard, and the local exposure plan.

2. Persons conducting the training must be knowledgeable in the subject matter.
3. The training program must:
    - Explain regulatory text and make a copy available
    - Explain the epidemiology and symptoms of bloodborne diseases
    - Explain the modes of transmission of bloodborne pathogens
    - Explain the employer's written exposure control plan

TABLE 8.1 Biohazard Labeling Guidelines

Item Description	No Label Needed if Universal Precautions Are Used and Specific Use of Container Is Known to All Employees	Biohazard Label		Red Container
Regulated waste container (e.g., contaminated sharps containers)		X	or	X
Reusable contaminated sharps container (e.g., surgical instruments soaking in a tray)		X	or	X
Refrigerator/freezer holding blood or other potentially infectious material		X		
Containers used for storage, transport, or shipping of blood		X	or	X
Blood/blood products for clinical use	No labels required			
Individual specimen containers of blood or other potentially infectious materials remaining in the facility	X or	X	or	X
Contaminated equipment needing service (e.g., dialysis equipment, suction apparatus)		X plus a label specifying where the contamination exists		
Specimens and regulated waste shipped from the primary facility to another facility for service or disposal		X	or	X
Contaminated laundry	* or	X	or	X
Contaminated laundry sent to another facility that does not use universal precautions		X	or	X

* Alternative labeling or color coding is sufficient if it permits all employees to recognize the containers as requiring compliance with universal precautions.

Labeling—Containers of regulated waste, refrigerators or freezers containing blood or other potentially infectious materials, and other containers used to store, transport, or ship blood or other potentially infectious materials must be labeled with a fluorescent orange or orange-red biohazard warning label. The warning label must contain the biohazard symbol and must include the word BIOHAZARD.

- Describe the methods to control transmission of hepatitis B virus and HIV
- Explain how to recognize occupational exposure
- Inform workers about the availability of free hepatitis B vaccinations and vaccine efficacy, safety, benefits, and administration
- Explain emergency procedures and procedures for reporting exposure incidents
- Inform workers of post-exposure evaluation and follow-up available from healthcare professionals
- Describe how to select, use, remove, handle, decontaminate, and dispose of personal protective clothing and equipment
- Explain the use and limitations of safe work practices, engineering controls, and personal protective equipment
- Explain the use of labels, signs, and color coding required by the standard
- Provide a question-and-answer period with each training session

E. HEPATITIS

Five types of viral hepatitis have been identified, and each one has a different prevalence and outcome. Refer to Table 8.2 for additional information.

Hepatitis A (Infectious Hepatitis)—This disease is primarily transmitted by person-to-person contact. It can also be contracted from contaminated uncooked shellfish, fruits, or vegetables and contaminated water.

- Pre-exposure prophylaxis is immunoglobulin given before exposure, especially for certain international travelers.
- Accounts for almost 50% of reported cases of hepatitis in the United States.

Hepatitis B Virus (HBV)—HBV is the world's most common bloodborne viral infection; it currently infects over 300 million people worldwide. The virus that causes hepatitis B is found in blood and other body fluids, including semen, vaginal secretions, urine, and even saliva.

- Hepatitis B can be very serious and even fatal, but most infected individuals either display no symptoms or have non-specific flu-like symptoms. About 25% will become ill with jaundice.
- Most people recover from HBV infection, but up to 10% become chronic carriers. These chronic carriers are capable of spreading the disease to others for an indefinite period of time and are at high risk for long-term complications such as cirrhosis of the liver and primary liver cancer.
- HBV is transmitted more easily than HIV and is commonly spread through sexual contact, sharing needles by drug abusers, injuries caused by contaminated objects including needles, and infected blood and blood products that enter through the eyes, mouth, or a break in the skin.
- While virtually anyone can get hepatitis B, some populations are at higher risk for becoming infected.
- High-risk occupations with exposure to blood or other body fluids include physicians, dentists, nurses, law enforcement officers, and fire fighters.

TABLE 8.2   Types of Viral Hepatitis

Type of Hepatitis	Mode of Transmission	Incubation Period	Serological Tests	Complications
A	Fecal/oral	15–50 days	Available	Rapid, intense hepatitis Relapse
B	Parenteral Sexual Perinatal	40–180 days	Available	Rapid, intense hepatitis Chronic liver disease Cirrhosis Primary hepatocellular carcinoma
C	Parenteral	35–75 days	Available	Chronic liver disease Cirrhosis Primary hepatocellular carcinoma
D	Parenteral Sexual Perinatal	21–49 days	Available	Chronic liver disease Rapid, intense hepatitis
E	Fecal/oral	28–42 days	Not widely available	High mortality in pregnant women Fetal death

- HBV can cause either acute or chronic outcomes. A healthy adult often produces an antibody and the disease is self-limiting. When this happens, the liver cells that contain the virus are destroyed, the virus is eliminated from the body, and the person has a lifetime immunity against re-infection. About 6 to 10% of adults cannot clear the virus from their liver cells and become chronic HBV carriers. These carriers are at high risk for developing chronic persistent hepatitis and other liver diseases.
- In cases of acute infection, one-third show no symptoms, one-third have a relatively mild case of flu-like illness, and one-third have more severe responses, including jaundice, dark urine, extreme fatigue, anorexia, nausea, abdominal pain, joint pain, rash, and fever.
- Death may occur in 1 to 2% of these cases. In cases of chronic infection, these symptoms recur and may be accompanied by symptoms of other more serious diseases.

1. **Chronic HBV**—Persistence of viral infection (the chronic HBV carrier state) occurs in 5 to 10% of persons following acute hepatitis B and occurs more frequently after initial anicteric hepatitis B than after initial icteric disease.
   - Consequently, carriers of hepatitis B surface antigen (HBsAg) frequently give no history of having had recognized acute hepatitis.
   - It has been estimated that more than 170 million people in the world today are persistently infected with HBV.
   - The CDC estimates that there are approximately 750,000 to 1 million chronic carriers of HBV in the United States. Chronic carriers represent the largest human reservoir of HBV.
   - Although the vehicle for transmission of the virus is often blood and blood

products, viral antigen has also been found in tears, saliva, breast milk, urine, semen, and vaginal secretions.

- HBV is capable of surviving for days on environmental surfaces exposed to body fluids containing HBV.
- Infection may occur when HBV, transmitted by infected body fluids, is implanted via mucous surfaces or percutaneously introduced through accidental or deliberate breaks in the skin.

2. *Hepatitis B Vaccination Requirements*—The hepatitis B vaccination series must be made available within ten working days of initial assignment to every employee who has occupational exposure.
   - The vaccination must be made available without cost to the employee, at a reasonable time and place for the employee, by a licensed healthcare professional, and according to recommendations of the U.S. Public Health Service, including routine booster doses.
   - The healthcare professional designated by the employer to implement this part of the standard must be provided with a copy of the bloodborne pathogens standard.
   - The healthcare professional must provide the employer with a written opinion stating whether the hepatitis B vaccination is indicated for the employee and whether the employee has received such vaccination.
   - Employers are not required to offer hepatitis B vaccination:
     a. To employees who have previously completed the hepatitis B vaccination series
     b. When immunity is confirmed through antibody testing
     c. If vaccine is contraindicated for medical reasons
     d. Following participation in a pre-screening program
     e. To employees who decline the vaccination, although these employees may request and obtain it at a later date, if they continue to be exposed
   - Employees who decline to accept the hepatitis B vaccination must sign a declination form indicating that they were offered the vaccination but refused it

3. *Hepatitis B Vaccine Information*—Current vaccines are recombinant or a non-infectious viral vaccine derived from HBsAg produced in yeast cells. A portion of the HBV gene is cloned into yeast, and the vaccine for hepatitis B is produced from cultures of this recombinant yeast strain.
   - The antigen is harvested and purified from fermentation cultures of a recombinant strain of the yeast.
   - The HBsAG protein is released from the yeast cells by cell disruption and purified by a series of physical and chemical methods.
   - The vaccine contains no detectable yeast DNA but may contain no more than 1% yeast protein.
   - Vaccines prepared from recombinant yeast cultures do not contain human blood or blood products.

4. *Dosage*—10 µg (1.0 ml)
   - First dose is at target date.

- Second dose is 30 days later.
- Third dose is 180 days after first dose.
- The recombinant vaccine is injected intramuscularly. The deltoid muscle is the preferred site in adults. Data suggest that injections given in the buttocks frequently enter fatty tissue instead of muscle. Such injections have resulted in a lower sero-conversion rate than expected.
- Contraindication is hypersensitivity to yeast or any component of the vaccine.
- Patients who develop symptoms suggestive of hypersensitivity after an injection should not receive further injections of the vaccine.
- Because of the long incubation period for hepatitis B, it is possible for unrecognized infection to be present at the time the vaccine is given. The vaccine may not prevent hepatitis B in such patients.
- As with any percutaneous vaccine, epinephrine should be available for immediate use should a reaction occur.
- Any serious active infection is reason for delaying use of the vaccine except when, in the opinion of the physician, withholding the vaccine entails a greater risk.
- Caution and appropriate care should be exercised in administering the vaccine to individuals with severely compromised cardiopulmonary status or those susceptible to febrile or systemic reactions.

Hepatitis C (Non-A, Non-B)—This parenterally transmitted virus is traditionally associated with blood transfusions.

- Parenteral drug users and dialysis patients are also considered at-risk groups.
- Accounts for about 30% of acute viral hepatitis in the United States.

Hepatitis D (Delta Agent Hepatitis)—This type may cause infection only in the presence of active HBV infection.

- Co-infection intensifies the acute symptoms of hepatitis D.
- Hepatitis D is dependent upon HBV for replication.

Hepatitis E (Enterically Transmitted Non-A, Non-B)—This type was first identified through waterborne epidemics in developing countries.

- Sporadic cases also can occur.
- The disease is mild except for women in the last three months of pregnancy.

F. HUMAN IMMUNODEFICIENCY VIRUS (HIV)

HIV affects the immune system, rendering the infected individual vulnerable to a wide range of disorders. Infections typically lead to death of the patient.

Symptoms—

- Symptoms can occur within a month after exposure and typically include fever, diarrhea, fatigue, and rash.

- Later, the person may develop antibodies and be without symptoms for a period of months to years.
- Finally, the infected person may develop a wide range of symptoms, dependent on the opportunistic infection(s) against which the person's immune system cannot defend.

Exposure and Transmission Routes—HIV is spread through:

- Contact with blood, semen, vaginal secretions, and breast milk
- Sexual intercourse and using needles contaminated with the virus
- Contact with HIV-infected blood under the skin, on mucous membranes, or non-intact skin
- Mother to child contact at the time of birth
- Blood transfusions or organ transplants

Workplace Transmission—HIV and other bloodborne pathogens may be present in:

- Body fluids such as saliva, semen, vaginal secretions, cerebrospinal fluid, synovial fluid, pleural fluid, peritoneal fluid, pericardial fluid, amniotic fluid, and any other body fluids visibly contaminated with blood
- Saliva and blood contacted during dental procedures
- Unfixed tissue or organs other than intact skin from living or dead humans
- Organ cultures, culture media, or similar solutions
- Blood, organs, and tissues from experimental animals infected with HIV or HBV

Means of Transmission—

- An accidental injury with a sharp object contaminated with infectious material, such as needles, scalpels, broken glass, and anything that can pierce the skin
- Open cuts, nicks, skin abrasions, dermatitis, acne, and mucous membranes
- Indirect transmission, such as touching a contaminated object or surface and transferring the infectious material to the mouth, eyes, nose, or open skin

G. TUBERCULOSIS

Mycobacterium tuberculosis is a slow-growing bacteria that infects the respiratory tract of humans. The TB bacteria is carried in airborne particles, called droplet nuclei, which can be generated when persons with pulmonary or laryngeal TB sneeze, cough, or speak. The bacteria is very small, measuring only 1 to 5 microns, and can spread through normal air currents in a room or building. Healthcare safety personnel must work closely with infection control personnel to properly control TB risks.

Transmission—Transmission occurs by inhalation. Two to ten weeks after infection, the body's immune system limits further growth and spread. Some of the bacteria can remain in the body in a dormant state for years. This is known as latent TB infection.

- There are no visible signs of latent TB and the person is not infectious. About 10% of people with latent TB are at risk for developing the disease later in life.
- OSHA will enforce the CDC's 1990 guidelines for TB control under its general duty clause until a standard is adopted using CDC guidelines.

OSHA TB Exposure Enforcement Guidelines—The United States has seen a significant (18%) increase in new TB cases since 1985. The CDC published its "Guidelines for Preventing the Transmission of Tuberculosis in Health-Care Settings" in 1990. These guidelines address patient and worker testing, source control methods, decontamination techniques, and prevention of TB-contaminated air. In 1993 OSHA issued its "Enforcement Policy and Procedures for Occupational Exposure to Tuberculosis." This enforcement policy uses the 1994 CDC guidelines and the OSHA general duty clause. OSHA will continue to enforce selected aspects of the CDC guidelines until the OSHA standard is published. Currently, OSHA inspections can be done in response to complaints and during routine compliance visits in the following workplaces:

- Healthcare settings
- Correctional institutions
- Homeless shelters
- Long-term care facilities
- Drug treatment centers

OSHA Citations—Citations can be issued to employers in the categories described above if employees are exposed as described below:

- Exposure or potential exposure to exhaled air of a suspected or confirmed case of TB
- Exposure to a high-hazard procedure performed on an individual with suspected or confirmed TB

OSHA Abatement Methods—

- Early identification of persons with active TB
- Medical surveillance at no cost to the employee, including pre-placement evaluation, TB skin tests, annual evaluations, and twice yearly exams for those who have been exposed
- Evaluation and management of workers with a positive skin test
- Utilization of acid-fast bacilli isolation rooms for those with active or suspected TB infection, where such rooms are to be maintained under negative pressure and have outside exhaust or high-efficiency particulate air (HEPA) filtered ventilation
- Employee information and training program

OSHA TB Respiratory Protection Policy—OSHA recently updated its enforcement policy for respiratory protection for occupational exposure to TB. OSHA has used existing CDC guidelines and requires respirators to:

- Have the ability to filter particles at 1 μm in size in the unloaded state with a filter efficiency of 95% at flow rates of up to 50 liters per minute

- Have the ability to be quantitatively or qualitatively fit tested in a manner that ensures a face seal leakage of less than 10%
- Be made available to workers in at least three difference sizes
- Have the ability to be effectively checked for facepiece fit by the worker each time the respirator is donned

The CDC recently revised respirator certification criteria to recommend that filter materials be tested at a flow rate of 85 liters per minute for penetration by particles with a mean aerodynamic diameter of 0.3 microns. Three classes of filters (N, R, P) will be certified with three efficiency levels, resulting in a total of nine classes. The National Institute of Occupational Safety and Health (NIOSH) has determined that any of the three filter levels described below will meet CDC criteria for TB protection:

Type 100   99.97% efficient
Type 99   99% efficient
Type 95   95% efficient

The minimum acceptable respirator is the N-95. All respirators must be certified by NIOSH and meet all four criteria listed previously. The classes of air-purifying particulate respirators to be certified are described in 42 CFR Part 84, Subpart K. Until these are available on the market, OSHA states that the minimum acceptable respiratory protection must meet the HEPA certification criteria. Respiratory protection for workers must be provided under the following circumstances:

- When workers enter rooms housing people with confirmed or suspected infectious TB
- When workers are present during performance of high-hazard procedures on persons with confirmed or suspected infectious TB
- When emergency medical response personnel transport in a closed vehicle a person with confirmed or suspected infectious TB

Until a TB standard is published, OSHA requires employers with workers exposed to infectious TB to develop a written respirator program that meets the requirements in 29 CFR 1910.134. NIOSH published "A Respiratory Protection Guide for Health Care Workers" (NIOSH Publication 96-102) in December 1995, which is an excellent resource and training publication.

1994 CDC Guidelines for TB Control—The CDC recently issued new guidelines for prevention and control of TB in healthcare settings. These updated guidelines emphasize the importance of control measures, including administrative and engineering controls and personal respiratory protection. The use of risk assessments should be considered when developing a written TB control plan. Early identification and management of persons who have TB, TB screening programs, worker training and education, and the evaluation of TB infection control programs are crucial to controlling the risks.

- The risk for exposure to TB in a given area depends on the prevalence of TB in the population served and the characteristics of the environment, including the community.
- Designated personnel at each facility should conduct a formal risk assessment for

the entire facility, each area, and each occupational group to determine the risk for nosocomial or occupational transmission of TB.

- The CDC defines specific elements that comprise the risk assessment, which include review of the community TB profile from public health department data and an analysis of purified protein derivative (PPD) skin test results of healthcare workers.
- Using the results of the risk assessment, one of five categories of risk is assigned to the facility, the specific area, or the specific occupational group.
- The facility should then implement an appropriate TB infection control program, based on the risk classification (minimal, very low, low, intermediate, high).
- The risk classification should be based on the profile of TB in the community, the number of infectious TB patients admitted, or the estimated number of infectious TB patients to whom healthcare workers may be exposed. Also considered are results and analysis of healthcare workers' PPD test conversions.

Fundamentals of TB Infection Control—The result of risk assessment should drive the extent of a healthcare facility's TB infection control program. The program should be based on a hierarchy of control measures.

1. *First Level*—Administrative procedures to reduce the risk of exposing uninfected individuals to those with active TB. The procedures include:
   - Developing and implementing written procedures to ensure rapid identification, isolation, evaluation, and treatment of persons likely to be infected
   - Implementing effective work practices among healthcare workers
   - Educating, training, and counseling workers about TB
   - Screening workers for TB infection
   Note—All facilities must implement Level 1 measures regardless of risk.

2. *Second Level*—Use of engineering controls to prevent the spread and reduce the concentration of droplet nuclei. Controls include:
   - Direct source control using local exhaust ventilation
   - Controlling direction of airflow to prevent contamination of adjacent areas
   - Diluting and removing contaminated air through the use of general ventilation
   - Air cleaning through the use of filtration devices or ultraviolet germicidal irradiation

3. *Third Level*—Use of personal respiratory protective equipment. This control measure is to be used in rooms with patients with known or suspected (active) TB and also in areas in which cough-inducing or aerosol-generating procedures are performed on such patients.

TB Exposure Control Program—The extent of the facility's TB exposure control program may range from a simple program which emphasizes administrative controls to a comprehensive program which includes not only administrative controls but also engineering controls and respiratory protection. The CDC guidelines include a chart to assist facilities in defining the specific elements of an infection control program for each risk classification.

- For the purposes of TB control and prevention, the CDC has retained the following definition of healthcare workers: all paid and unpaid persons working in healthcare facilities, including, but not limited to, physicians, nurses, aides, technicians, students, part-time personnel, temporary staff not employed by the facility, volunteers, dietary, housekeeping, maintenance, and clerical staff.
- Patients and/or healthcare workers with suspected or confirmed TB should be reported immediately to the appropriate public health department so that standard procedures for identifying and evaluating can be initiated.

New Guideline Highlights—The risk assessment procedures have been expanded by the addition of two new categories—very low risk and minimal risk. According to the CDC, this was done to accommodate facilities that rarely or never provide services to patients with TB.

- *High-Risk Settings*—The CDC also redefined a high-risk setting as an outbreak setting in which there is evidence of transmission of TB. The CDC clarified that in order to determine if cluster skin testing conversions represent nosocomial transmission of TB, such a cluster would have to be investigated. The cluster conversions would only be classified as high risk if the evaluation supported a conclusion that nosocomial transmission had occurred.
- *Isolation Room Requirements*—The CDC specifically clarified that nursing homes do not need TB isolation rooms if they do not provide care to TB patients. However, such facilities must have a written protocol for referral and periodic (annual) risk assessments as well as a written infection control plan that is periodically reviewed. Except for those acute care in-patient facilities that are determined to be at minimal and very low risk, the CDC recommends that all acute care facilities have at least one TB isolation room. This should ease the burden on nursing facilities that do not accept or treat infectious TB patients in locating a hospital with which to establish a referral (transfer) protocol specific to infectious TB patients.
- *Written Protocols*—The guidelines clarify that facilities that do not have isolation rooms for TB and do not perform cough-inducing procedures on patients who may have TB may not need to have a respiratory protection program for TB. However, such facilities should have written protocols for referral and periodic (annual) risk assessments as well as a written infection control plan that is periodically reviewed. The guidelines recommend that these protocols be regularly evaluated and revised as needed.
- *Patient Status*—The CDC recommends that patients who are infectious at the time of discharge should only be discharged to a facility that has isolation capability or to their homes. Facilities that accept and treat TB patients should have an engineer on staff or on a consulting basis to provide guidance in ventilation.
- *Performance Criteria*—The performance criteria for respiratory protection were not changed in the new guidelines. The CDC did remove details on specific respirators such as dust-mist and dust-fume-mist. The CDC will use the new NIOSH certification process in determining appropriate respiratory protection. This will allow healthcare facilities to choose from a broader range of and less expensive certified masks.

- **Testing**—The CDC recommended that all personnel not employed by a facility but working in the facility also receive skin testing at appropriate intervals. Healthcare workers with potential for exposure, including those with a history of BCG vaccination, should have baseline PPD testing. Those with a negative test should repeat the procedure at regular intervals as determined locally by the risk assessment. The two-step skin testing procedure is not necessary if a healthcare worker had a documented negative PPD result in the past 12 months or if the facility determines that boosting is not common in its population.

Worker Training and Education—An effective TB training program should include the following:

- Basic concepts of transmission, pathogenesis, and diagnosis
- Explanation of the difference between latent and active TB
- Signs and symptoms of active TB
- Increased risk for those infected with HIV
- Potential for occupational exposure
- Information about prevalence of TB in the community
- Situations that increase the risk of exposure
- Principles of infection control
- Importance of skin testing and significance of a positive test
- Principles of preventive therapy for latent TB
- Drug therapy procedures for active TB
- Importance of notifying the facility
- Information about medical evaluation for symptoms of active TB

Engineering Controls—Engineering controls are critical in preventing the spread of TB within a facility. The CDC guidelines recommend exhausting air from possibly infected areas to the outside. Healthcare facilities should have isolation rooms with negative pressure. Six air changes per hour is recommended, with new construction requiring 12 air changes per hour. Some facilities are also using germicidal lights that use ultraviolet irradiation to supplement ventilation and isolation efforts.

H. SUMMARY OF EXPOSURE CONTROL MANAGEMENT ELEMENTS (29 CFR 1910.1030)

Universal (Standard) Precautions—The single most important measure to control transmission of HBV and HIV is to treat all human blood and other potentially infectious materials as if they were infectious for HBV and HIV. Application of this approach is now referred to as universal precautions. *Blood and certain body fluids from all acute care patients should be considered as potentially infectious materials.* These fluids cause contamination, defined in the standard as "the presence or the reasonably anticipated presence of blood or other potentially infectious materials on an item or surface."

- Puncture-resistant, leak-proof containers should be used to collect, handle, process, store, transport, or dispose of potentially infectious materials. These specimens should be labeled if shipped outside the facility.
- When gloves are removed, hands should be washed as soon as possible after contact with blood or other potentially infectious materials.
- A mechanism for immediate eye irrigation should be made available in the event of an exposure incident.
- Workers should not eat, drink, smoke, apply cosmetics, or handle contact lenses in areas of potential occupational exposure.
- Red or biohazard labels should be affixed to containers when storing, transporting, or shipping blood or other potentially infectious materials.

Personal Protective Equipment—Personal protective equipment is specialized clothing or equipment used by employees to protect against direct exposure to blood or other potentially infectious materials.

- Protective equipment must not allow blood or other potentially infectious materials to pass through to workers' clothing, skin, or mucous membranes.
- Equipment includes, but is not limited to, gloves, coats, face shields or masks, and eye protection.
- The employer is responsible for providing and assuring the proper use of personal protective equipment.
- The employer is responsible for ensuring that workers have access to protective equipment, at no cost, in proper sizes and types, including taking allergic conditions into consideration.

Housekeeping Procedures—Contaminated work surfaces must be disinfected upon completion of procedures or when contaminated by splashes, spills, contact with blood or other potentially infectious materials, and at the end of the work shift.

- Broken glass should be cleaned up with a brush or tongs; it should never be picked up with hands, even when wearing gloves.
- Special precautions are necessary when disposing of contaminated waste.
- Contaminated wastes should be disposed of in closable, puncture-resistant, leak-proof, red or biohazard-labeled containers.
- If outside contamination of the regulated waste container occurs, it should be placed in a second container that is closable, leak-proof, and appropriately labeled.

Exposure Incidents—An exposure incident is the specific eye, mouth, mucous membrane, non-intact skin, or other parenteral contact with potentially infectious material that results from the performance of an employee's duties.

- Employees should immediately report exposure incidents to enable timely medical evaluation and follow-up by a healthcare professional.
- The employer can request testing of the source individual's blood for HIV and HBV. The source individual is any patient whose blood or body fluids are the source of an exposure incident to an employee.

- At the time of the exposure, the exposed employee must be directed to a healthcare professional.
- The employer must provide the healthcare professional with a copy of the bloodborne pathogens standard and a description of the employee's job duties as they relate to the incident.
- The employer also must provide a report of the specific exposure, including route of exposure, relevant employee medical records (including hepatitis B vaccination status), and results of the source individual's blood tests, if available.
- A baseline blood sample should be drawn if the employee consents. If the employee elects to delay HIV testing of the sample, the healthcare professional must preserve the employee's blood sample for at least 90 days.
- Testing the source individual's blood does not need to be repeated if the source individual is known to be infectious for HIV or HBV.
- Testing cannot be done in most states without written consent.
- The results of the source individual's blood test are confidential, but the results must be made available to the exposed employee through consultation with the healthcare professional.
- The healthcare professional will provide a written opinion to the employer. This opinion is limited to a statement that the employee has been informed of the results of the evaluation. It also recommends further evaluation or treatment as necessary.
- The employer must provide a copy of the written opinion to the employee within 15 days. This is the only information shared with the employer following an exposure incident. All other employee medical records are confidential.
- All evaluations and follow-up visits must be available at no cost to the employee. They must take place at a reasonable time and place. Evaluations and follow-up visits must be performed by or under the supervision of a licensed physician or another licensed healthcare professional.
- All evaluations must follow the U.S. Public Health Service guidelines current at the time.
- All laboratory tests must be conducted by an accredited laboratory and at no cost to the employee.

Note—Table 8.3 provides information on exposure and HBV vaccination.

Recordkeeping—The OSHA bloodborne pathogens standard requires that medical and training records be maintained.

- A medical record must be established for each employee with occupational exposure. *This record is confidential and separate from other personnel records.*
- This record may be kept on-site or may be retained by the healthcare professional(s) who provides services to employees.
- The medical record contains the employee's name, social security number, and hepatitis B vaccination status, including the dates of vaccination and the written opinion of the healthcare professional regarding the hepatitis B vaccination.
- If an occupational exposure occurs, reports are added to the medical record to document the incident and the results of testing following the incident.

TABLE 8.3   Bloodborne Exposure Evaluation Actions

Employer	Employee	Healthcare Professional
Provides copy of standard to healthcare professional		Receives a copy of the standard (from employer)
Provides training to employee	Receives training	
Offers vaccination (within 10 days)	Vaccination offered • If accepted, sent to healthcare professional • If declined, employee signs a declination form • If employee changes mind later and accepts vaccination, sent to healthcare professional	• If employee accepts the vaccination offered, receives referred employee • Receives referred employee after he or she has declined once but changes mind later and accepts the vaccination offered
Receives record of healthcare professional's written opinion	Receives copy of healthcare professional's written opinion from employer	The following information is confidential: • Establishes medical record • Evaluates employee for contraindications to vaccination or prior immunity • Vaccinates employee or discusses contraindications/immunity with employee
Provides a copy to employee within 15 days		• Records written opinion (i.e., whether vaccine is indicated and whether vaccine was received) • Provides copy of written opinion to employer

- The post-evaluation written opinion of the healthcare professional is also part of the medical record.
- The medical record must document what information has been provided to the healthcare provider.
- Medical records must be maintained 30 years past the last date of employment of the employee.
- Emphasis is on confidentiality of medical records. No medical record or part of a medical record should be disclosed without direct written consent of the employee or as required by law.
- Training records documenting each training session are to be kept for three years.
- Training records must include the date, content outline, trainer's name and qualifications, and names and job titles of all persons attending the training sessions.
- If the employer ceases to do business, medical and training records are transferred to the successor employer. If there is no successor employer, the employer must notify the director of NIOSH for specific directions regarding disposition of the records at least three months prior to disposal.

- Upon request, both medical and training records must be made available to the Assistant Secretary of Labor for Occupational Safety and Health. Training records must be available to employees upon request. Medical records can be obtained by the employee or anyone who has the employee's written consent.

I. INFECTIOUS WASTE MANAGEMENT

Medical or infectious waste has been defined in 40 CFR 259.10 as any solid waste that is generated in the diagnosis, treatment, or immunization of human beings or animals or in related research, biological production, or testing. This topic is covered more fully in Chapter 7.

Basic Handling Guidelines—

- Infectious waste should be segregated from other waste at the point of generation within the facility.
- Infectious waste must be packaged to protect waste handlers and the public from exposure to the waste.
- Packaging should provide for containment of the infectious waste from the point of generation up to the point of proper treatment or disposal.
- Packaging must be selected and utilized for the type of infectious waste and how it will be handled prior to disposal.
- Contaminated sharps should be placed directly in rigid, leak-proof, puncture-resistant containers.
- All containers used for disposal of infectious waste must be conspicuously identified.
- Infectious waste must be handled and transported so as to ensure integrity of the packaging.
- Plastic bags containing infectious waste must not be transported by chute, dumb-waiter, conveyor belt, or similar device.
- Infectious waste must be stored in a manner which inhibits rapid microbial growth and minimizes exposure potential.

Containing Spills—In the case of a spill, ruptured packaging, or other incident involving potentially infectious materials:

- Isolate the area from the public and all non-essential personnel.
- When practicable, repackage all spilled waste and containment debris.
- Disinfect all containment equipment and surfaces appropriately.
- Complete the appropriate incident report.
- Consult written management procedures to ensure all actions are accomplished.

SUMMARY

Healthcare workers are exposed or potentially exposed to a number of infectious agents, diseases, and substances. Healthcare organizations must educate and train workers in how to protect themselves, their patients, and others from exposure to biohazards. The development and spread of pathogenic diseases resulted in the establishment and use of universal precautions and body isolation procedures. In May 1996, the Centers for Disease Control and Prevention issues a new control guideline. The new guideline continues to stress the importance of using universal-type precautions to prevent the spread of pathogenic diseases. However, the new guideline stresses the importance of following disease-specific transmission precautions in addition to following standard (universal) precautions. The new guideline also stresses the importance of expanding traditional infection control programs beyond the scope of isolation procedures and the spread of infectious diseases. This chapter placed special emphasis on protection against the spread of AIDS, hepatitis, and tuberculosis in the healthcare environment.

REVIEW QUESTIONS

1   Some environmental services and cleaning professionals consider _____ to be the safest category of cleaning disinfectants.
    a.   synthetic phenols
    b.   quaternary ammonia compounds
    c.   alcohols
    d.   chlorine (household bleach)

2   The process that renders a surface completely free from bacteria is known as:
    a.   sanitizing
    b.   disinfecting
    c.   sterilization
    d.   all of the above

3   The term that describes the process of preventing the spread of a communicable disease within a healthcare facility is known as:
    a.   contact isolation
    b.   universal precautions
    c.   strict isolation
    d.   quarantine

4   The *best* protection against individual needlesticks is to:
    a.   never recap a used needle
    b.   place used needles in approved sharps containers
    c.   wash hands immediately after completing the procedure
    d.   wear disposable gloves

5   Healthcare workers covered by the OSHA bloodborne pathogens standard must be:
    a.   trained when assigned to any position with the potential for exposure
    b.   retrained within two years of initial training
    c.   required to take the hepatitis B vaccination within ten days of hire
    d.   all of the above

6 Tuberculosis can be contracted:
   a. through casual contact with an infected person
   b. by contact with infected blood
   c. by ingesting a droplet nuclei from an infected person's cough
   d. by being stuck with a used needle

7 The scope of a healthcare facility's TB exposure control program should be based on:
   a. the results of the local risk assessment
   b. national trends published in CDC guidelines
   c. the number of workers with positive TB skin test conversions
   d. the number of TB-infected patients

8 Which of the following statement(s) is/are true concerning hepatitis?
   a. HBV is transmitted more easily than HIV
   b. some populations have a greater risk of HBV infection
   c. HBV can cause either acute or chronic outcomes
   d. all of the above

Answers: 1–b   2–c   3–a   4–a   5–a   6–c   7–a   8–d

CHAPTER 9

SAFETY IN PATIENT CARE AREAS

A. HEALTHCARE LIFTING HAZARDS

Back-Related Injuries—It is estimated that only about 5% of back injuries result from a single incident; most are the result of long-term wear and tear on the back. This can lead to disc degeneration and nerve damage. Muscles and ligaments can also become damaged through continuous or repetitive pulling, tugging, or pushing. Reduction efforts in the healthcare environment should focus on the following:

- Education on the back and proper body mechanics
- Recurring training on patient transfer techniques
- Exercise routines for those involved in lifting
- Formation and required use of lifting teams
- Ergonomic evaluations to detect problem areas
- Effective housekeeping procedures
- Lift and patient assist equipment

Healthcare Lifting Evaluation—The National Institute of Occupational Safety and Health (NIOSH) revised its "Work Practices Guide for Manual Lifting" in 1991. This well-known quantitative method really has difficulty when applied to nursing care lifting tasks. Lifting or moving patients is much different than lifting a static object. Healthcare lifting may result in trunk twisting, and the small space requirements in many healthcare situations add difficulty to the task. NIOSH is currently evaluating lifting tasks in healthcare facilities. This research could eventually contribute to the development of industry-wide safety standards and protocols.

Back Injury Prevention—Back injury management and prevention efforts in the healthcare environment should focus on the following:

- Study lifting requirements and eliminate lifts wherever possible.
- Provide patient handling, transfer, and lifting equipment.

- Keep equipment in good repair. Ensure wheelchairs and carts can be moved without excess strain.
- Establish patient lift guidelines to help workers safely assess patient handling situations.
- Redesign the workplace to increase efficiency and decrease the potential for injuries.
- Educate workers about back anatomy and personal back care responsibilities.
- Provide recurring education and training on proper body mechanics and patient transfer techniques.
- Require employees to participate in exercise and/or stretching routines before lifting.
- Establish and train two-person lift and transfer teams.
- Use physical or occupational therapy professionals to instruct workers in patient handling techniques.
- Investigate all accidents and make changes to prevent recurrence.
- Assign a case management worker to oversee medical treatment and return-to-work efforts.

Basic Safe Lifting Tips—

- Never move or lift from side to side.
- Keep items close to the body when reaching, carrying, or lifting.
- Plan the lift and size up the load to better reduce spine movement.
- Keep the patient load as close to the body as possible. *Note*—Ten pounds at waist height equates to 100 pounds force on the back with arms extended away from the body.
- Bend at the knees when lifting loads from floor level. *Note*—Ten pounds at floor height with bent knees is equal to 100 pounds of force when bending at the waist with legs straight.
- Avoid any twisting motion and be sure to pivot the feet to turn.
- Always push rather than pull loads. Pushing reduces the force necessary to move an object by 50%.
- Use lifting equipment and devices such as chair lifts, mechanical lifts, transfer boards, and gait belts.
- Keep beds at proper heights.
- Keep the back straight and maintain correct posture with head up and stomach tucked in.

1. *Safety During Transfers*—
   - Communicate the plan of action to the patient and other workers to ensure that the transfer will be smooth and without unexpected moves.
   - Position equipment and furniture effectively. Be sure to remove obstacles.
   - Ensure good footing for the staff and patient. Patients should wear slippers that provide good traction.
   - Maintain eye contact, communicate with the patient, and be alert for trouble signs.

- Record any problems on the patient's chart so that other shifts will know how to cope with difficult transfers. Also note the need for any special equipment.

2. *Mechanical Lift Safety*—
   - Most mechanical devices require two persons to accomplish the lift—one to operate the lift and the other to stabilize the patient.
   - Nursing personnel must be trained on lifting equipment and proper procedures before using mechanical lifting devices.
   - Always explain the lift to the resident or patient before beginning the procedure.
   - Ensure the resident or patient is positioned correctly in the sling before continuing the lift procedure.
   - One person must ensure that the patient remains stable during the entire lifting procedure.
   - Never allow the sling to swing and never leave a patient or resident suspended in the sling.
   - Never use lift devices to transport patients unless the equipment has been designed to do so.

3. *Lifting Teams*—Some healthcare organizations have trained and organized lift teams to:
   - Eliminate uncoordinated lifts
   - Prevent unprotected personnel from performing lifts
   - Reduce weight and/or height differences between partners
   - Not allow tired or exhausted workers to perform lifting tasks
   - Eliminate personnel recovering from a back injury from performing lifting tasks
   - Prevent untrained personnel from lifting
   - Encourage the use of lifting equipment when possible

Back Care Safety Tips for Nursing Personnel—

1. *Bed-to-Stretcher Transfer*—
   - Best accomplished with two persons.
   - Use a plastic bag beneath the draw-sheet.
   - Adjust bed to the level of the stretcher.
   - Bend knees during transfer to reduce back strain.
   - Accomplish transfer in two stages and move legs first if possible.

2. *Bed-to-Wheelchair Transfer*—
   - Lock wheelchair and adjust height of bed
   - Remove wheelchair armrest nearest the bed.
   - Support patient's knees between legs.
   - On a signal, move patient to standing position using a rocking motion.
   - Keep knees slightly bent and back balanced.
   - If patient must hold onto person assisting, ensure the neck is not held.
   - Use a transfer belt on the patient to prevent twisting and provide support.

3. ***Pulling Patient Up in Bed—***
   - Ensure bed height is below the waist.
   - Work from side of bed and place feet in direction patient is to be moved.
   - Placed hands under patient's shoulders and back and slide—do not lift.
   - Keep feet apart and knees bent.

4. ***Turning Patient Over—***
   - Ensure bed height is to mid or upper thigh.
   - Slide patient in opposite direction he or she will be rolling.
   - Place bed rails down if safe for patient.
   - Place knee on bed near patient's shoulder and use the entire body, keeping knees bent and back balanced.
   - If possible, always turn the patient toward you.

5. ***Wheelchair-to-Toilet Transfer—***This maneuver can be difficult because of small bathrooms and the potential for patient falls.
   - If patient has weak leg or cast, ensure that strong leg is closest to toilet.
   - Lock wheelchair.
   - Instruct patient to use grab bar and wheelchair arm for support.
   - Bend knees and maintain a natural back curve.
   - Get help for difficult transfers.

6. ***Falling Patients—***
   - Never try to prevent a fall, but try to guide the patient safely and gently to the floor.
   - Stay close to the patient.
   - This should be mentally practiced before an incident occurs.
   - Get help to move the patient from the floor.

7. ***Wheelchair to X-Ray Table—***
   - Lock chair in position with patient's good leg toward table.
   - Help patient stand up and pivot by providing support to bad leg.
   - Bend at the knees and maintain a natural back position.
   - Have patient pivot first onto the table footrest and then onto the table.
   - Get help for difficult situations.

General Back Care Safety Tips for Support Personnel—

1. ***Bending—***
   - Kneel down on one knee (do not bend from the waist when cleaning).
   - Bend knees and hips and not the back.
   - When leaning forward, move the entire body and not just the arms.

2. ***Twisting—***
   - Kneel down on one knee.
   - Position yourself so that you can have the best possible leverage.
   - Use arms and legs, and not the back, to do the task.

3. ***Repetitive Motions—***
   - Keep loads small.

- Turn the whole body and do not twist.
- Get the body close to the load.
- Do not reach and lift.
- Lift with arms and legs.
- Change position frequently.

4. *Reaching*—
   - Reach only as high as is comfortable; never strain.
   - Never reach above shoulder level; use a stool.
   - Test the weight of the load.
   - Use your arms and legs to do the work.
   - Tighten stomach muscles as you lift.

5. *Pushing and Pulling*—
   - Stay close to the load, but do not lean forward.
   - When possible, push instead of pull.
   - Use both arms.
   - Tighten stomach muscles when pushing.
   - Use a carrier with wheels.

B. WHEELCHAIR SAFETY

Hospital patients or nursing home residents may require the use of wheelchairs, which can be a source of injury from falls or improper positioning. The major considerations when selecting a wheelchair for a patient are patient safety, proper alignment, comfort, and maximum independence.

Types of Wheelchairs—There are two basic types of wheelchairs; one has large wheels in the rear and the other has large wheels in the front. The one that has large rear wheels is better for general use because it makes transfer easy, encourages good positioning, and wheels easily over difficult surfaces. Long-term care facilities also use a wheelchair referred to as a "geri-chair." A geri-chair looks like a recliner on wheels. It is equipped with a high back and has a tray that fits in front. A geri-chair should be considered to be a restraining device.

Basic Safety Considerations—

- Ensure that wheelchairs are maintained in top mechanical condition.
- Keep wheelchairs clean of all food and body waste by disinfecting on a regular basis.
- When moving a patient down a ramp or entering an elevator, always back in, pulling the chair toward you.
- Never allow wheelchairs to block hallways or exits.
- The brakes should be locked when the patient is being moved and/or is to remain seated for a length of time.
- Caster locks are desirable to keep the casters from turning when the patient moves into or out of a wheelchair.

- Hand-rim projections on chair wheels will facilitate better control and safer use of the wheelchair.
- High-density foam rolls, called lateral stabilizers, slip easily over the backrest of the wheelchair to help distribute weight evenly over the patient's back. They may also be applied over the armrests for lower trunk support.
- Side slumping can be prevented by using stabilizers slipped over the metal frame of the wheelchair backrest.
- A thick foam roll slipped over the armrest of the affected side provides rehabilitative positioning of the affected upper extremity.
- Wheelchair lap boards can provide rehabilitative support and encourage use of the affected upper extremity. *Note*—An occupational and physical therapist should be consulted to recommend the optimal positioning device for each patient.
- Forward-sliding patients increase the amount of weight bearing on their sacrum, which can result in pressure sores. Place pillows under the buttocks to raise the hips above the knees.
- Seat belts secured around patients' hips remind them to keep their buttocks back in the wheelchair.
- Proper preventive positioning is critical. A wedge cushion can be used as a "fill-in" on sling wheelchairs, which can help prevent adduction of the hips.
- Maximal-assist patients with conditions such as cerebral palsy or traumatic brain injury cannot control their bodies when the trunk suddenly extends. This results in leg crossover, which causes slides out of the wheelchair.
- A wedge cushion helps keep a patient in the seat by positioning the hips in abduction and external rotation.
- A critical aspect of nursing care is proper patient positioning in wheelchairs. To meet this standard of quality care, nursing personnel should be instructed in basic positioning principles.

C. PREVENTING SLIPS AND FALLS

Fall Prevention Management—Slips, trips, and falls are some of the most common patient-related occurrences in healthcare facilities. Many healthcare workers are also involved in slip, trip, and fall incidents each year. The safety director should manage the program and the safety committee should approve and review all policies. Trends and problem areas will be determined by analyzing data from the risk manager and by using workers' compensation data available from the claims handling agency. During 1992, falls were the fourth leading cause of accidental death on the job. Falls happen because people and conditions do not remain static. Slip, trip, and fall prevention programs must be continuously monitored to be effective in preventing and reducing loss. Healthcare costs for falls totaled almost $40 billion in 1992; only motor vehicle injuries had higher costs.

1. *Objectives*—An effectively managed program should seek to:
   - Identify, evaluate, and correct safety hazards that could contribute to slipping, tripping, and/or falling of employees, patients, or visitors

- Communicate to all staff members the physical hazards and behavioral aspects of slip, trip, and fall prevention
- Provide all employees with training related to slip, trip, and fall prevention
- Establish procedures to better manage and analyze the trends and problems within the facility

2. *Management Elements*—The following must be considered crucial to the program.
   - Hazard surveillance program for physical hazards
   - Regular inspection of all areas, especially patient and visitor areas
   - Special emphasis on high-risk areas as determined by a statistical analysis of incident and accident data
   - Analysis of accidents/incidents to determine trends
   - Involvement of safety committee in approving and/or reviewing all policies
   - Awareness training for employees, staff, and students
   - Training for those responsible for facility maintenance and housekeeping
   - Quarterly assessment of program effectiveness
   - Annual audit

3. *Patient Fall Prevention*—Nursing personnel must be actively involved in fall prevention programs. An effective program would strive to increase awareness of fall hazards located in the unit or department. The nursing care plan should contain information on each patient that would help minimize the chance of falls. Basic patient fall prevention measures are as follows:
   - Keep patients' belongings and call buttons close to beds so that patients do not have to reach too far.
   - Place call buttons close to commodes and showers.
   - Respond to trends in falls with new programs and patient fall prevention in-service education.
   - Keep lighting bright in corridors and public areas.
   - Encourage patients to go to the bathroom before they go to bed. Even this simple measure can reduce nighttime falls.
   - Re-evaluate drug treatments that contribute to frequent falls and constant disorientation.
   - Where possible, help nursing home residents to improve physical disabilities through physical therapy and walking aids.
   - Increase nursing coverage during high-risk activities such as getting out of bed and going to the bathroom.
   - Work to subdue hostile patients. Anger and resentment have been shown to contribute to falls.
   - Introduce softer surfaces for floors or furniture, and provide grab bars and handrails throughout the facility.
   - Use safety devices such as non-slip surfaces and footstools with rubber feet.
   - Repair torn carpet, and remove other obstructions from patient areas.
   - To prevent a fall caused by overreaching, place bedside tables, water pitchers and glasses, toiletries, footstools, the nurse call button, and other often-needed items close enough for the patient to grasp them without overextension.

- Remove temptation by placing articles that are to be used only with assistance or supervision out of the patient's sight to eliminate the possibility of reaching for them from bed.
- Footstools should have non-slip surfaces and rubber feet caps. Preferably, they should also be attached to the bed.
- Use bedrails when necessary and gentle restraints only when authorized and countersigned by the attending physician.
- Correctly position the patient in bed and be aware that extended pressure on a certain area of the body may cause various injuries.
- Provide sufficient support to prevent the patient from sliding into a vulnerable position.
- Keep beds and bed areas in good repair. Promptly repair loose or broken tiles and frayed or loose carpet. Repair or replace all worn or poorly functioning self-help appliances.
- Train employees in correct patient carrying, lifting, and moving procedures. Make certain all personnel responsible for these lifts and carries are trained in safety precautions.
- Get patients involved in their own safety. Most patients will try to avoid hazards if they understand the possible danger.
- Make certain handrails are sturdy. Adequate hand grips and armrests should surround bathtubs.
- Bathroom floors, tubs, and showers should have slip-resistant surfaces.
- Answer patient calls promptly. Some patients may be particularly impatient and may attempt activities beyond their capacity.
- Never block doorways or elevator entrances.
- Stair treads should have a slip-resistant surface, and rises and threads should be correctly sized. Keep landings and stairways free of all obstructions. Do not use landings as storage areas.
- Do not use throw rugs or mats, which can be very hazardous, especially around wheelchairs.

4. ***Long-Term Care Fall Prevention***—The incidence of falls increases with the age of the patient for a variety of reasons. Slowed reaction time makes it difficult to regain balance once balance is lost. Changes in balance are caused by:
   - Tilting the head back to look up
   - Posture changes that shift the center of gravity
   - Dizziness caused by medications
   - Joint pain or stiffness
   - Muscle weakness
   - "Drop attacks" (sudden loss of muscle tone)
   - Foot problems and improper shoes
   - Fractures that cause falls
   - Sensory changes
   - Low blood pressure
   - Environmental hazards such as slippery floors and faulty equipment
   - Improper use of wheelchairs, walkers, and canes

5. ***High-Risk Nursing Home Patients***—Some characteristics or signs of a high-risk fall potential include patients and residents who:
   - Experience high stress and anxiety for any reason
   - Are weak or unsteady
   - Use assistance devices
   - Take certain medications
   - Are known to have low blood pressure
   - Experience arthritis or painful joints
   - Are not able to see very well
   - Are known to have a recent change in health status
   - Have a history of falls
   - Have had limited activity due to illness

6. ***Lowering the Potential for Falls***—Healthcare personnel should understand that:
   - Most falls occur between 6 and 9 p.m. near the bed.
   - Falls may be a symptom of a developing illness.
   - Persons who are restrained fall more than those who are not.
   - Carpets decrease the severity of injuries from falls.
   - Identification of high risks results in planning appropriate care.
   - Planning exercises for the patient can increase or maintain strength.
   - Decreasing medications to the lowest effective dosage can be beneficial in lowering risks.
   - Patients must be taught to ask for help and rise slowly.
   - Patients must be educated in the use of wheelchairs, canes, and walkers.
   - Keeping necessary items within reach can lower the risk of falls.
   - Allowing residents to proceed at their own pace can reduce falls.

7. ***Investigating Patient and Resident Falls***—When the hazard control or risk manager reviews incident and accident reports, the following items should be noted to help determine trends:
   - Location, time, day, and shift
   - Level of nursing staff on duty
   - Use and position of bedrails and/or restraints
   - Condition of patient at the time of the fall
   - Medications that could have contributed to the fall
   - Environmental factors such as floor conditions and lighting
   - Whether the patient requested nursing assistance
   - Availability of physician orders concerning patient activity
   - Violation of safety procedures or organizational policies

Correcting Environmental Hazards—The healthcare staff must be sure the environment is safe, needs are met, and patients have the necessary information to assist with their own safety. Hazard assessments should be done regularly and should include:

1. ***Lighting***—
   - Switches should be accessible to the patient upon entering the room and from the bed.

- Lights should be bright enough to compensate for limited vision and appropriate for the activities likely to be performed in that room.
- Night-lights should be available in patient rooms, bathrooms, and hallways.
- Lighting should be adequate on stairs and hallways. Floor-level lighting will reduce glare.

2. *Stairs*—
   - Secure handrails should be provided on both sides of the staircase.
   - Steps can be painted or outlined for increased visibility and covered with non-slip material.
   - Stairs should be clutter-free and well maintained.

3. *Floors*—
   - Floors should have non-glare and non-skid wax surfaces.
   - Throw rugs should have non-slip backing.
   - Carpet edges should be taped or tacked down. Note that some carpet patterns impair perception and may cause falls.
   - Doorways should not have surface drops between carpet and other flooring.
   - Spills and liquids should be wiped up immediately.
   - Physical and/or visual warnings should be used to alert patients to flooring changes and hallway turns.

4. *Patient and Nursing Home Resident Rooms*—
   - Pathways from bed to bathroom should be unobstructed.
   - Low-level furniture such as coffee tables and ottomans may create tripping hazards.
   - Floors must be kept clear and free of clutter.
   - Chairs, tables, nightstands, and over-bed tables should be secure and tip-resistant in case they are used for support.
   - Furniture should not have sharp edges or corners in case of a fall.

5. *Beds*—
   - Beds should have the capacity to be lowered to 18 inches.
   - Bed height should be adjustable to allow for transfers.
   - Mattresses should be firm on the sides to allow seating.
   - Bed wheels should lock to prevent sliding when leaned against.
   - Side rails should be appropriate for the patient and the situation.
   - Side rails should prevent the patient from slipping through the gaps and becoming entangled.
   - Call buttons or bells should be easily accessible from the bed.

6. *Bathrooms*—
   - Doors should be wide enough to allow easy wheelchair or walker passage.
   - Floors should not be slippery, especially when wet.
   - Tubs, showers, and floors should have non-skid strips or mats.
   - Grab bars should be securely attached to walls and low enough for easy reach, especially near tubs, showers, and toilets.
   - Toilet seats should not be elevated.
   - Restraints should never be used on toilets.

7. *Seating*—
   - Chairs should be the appropriate height and size.
   - Chairs with armrests should provide maximum leverage during transfers.
   - Appropriate cushion, wedge, pommel, or wheelchair foundation should be available if needed.
   - All lap belts should be placed low over the pelvis and be snug enough to prevent the patient from sliding under the belt.
   - When the care plan calls for a seat belt, a determination should be made as to whether the belt should be self-release or assisted-release.

8. *Elevators*—
   - Doors should be time-delayed with very sensitive pressure door recoils.
   - Floor buttons should be easily accessible, visible, or marked in braille.
   - Elevator floors should be slip-resistant and level with the landing.
   - Emergency buttons and phones must be well marked and easily accessible.
   - Elevators should provide room for walkers and wheelchairs.

D. MEDICATION SAFETY

Medication errors receive widespread publicity. Many of the errors contribute to patient deaths. Organizations accredited by the Joint Commission on Accreditation of Healthcare Organizations must have a procedure to ensure the safe use of medication. Healthcare safety professionals must emphasize medication safety during orientation and training sessions. Nursing personnel must understand the importance of following established procedures when administering medications.

Voluntary Reporting—The American Society of Hospital Pharmacists promotes a voluntary reporting system called the USP-ISMP Medication Errors Reporting Program, which is operated by the United States Pharmacopeia Convention and the Institute for Safe Medication Practices. This voluntary reporting system encourages compliance from healthcare professionals and provides a framework to protect the public.

Prevention Guidelines—Listed below are some general guidelines for preventing medication errors:

- Check each medication against the order book to be certain that there is an order for the medication and that there have been no changes in the order.
- Store all drugs in a locked storage cabinet or refrigerator. Drugs marked "for external use only" should be stored in separate locked cabinets or containers and kept away from ingestible medications and foodstuffs.
- Keep all medications in their original containers and return them to the pharmacy for relabeling if the label becomes illegible.
- Return to the pharmacy any containers not clearly labeled with the prescription number, drug name, strength, and quantity. The container must also have the patient name, prescribing physician, pharmacy, and date of issue.
- When a patient brings a medication from home, it should be sent to the pharmacy

for analysis and relabeling. These drugs should only be administered if ordered by the attending physician.

- Wash hands before starting to prepare medications.
- Nursing personnel should read each label when removed from the container, before measuring, and when replacing the container in the storage cabinet.
- Nursing personnel must be able to calculate dosages in both the metric and apothecary systems of measurement.
- Use small disposable cups to dispense individual doses. Cups should be clearly labeled or drug information placed on a medicine card.
- Labels and cards must include the name of the patient, his or her room number, the name of the drug, the prescription number, the dosage, and the exact time the drug is to be given.
- When unit dose systems are used, medications should be given as they are poured.
- Nursing personnel must be aware of drug interactions in order to prevent synergistic effects or overdosing.
- Maintain a time schedule for administering medications. Each new drug order should be added to the list.
- Identify a patient by at least two means before giving a medication. Ask the patient his or her name if possible and check the patient's bed chart or armband.
- Stay with the patient until certain all the medication has been taken.
- When in doubt about the dosage, time, content, or method of administration, do not administer the drug. Return the medication to the pharmacy for verification and discuss with the nursing supervisor.
- Never re-use intravenous bottles, because previous contents of the bottle could be administered inadvertently.
- Do not measure drugs with a syringe, because the contents could be administered inadvertently.
- Report medication errors to the nursing supervisor or physician. The location and phone number of the nearest Poison Control Center should be posted at the nursing station telephone.
- Nursing staff should be instructed in measures to take in the event of a serious error.

Categories of Medication Errors—Hazard control personnel investigating medication errors categorize them in a number of ways. Some of the most common categories are listed below:

- Failure to administer medication when required or as prescribed
- Administering medication at the wrong time or using incorrect route of administration
- Administering the wrong dosage or concentration of a drug
- Administering the wrong medication (*Note*—This is becoming a problem because some medications having similar names.)
- Misunderstanding verbal or written medication orders, including transcription mistakes

- Administering medication to the wrong patient
- Failure to read labels on vials and containers
- Using an improper injection technique

Investigating Medication Errors—The hazard control officer, risk manager, or nursing services manager should conduct a thorough investigation to document all the facts. Investigations should seek to determine:

- Location and type of unit in which the error or misadministration occurred
- Time, date, and shift
- Staff levels at the time the incident occurred
- Other departmental or unit events that could have contributed to the incident
- Legibility and accuracy of physician orders
- If failure to follow safety precautions or other medication procedures contributed to the event
- If communication failure among hospital staff or with the physician contributed to the event

Other Considerations—

- *Evaluating Facts*—Ensure that all data and information surrounding the incident are analyzed and evaluated by nursing, management, and pharmacy personnel.
- *Trends*—Ensure that any trends or patterns in medication errors are reported to the appropriate committee for discussion and implementation of corrective actions.

Chemotherapeutic Drug Safety—All hazardous chemotherapeutic agents dispensed by the pharmacy should be issued in a plastic bag and labeled as being potentially dangerous. Chemotherapeutic drugs are also covered in Chapter 7 under antineoplastic drugs.

1. *General Recommendations*—
   - Special handling and administering instructions should also be on the label.
   - All supplies and equipment that come into contact with hazardous cytotoxic agents should be placed in a red plastic container provided for the disposal of biohazards. This includes all empty vials, syringes, needles, intravenous bottles and tubing, gloves, disposable gowns, and any other items that may be contaminated.
   - The facility should establish special handling procedures to ensure proper handling and disposal.

2. *References*—Sources of information about working with cytotoxic drugs include:
   - *Safe Handling of Cytotoxic Drugs—Study Guide,* published by the American Society of Hospital Pharmacists
   - "Work Practice Guidelines for Personnel Dealing with Cytotoxic (Antineoplastic) Drugs," OSHA Instruction Publication 8-1.1

E. GENERAL SAFETY IN PATIENT AREAS

Equipment Safety, Electrical Safety, and Burn Prevention—Many hazards found in a healthcare facility can pose a threat to patients and staff members. Burn prevention and electrical safety should receive daily emphasis from unit or department supervisors.

Equipment Safety—Injuries caused by the use of equipment can be greatly reduced if healthcare staff members follow some basic safety rules:

- Always use equipment properly.
- Never use damaged, defective, or improperly working equipment.
- Never use equipment unless properly trained and authorized to do so.
- Always follow safety policies and procedures when using equipment.
- Report defective equipment to the biomedical engineering department.

Electrical Safety—

- Electrical pads should kept dry and sharp objects should never be allowed to puncture them.
- Heating appliances should not exceed a temperature of 120°F or be placed directly against the skin.
- Patients and nursing home residents should be closely supervised when permitted to use electrical appliances.
- The use of electrical appliances should be restricted and appliances tested by maintenance personnel before they are brought into the facility.
- Maintenance should regularly check electric beds for damaged cords and plugs and ensure ground connections are attached and working.
- Electric beds should be repaired immediately and placed back in use when certified to be in safe working condition. *Note*—All beds should conform to Underwriters Laboratories UL-544, Hospital Equipment Standard.
- Smoking rules and areas for staff members should be strictly enforced. Visitors should be reminded of smoking regulations and areas.
- Patients, visitors, and workers should be prohibited from using ungrounded coffeepots, radios, cooling fans, portable heaters, or other appliances.
- A program to check all electrical equipment and connections in all nursing units and patient care areas should be implemented.
- Microwave ovens should be cleaned regularly, checked periodically for proper door closure and seal, and used only in designated areas.
- Electrical equipment such as radios, televisions, and lamps that have been checked and approved for electrical safety may be used in a room where oxygen is being administered.
- Under no circumstances should electrically operated equipment be used inside of a canopy when oxygen is being administered.
- Patients or visitors may not utilize personal electrical appliances within the facility without written approval from the biomedical department. If approved, a label denoting the approval date should be affixed to the appliance.
- The use of extension cords should be prohibited or strictly regulated.

- Adapters that convert three-prong outlets to two-prong outlets should be prohibited.
- Broken or cracked receptacle covers should be reported to the maintenance department immediately and should not be used until repaired.
- Only one electrical appliance should be plugged into a single electrical outlet.
- Any outlet or switch that smokes or emits an odor should not be used and should be reported immediately to the maintenance department for repair.
- Electrical equipment with frayed cords, damaged plugs, loose knobs, or not in proper operating order should be removed from service and reported immediately to the biomedical engineering department.
- Multiple outlet strips should be prohibited from use in patient areas.

Electric Heating Pad Hazards—The Food and Drug Administration (FDA) and the Consumer Product Safety Commission receive numerous reports of injury and death relating to burns, fires, and electric shock caused by the use of heating pads. Heating pads pose a danger for patients with decreased temperature sensation, diabetes, and spinal cord injury. Stroke victims and those taking medication for pain/sleeplessness are also at risk. Prolonged use on a single area of the body can cause a severe burn even at low temperatures. Most hospitals today use a circulating hot water pad or a hypo/hyperthermia machine which allows temperatures to be controlled by a thermostat. However, nursing home and home health personnel may encounter heating pad use. The following precautions should be observed when using electric heating pads:

- Inspect heating pad before each use and discard if it looks worn or if the cord is frayed.
- Ensure the removable cover is on during use.
- Place pad on top and never under the body part needing heat.
- Always unplug the pad when not in use.
- Read and follow all manufacturer instructions prior to use.
- Never use on an infant or person who is not sensitive to temperature changes.
- Never use in an oxygen-enriched environment or near equipment emitting oxygen.
- Do not use on a sleeping, unconscious, or paralyzed person.
- Do not fold or crush pad during storage and never unplug a pad by the cord.

F. BED SAFETY

Beds can cause patient and employee accidents and injuries. Consider the following when evaluating bed safety.

Bed Height—The bed height is proper when the person can sit on the edge of the mattress with knees flexed at 90 degrees and both feet can be planted firmly on the floor.

- Adjustable height beds should be used to obtain the desired mattress height. If the bed is still too high, proper height can be achieved by replacing the mattress with one that is thinner.

- Because bed height may inadvertently be altered during routine care, it should be checked regularly.

Headboard Safety Tips—

- For persons who use the headboard or footboard for assistance when transferring in or out of bed, these surfaces should be slip-resistant and easy to grab.
- The application of non-slip adhesive tape along the top length of the headboard and footboard will help prevent a person's hands from slipping.

Mattresses—Mattress edges must be firm enough to support a person securely when seated in an upright position. Mattress edges that are rolled offer a good grasping surface and provide stability when transferring.

Bed Wheels—

- Bed wheels that roll or slide away during transfers are a particular hazard. Beds, even those with adequate wheel locking systems, can be unsteady. A combination swivel and wheel brake provides the most stability.
- Properly locked wheels may slide on slippery linoleum or tiled floors. Non-slip adhesive strips or decals placed beneath the wheels can minimize sliding.
- Beds equipped with immobilizer legs (wheels recess when legs are on the floor) provide the highest level of stability.

Safety of Bed Side Rails—Since 1990, the FDA has received more than 100 reports of head and body entrapments in side rails. Most incidents involved elderly patients, with 68 deaths reported. The FDA recently released a safety alert concerning the operation of hospital beds. The following safety precautions should be observed:

- Inspect all bed frames, side rails, and mattresses as part of the preventive maintenance program.
- Ensure that the bed is properly aligned and that no gap is wide enough to entrap a patient.
- Never replace mattresses and side rails with dimensions different than the original equipment supplied by the manufacturer.
- When purchasing side rails and mattresses separate from the frame, ensure that all components are compatible.
- Check bed side rails for proper installation using manufacturer's instructions. Avoid bowing and make sure proper headboard to footboard distance is maintained.
- Establish safety rules/procedures for patients considered at high risk for entrapment.
- Use bed side rail protectors to close off open spaces which could lead to entrapment.
- Never use bed side rails as a patient protective restraint. *Note*—Use of restraints requires frequent monitoring and compliance with local, state, and federal regulations.

- The Safe Medical Device Act of 1990 requires healthcare organizations to report bed-related incidents resulting in death or injury.

Slippery Floors Near Beds—

- The application of non-slip adhesive strips on the floor along the length of the bed provides a slip-resistant surface. The color of strips should blend with the floor color so persons with altered depth perception do not misinterpret the strips as hazards to avoid.
- Additional measures to avoid slippery floor surfaces include the application of anti-skid acrylic coating to linoleum tiles and traction-soled socks or slippers.

Bed Alarms—Bed alarms allow normal bed activity but an alarm sounds by the bedside and/or nursing station if the person is about to transfer unsafely from the bed. Nurses who have used bed alarm devices find them to be "user friendly." They reduce the risk of falls; are easy to install, operate, and maintain; and are safe for patients.

G. SPECIAL NURSING CONCERNS

Ethyl Alcohol—Most hospitals utilize 70% ethyl alcohol as a topical application in local skin disinfection. Ethyl alcohol, according to NFPA 325M, is flammable in all dilutions where vapor may come in contact with an ignition source. The flash point of a 70% solution is approximately 70°F and is considered a possible fire hazard. Refer to Chapter 7 for additional information on ethyl alcohol.

 Safe Handling and Storage Procedures—Healthcare facilities must utilize reasonable care in the use, handling, and disposal of ethyl alcohol.
- The pharmacy, if equipped with a flammable storage cabinet, should be the only location of 100% ethyl alcohol.
- Volumes greater than 70% should not be stored on patient units.
- Volumes over 70% should be maintained in a flammable storage cabinet, away from patient care areas.
- When used topically, alcohol enhances the drying of the skin. Care should be exercised to avoid dermatitis in its use.
- Ethyl alcohol is highly miscible in water; therefore, it should be disposed of through dilution with water in an area where adequate ventilation is provided.
- A type BC fire extinguisher should be used to extinguish alcohol fires.

Aerosol-Delivered Drugs (Ribavirin/Pentamidine)—

- Visitors who are pregnant, breast feeding, or attempting to become pregnant should be discouraged from visiting while drug is in use.
- Workers who are pregnant, breast feeding, or attempting to become pregnant must be advised of the manufacturer's guidelines and should not be required to render primary care to patients during the period when drug is being administered.

- Staff should document the medical record if a visitor elects to enter the patient's room.
- Symptoms thought to be associated with overexposure should be reported according to administrative policy, including notification of the employee health nurse. Symptoms such as sore, itchy, burning eyes; headaches; asthma-like respiratory difficulties; chest pain; rashes; or coughing should be reported.
- A separate room or critical care unit is required, but an isolation room is not necessary. Special isolation ventilation is not necessary.
- Contact lens damage has been reported; individuals who wear contact lenses should consider wearing glasses while caring for these patients.
- In the event of an accidental spill of medication, the housekeeping department should be notified. Housekeeping should clean the spill according to the facility's policies and procedures.
- A warning sign should be posted outside the patient's room by respiratory therapy while the drug is in use.

Call System Operation—Call systems are tools for safety. To promote legitimate use of call systems, patients and nurses need to be educated. Nurses must know that call systems are designed to encourage patients not to do things for which they need help.

- Even though a call for help may seem to be a nuisance, nurses should be encouraged to consider what might happen if a patient attempted a potentially dangerous activity without assistance.
- Nursing personnel should explain to each patient the proper use of the call system and specify what the patient can do and should not attempt to do without assistance.
- Responding promptly and courteously to patient calls will encourage patients to use the system rather than attempt a dangerous activity.
- Call buttons should be placed within easy reach of the patient's bed. Ensuring that the call button is as close as possible to the patient prevents a major fall when the patient attempts to use a button that is out of reach.

Accurate Charting—Correct and accurate charting is crucial to patient care and safety. The following should be considered as important to the charting process:

- Each nursing action should be recorded as soon as possible in chronological order.
- All information and education given to the patient, family members, or other care providers should be recorded.
- Accurate charting actually catalogs the patient's medical care history.
- All charting entries should be thorough, accurate, and in clinical terms.
- Charting also involves documenting all patient-related observations made during care activities.
- Nursing personnel must keep detailed records of all nursing intervention activities as they occur.

Completing an Incident Report—Healthcare organizations collect incident data for a number of reasons. Nursing and patient care personnel must follow correct incident reporting procedures. Reports should be completed promptly so that all information is accurate. Incident reports protect patients, visitors, and employees. The report should never be used as an evaluation form and should not place blame or point a finger at others. Reports must accurately document information and the facts of an incident. Forms should be designed so that they can be completed quickly using ballot boxes. Healthcare incident reports:

- Provide a means to assess organizational problems
- Uncover areas or topics that require training
- Reduce patient risks by identifying care-related problems
- Evaluate the need for policy revisions
- Provide information on individual healthcare providers

No-Smoking Requirements—The Joint Commission's 1995 Environment of Care standards require hospitals to establish and maintain a no-smoking facility. The Joint Commission standard applies to all staff, medical professionals, contractors, visitors, and employees.

1. *Exceptions to the Policy*—
   - A licensed independent medical professional authorizes patient smoking
   - Patients who are mentally ill and/or in long-term nursing units. (*Note*—No exceptions can be made for employees, volunteers, physicians, contractors, or volunteers.)

2. *Conditions When Exceptions Are Not Honored*—
   - When the area is undergoing major construction or renovation
   - When the facility is operating under interim life safety rules
   - When the building has been classified as Level 3 and interim life safety measures have been implemented

3. *Enforcing No-Smoking Rules*—Facilities must establish procedures to ensure that no-smoking rules are followed. No-smoking signs should be posted. Some facilities take disciplinary action against employees who violate the policy. Smoking areas should be located in areas segregated from the rest of the facility. A no-smoking policy reduces the risk of fire and reduces indoor smoke pollution.

Ethylene Oxide Emergency Response—The following procedures should be carried out in the event of a major gas leak, suspected leak, or gas spill, such as but not limited to equipment failure, defective cylinders or canisters, or ventilation alarm malfunctions. Refer to Chapter 7 for additional information on ethylene oxide.

- Alert all personnel in the immediate area of the ethylene oxide release and evacuate the area immediately. DO NOT RE-ENTER THIS AREA UNTIL THE HOSPITAL SAFETY OFFICE PERMITS RE-ENTRY.
- Close doors to the affected area and report the incident by calling the emergency number.
- Wait for responding personnel to arrive and identify the location of the leak.
- Proceed to the emergency room (accompanied by security personnel) for first aid and medical attention.

H. DEALING WITH PATIENT EMOTIONS

Recognizing Patient Anger and Irritation—Nursing personnel must recognize that irritation, aggravation, and rage are the three phases of anger. Because they work closely with patients, nursing personnel are in a good position to observe and evaluate human emotion and recognize the first signs of anger.

- *Anger*—Patient can become fidgety, verbally abusive, or withdrawn. Attempt to establish a conversation with the person by asking about what is causing the anger. Showing concern can be a very effective way to prevent an explosion later.

- *Aggravation*—If the situation has progressed to aggravation, try to remove the patient from the aggravating situation or person. Speak in a reassuring tone. Ensure other staff members are aware of the escalating anger.

- *Rage*—Aggressive acts occur when rage has been reached; often the cause of the event may seem unimportant. Rage can build slowly and the anger can be directed toward a staff member. At this point, unwilling transport or restraint may be necessary.

- *Aggressive Behavior*—Many hospitals and most long-term care workers admit that destructive behavior by patients is a major concern. Many employees are injured when patients act out by biting, kicking, scratching, slapping, or punching. There are four possible outcomes of an acting out incident:
  1. The employee may be hurt
  2. Another patient may be harmed
  3. The aggressor may be injured
  4. The situation may be resolved peacefully

- *Prevention*—This is the best course of action because violence occurs in a cycle that leads to physical aggression. Healthcare workers need to feel that they can intervene during the anxiety stage of the cycle. Try to calm the patient by listening and addressing his or her concerns.

- *Options*—During verbal attacks, give the patient some well-defined options with enforceable limits. The most important goal during this phase of the aggression cycle is to help the patient return to a state of control.

- *Security Assistance*—Some organizations require that nursing personnel call for security assistance before the patient resorts to physical aggression.

- *Non-Verbal Clues*—When evaluating a patient's mental condition and situation, look for non-verbal clues. Pay attention to the eyes; they often provide information about what the patient is about to do. Be prepared for physical aggression if verbal communication stops abruptly.

Hazards of Protective Restraint Devices—Due to an increasing number of reports of injury and death associated with the incorrect use of patient restraints, the FDA is warning health professionals to make sure these devices are used safely. Devices include safety vests, lap belts, wheelchair belts, and body holders. Incorrect use of these devices has involved using the wrong size for a patient's weight, errors in securing restraints, and inadequate patient monitoring. Such mistakes have resulted in fractures, burns, and strangulations. Injuries and deaths have been reported in healthcare facilities as well as patients' homes.

1. *Types of Restraints*—
   - **Safety Bars**—Can be applied to wheelchairs to prevent falls. They are less restrictive than soft belts.
   - **Soft Belts**—Restraints similar to seat belts to prevent falls from beds and wheelchairs. They can be applied over clothing and around the waist.
   - **Safety Vest**—Provides more support than a belt in preventing falls from a chair or bed. This sleeveless restraint should be applied across the front of the body.
   - **Wrist Restraint**—A limb-holding restraint that prevents the patient from removing tubes or bandages. It is applied around the wrists and secured to the bed or chair. The restraint must be physically checked every 15 minutes to ensure it does not interfere with circulation.
   - **Mitt Restraint**—Looks like a big mitten without the thumb. It restricts finger movement but permits movement of the arm and wrist.

2. *Safety Recommendations*—The FDA recommends the following steps to reduce the risk of injury or death:
   - Follow good nursing and basic patient care practices.
   - Monitor patients frequently and remove restraints often.
   - Apply and adjust devices properly to maintain body alignment and patient comfort.
   - Allow the use of restraints only by prescription and for a strictly defined period.
   - Define and communicate a clear institutional policy on the use of restraints. The policy should cover such issues as appropriate conditions and length of time for use of restraints.
   - Display user instructions for restraints in a highly visible location and in foreign languages as necessary.
   - Keep accurate patient records on the use of restraints, including the reason for use, the type selected, and the length of time it was used.
   - Follow local and state laws regarding these devices.
   - Explain to patients and their families why the restraint is necessary, in order to facilitate understanding and cooperation.

3. *Other Recommendations*—
   - Select the appropriate restraint for the patient's condition.
   - Use the correct size.
   - Note front and back of the device and apply it correctly.
   - Tie knots with easily released hitches.
   - Secure bed restraints to the springs or frame, not the mattress or bed rails.
   - If the bed is adjustable, secure the restraints to the parts of the bed that move with the patient to avoid constricting the patient.

- Never use a sheet as a restraint.
- Visually check restraints every 30 minutes.
- Remove restraints every two hours to provide exercise and skin care.
- Document type of restraint and time it was applied.
- Never restrain a patient to a toilet or portable commode.

4. *Reporting Restraint Injuries*—Healthcare professionals should report serious injuries or deaths associated with the use of physical restraints to the FDA Problem Reporting Program operated by the U.S. Pharmacopeia. The toll-free number is 800-638-6725. Additional questions about how to prevent safety problems with protective restraints should be referred to the FDA, Center for Devices and Radiological Health, 5600 Fishers Lane, Rockville, MD 20857.

Safe Medical Device Act of 1990 (SMDA)—On December 15, 1995, the FDA published its final rule requiring healthcare facilities to report adverse events related to medical devices under a uniform reporting system. The new regulation was mandated by the 1992 congressional amendment to the SMDA of 1990. The FDA has also developed a standardized reporting form (FDA Form 33500 A) to assist facilities in filing reports with the manufacturer and the FDA. The new rule requires facilities to comply with reporting, recordkeeping, and training requirements effective April 11, 1995. It simplifies reporting requirements, identifies reportable events, and clarifies medical device user requirements. Key points of the major provisions are as follows:

- The FDA eliminated previously proposed training requirements which would have been costly to implement.
- Standardized forms for semi-annual reporting have been changed to streamline the process. The FDA also designated Form 3500 A to be used for reporting individual patient adverse device-related incidents.
- User facilities must only report information that is reasonably known rather than being required to investigate the incident.
- Incidents must now be reported within ten days of the event becoming known.
- Facilities are encouraged but not required to report malfunctions.
- User errors are considered a device-related event and are required to be reported.

1. *SMDA Reportable Events*—These include adverse events or problems that medical device regulations require to be reported. Medical device reportable events include patient deaths or serious injuries caused by devices or in which devices played a role. FDA Form 3500 A must be used for adverse incident reports. Semi-annual reports (FDA Form 3419) to the FDA are due by January 1 and July 1 for the previous six-month period (negative reports are not required).

2. *User Facility Report Number*—This is a unique number that identifies each report from a user. The number consists of three parts:
   - User facility's ten-digit Health Care Financing Administration (HCFA) number
   - The four-digit calendar year in which the report is submitted
   - The four-digit sequential number of reports submitted for the year, starting with 0001

3. **Facilities Required to Report**—Device user facilities are hospitals, outpatient diagnostic centers, outpatient treatment facilities, nursing homes, and ambulatory surgical facilities. Private offices of healthcare practitioners are not considered user facilities under the rule.

4. **Information to Be Reported**—
   * Patient information
   * Type of adverse event
   * Description of the event
   * Relevant laboratory/test data and patient history
   * Manufacturer and identification of suspected device
   * Initial reporter of the event
   * User facility name, address, and contact person
   * Event problem codes for the device and patient
   * Where and when the report was sent

5. **Written Program Requirements**—Regulated facilities must develop, implement, and maintain written procedures that address the following:
   * A mechanism to ensure timely and effective identification/evaluation of events
   * A standardized review process and methods for determining reportable events
   * Ways to ensure timely submission of completed reports

6. **User Documentation Requirements**—Device user facilities must maintain and provide FDA access to files that include:
   * Information related to the event
   * Reference to the location of such information
   * Documentation of reporting decisions and decision-making processes
   * Copies of all completed medical device reportable forms and other information submitted to the FDA and/or the manufacturer
   * Records related to an event for a period of two years

I. HELICOPTER SAFETY

Many acute care and trauma facilities are serviced by helicopters. Safety is paramount, and all staff who come in contact with the helicopter must be trained in helipad safety procedures. The following guidelines should be followed when handling helicopter-transported patients:

* The helipad should be restricted to authorized personnel. Personnel should not approach the landing zone until the craft has landed and the crew grants permission to approach.
* Hearing protectors must be worn while aircraft engines are running.
* Smoking is not allowed near the craft.
* Personnel should never shine lights directly in front of the aircraft during landing. Helipad landing lights should be adequate for landing.
* Approach the craft from the front or side as required, but never approach from the rear.

- Use a crouched stance when approaching.
- Be aware that the wind created by the rotor blades can result in a hazardous situation due to flying dust, litter, and loose clothing.
- Stethoscopes should be placed in pockets, and hats/scarves must be properly secured.
- The sheet must be taped to the stretcher.
- The portable oxygen bottle should be placed so that it does not extend beyond the cart and hit the craft when the stretcher is brought out for off-loading.
- The flight crew is responsible for opening and closing the aircraft doors.
- In most situations, the flight crew is responsible for off-loading the patient and will direct the hospital crew about proper placement of the transport stretcher.
- The flight crew will give directions about the departure from the helipad; not following instructions could result in an intravenous line being pulled out or changing the patient's skeletal alignment.
- Off-loading with the rotor blades turning requires great caution; all equipment is to be carried low to the ground.
- Fire extinguishers must be available and all personnel must be properly trained in their use.

J. MANAGING STRESS

Stress Management—Stress is the physiological response to a psychological event or a change.

Understanding Stress—

1. The effects of adaptation on the body are referred to as stress response. The stress response is also due to the buildup of small daily demands, difficulties, conflicts, threats, and frustrations. It is the body's general reaction to any change or demand from the environment.

2. Stress is also the body's response to anything the mind perceives or imagines to be threatening, frustrating, or demanding. This response has both physical and emotional components.

3. Working in healthcare facilities is stressful because of the nature of work. The National Institute of Mental Health has estimated that 25% of people age 25 to 44 suffer some type of psychological disorder. Some clinical disorders have been attributed to job-related stress:
   - Anxiety, depression, and job dissatisfaction
   - Maladaptive lifestyles and behavioral patterns
   - Substance abuse and dependency

Stress in Healthcare Organizations—Healthcare workers deal with life-threatening injuries and illnesses on a daily basis. Their work is complicated by overwork, under-staffing, paperwork, tight schedules, equipment breakdowns and malfunctions, demanding and dependent patients, and even death. Some healthcare workers feel that the bureaucratic and depersonalized nature of the healthcare organization leaves them feeling alone, isolated, angry, and frustrated.

- Stress contributes to worker apathy, lack of confidence, and absenteeism. Studies have shown that healthcare workers have a high rate of hospital admission for mental disorders. Suicide rates are also elevated for some healthcare professionals.
- Stress has been associated with loss of appetite, mental disorders, migraines, sleeping disorders, and emotional instability. It can increase the use of tobacco, alcohol, and drugs. Stress affects attitude, motivation, and behavior.

Stress and Patient Care—Stress not only affects the worker but also affects patient care. The following are some contributing causes of stress in healthcare workers:

- Understaffing and unbearable workload
- Inadequate resources to accomplish the task
- Working in an unfamiliar area, job, or role
- Rotating shift work
- No input or participation in planning and decisions
- No rewards for a good job
- Talents not used properly
- Exposure to biological and chemical hazards
- Lack of supervision and/or organization

Stress and Overtime—Employees who work overtime are at risk for accidents for a number of reasons. Working too many hours over a long period of time can cause fatigue, unnecessary stress, and morale problems among workers. Improper management techniques, insufficient numbers of workers, and poor job planning can result in unnecessary overtime. Supervisors must evaluate each situation and take corrective action as soon as possible.

Positive Stress—Not all stress is negative. A positive situation can be created when a person experiences short-term challenges that still leave time for relaxation. Positive stress can generate excitement and a positive attitude toward oneself and task accomplishment. A person in control can assess the situation, establish priorities, and then get things done. Asking others to help or delegating can be a positive step in helping control a stressful situation.

Negative Stress—This type of stress occurs when there is a constant uproar in one's life over a period of time. The person feels pressured and does not see any time for relaxation. Such stress can affect a person's physical and psychological well-being. Many people who experience out-of-control stress turn to cigarettes, caffeine, drugs, or alcohol as stress relievers. People who experience negative stress do not prioritize activities and often do not accomplish job-related tasks effectively. Stress can affect a person's human relations skills and often results in feeling isolated or deserted.

Healthcare Stressors—Some of the stressors experienced by healthcare workers are listed in Table 9.1.

TABLE 9.1   Stress Elements by Category

Category of Worker	Stressors
Nurses	Heavy workload, shift work, change in the organization, new technologies, and confronted daily with life-and-death issues
Medical Technicians	Not being appreciated, doing the same type of work every day, having to accomplish pulling/pushing tasks, and not being able to participate in the decision-making process
Administrators	Maintaining quality of care with smaller staffs and budgets, being involved in crisis management situations, maintaining competitiveness in a changing industry, and feeling unappreciated by healthcare staff and workers
Medical Staff	Having to make responsible decisions very quickly, dealing with sick and dying patients, having to face patients' families, and putting in long hours without proper rest

Shift Work and Sleep Deprivation—Supervisors and safety personnel should realize that shift work and sleep deprivation affect not only task accomplishment but also safety. People are diurnal and function best during daylight hours. Performance is decreased during periods of rapid eye movement (REM). This REM sleep is known as the dream period and normally occurs in the early morning hours. During this period, the body temperature is at its lowest. Humans have what is called "circadian rhythms" or a 24-hour body clock. This body clock can vary with individuals but can also be influenced by environmental factors. Organizations that have around-the-clock operations must realize that shift work can affect a worker or staff member's performance. Shift work and lack of sleep can contribute to health problems or the increased risk of health problems. Shift workers are also more prone to stress-related and family problems. Healthcare organizations should strive to make shift work safer and educate workers on adjustment strategies. Organizations should be fully staffed and workers should not be required to work double shifts or excessive overtime. Supervisors should be trained to evaluate workers and look for the signs of sleep deprivation, stress, and fatigue. Shift workers should be educated about the following:

- The importance of getting six to seven hours of uninterrupted sleep
- The necessity of sleeping in a dark room
- Tips on how to deal with noise during sleep periods
- The importance of eating a nutritious meal during a regularly scheduled "lunch" period
- Avoiding caffeine prior to completing a shift because it disrupts sleep patterns
- Exercising on a regular basis

Coping with Stress—Stress in the workplace can be reduced by following some simple suggestions:

- Conduct staff meetings and allow open communication.
- Implement a formal stress management program.

- Provide accessible counseling from a non-judgmental source.
- Promote flexibility and creativity within the department.
- Ensure adequate staffing and sufficient resources.
- Organize work areas and departments.
- Work to provide reasonable and flexible work schedules.
- Schedule rotation of unit assignments.
- Emphasize safety and health of workers.
- Conduct regular in-service education and training sessions.
- Provide group therapy on how to deal with patients.
- Implement a complaint and suggestion system.

K. VOLUNTEER SAFETY

Volunteers work in a variety of capacities in most healthcare organizations. Each volunteer needs to be trained for the work and risks to which he or she is exposed. Many organizations limit volunteers from patient contact tasks that could expose them to infectious agents, including bloodborne pathogens. Healthcare organizations that use volunteers should provide a comprehensive safety orientation program. Topics to consider include:

- Fire safety and emergency evacuation procedures
- Infection control precautions as necessary
- Universal precautions
- Patient care safety topics
- Radiation precautions
- Hazard identification and reporting procedures

L. HOME HEALTH SERVICES SAFETY

The home healthcare industry continues to grow, and many hospitals now offer home healthcare services. Hazards encountered can differ from those found in traditional healthcare settings. Home healthcare providers spend a great deal of time traveling and enter environments where they have little or no control. Many patients live in unsafe neighborhoods that expose providers to violence. Home healthcare providers are exposed to bloodborne pathogen and patient moving hazards much like their hospital counterparts. Nursing aides, nurses, and therapists often work alone and have no one to call to support them in a time of crisis. Home-based healthcare is unpredictable. The agency is legally responsible for worker safety.

General Training and Orientation—Orientation and regular in-service training for home healthcare workers should include safety-related topics. The following items should be covered:

- Age, cultural, economic, and social factors
- Disease manifestations

- Mental, emotional, and spiritual needs
- Personal security precautions including travel safety
- How to conduct a hazard assessment during a consultation visit
- Protection guidelines for bloodborne pathogens exposure
- Sharps and needle safety precautions
- Medical waste disposal procedures
- Back injury prevention techniques
- Safe lifting and transfer techniques
- Care plan development to identify risks

Safety in Home Care Environments—

- Ensure care plan identifies hazards and care requirements.
- Use well-lighted and common walkways when visiting patients.
- Instruct patients and family members on infection control.
- Always knock or ring doorbell before entering homes.
- Know injury and emergency reporting procedures.
- Document unsafe behaviors, threats, and menacing pets.
- Never run away from a threatening dog; back away slowly.
- Schedule "joint" visits in unsafe neighborhoods or homes.
- Request security escorts for night visits.
- Carry a nursing bag but never a purse.
- If threatened, scream, kick, and use chemical spray or a whistle.
- Keep automobile in top mechanical condition.
- Keep car locked and park near the patient's home.
- Look for slip, trip, fire, and electrical hazards.

Infection Control Guidelines—

- Use protective precautions and clothing.
- Follow universal precautions at all times.
- Wear disposable gloves and treat as medical waste.
- Dispose of all sharps and needles in an approved container.
- Take precautions when handling laboratory specimens.
- Segregate infectious waste at point of origin.
- Report potential exposures immediately.
- Follow bloodborne pathogens exposure protection requirements.

Note—Effectiveness of infection control management and training programs can be determined through field evaluations, written exams, and infection documentation audits.

Joint Commission Home Healthcare Infection Control Requirements—The Joint Commission requires home healthcare agencies to develop a formal infection control plan that addresses the following:

- Personal hygiene policies
- Equipment cleaning and sterilization procedures

- Staff health concerns
- Protection against transmitted diseases
- Isolation precautions to be followed
- Nursing aseptic procedures
- Documentation of infections relative to patient care
- Training on the personal responsibilities of home healthcare providers

M. OTHER PATIENT CARE AREAS

Surgical and Operating Room Safety—

1. *General Safety*—
   - Hazardous materials found in operating rooms include anesthetic gases, their vapors, and vapors of various solvents.
   - Anesthetic gases can pose both safety and health hazards, and testing for leaks should be done on a continuing basis.
   - The volume of anesthetic gases used should be properly noted, and records should be analyzed routinely as a check for leakage.
   - Nitrous oxide is the most commonly used anesthetic gas. The vapors of cyclopropane, halothane, and isoflurane are also considered to be gases that should be monitored.
   - The principal source of waste anesthetic gases in operating rooms is leakage from equipment, especially when administered by face mask.
   - The NIOSH criteria document on waste anesthetic gases provides a description of work practices for areas where anesthetic gases are used.
   - Only electrical equipment approved by the hospital engineering department should be used in operating rooms.
   - Equipment should be checked regularly to ensure that it is operating properly.
   - Flammable anesthetics should be stored in a separate, fire-resistant location that is vented to the outside.
   - The floors of operating rooms should be covered with an approved conductive material and tested regularly for conductivity.
   - Conductive clothing and footwear should be worn where required.
   - NFPA 56A, Standard for Inhalation Anesthetics and NFPA 70, National Electrical Code contain additional information.

2. *Compressed Gases*—Compressed gases used for anesthesia or other purposes in surgical suites include oxygen, nitrous oxide, ethylene oxide, and air. These gases may be piped in from a central storage area or used directly from cylinders in the surgical suite.
   - Hospital administrative personnel must ensure that cylinders of compressed gas are stored and used safely.
   - NFPA 56A contains recommendations for the storage and labeling of compressed gas cylinders and the use of regulators, valves, and connections.
   - NFPA 53M, Fire Hazards in Oxygen-Enriched Atmospheres provides additional safety guidance.

- Personnel should also reference Chapter 12 of NFPA 99, Health Care Facilities Handbook.
- The principal recommendations are to conduct proper inspections to ensure that the gas delivered is the same as that shown on the outlet label and to provide appropriate storage rooms for oxidizing gases such as oxygen and nitrous oxide.
- The National Fire Code, NFPA 56A, and NFPA 704 give a more detailed explanation of the recommendations and safety precautions.

3. *Scavenging*—Scavenging is the process of collecting and disposing of waste anesthetic gases and vapors from breathing system at the site of overflow. It is carried out to protect operating room personnel by preventing the dispersal of anesthetic gases into the room air. A scavenging system has two major parts: a collecting device or scavenging adapter to collect waste gases and a disposal route to carry gases from the room. Also refer to ANSI/Z 79.11-1982, Anesthesia Gas Scavenging Devices and Disposal Systems.

- The NIOSH publication "Development and Evaluation of Methods for the Elimination of Waste Anesthetic Gases and Vapors in Hospitals" contains information about control methods to establish and maintain low concentrations of waste anesthetic gas in operating rooms.
- This document includes techniques for scavenging, maintaining equipment, monitoring air, and minimizing leakage while administering anesthesia.
- Persons responsible for health and safety in the hospital surgical department should be aware of the availability of new products and new information on familiar products.
- Methyl methacrylate used in bone surgery has been recently investigated as a potentially hazardous substance.

4. *Worker Protection*—The following guidelines serve to help protect workers in surgical service:
   - Separate collection containers should be used for glass, empty ether cans, aerosol cans, disposables, etc. that will not be incinerated.
   - Sharp instruments, blades, and needles should be disposed of in designated puncture-resistant containers.
   - All supplies and instruments used should be accounted for to prevent their disposal in linens and materials that will be handled by hospital workers.
   - Towel clips and scissors should be closed when not in use.
   - Suction lines and electrical cords should be installed so as to minimize tripping hazards. Lines and cords should be suspended from the ceiling or placed under the floor whenever possible.
   - Personnel should be instructed to report defective equipment.
   - Warning signs should be posted where necessary and proper work practices enforced.
   - Workers should be instructed in proper lifting practices.
   - Safe work practices and health hazards should be discussed with each new worker as part of orientation and reviewed periodically.

Special Care Units and Dialysis—Electrical safety is a major concern in dialysis and special care units. The following are some practical safety precautions:

- A system of preventive maintenance should be established whereby all electrical equipment is inspected regularly and constantly maintained to prevent hazards.
- The Joint Commission requires specific inspection programs for patient-use and non-patient-use equipment.
- Electrical cords should not be frayed or exposed.
- The maintenance or engineering department should maintain compliance with NFPA 101, Life Safety Code.
- Dialysis machines should be maintained at the correct temperature range (35 to 40°C, 95 to 104°F) and conductivity (12 to 14 milliohms).

N. CENTRAL SUPPLY

Sterilization activities are normally the main function of this department. Many sterilization activities are very similar to small manufacturing plants or distribution centers. Central supply departments receive, package, process, sterilize, and distribute non-drug items for most medical and patient care units in the organization. These supplies include items such as glassware, gloves, surgical accessories, and intravenous solutions. Central supply also processes, inspects, and packages sterile linen for use in areas such as surgical and delivery areas. Traditionally, this department has been under the nursing services department, but many hospitals have now placed it under the control of a material management department.

Central Supply Hazards—Hazards in this department include those associated with material handling and sterilization processes.

- Sterilization should be conducted by a pressured steam process at a temperature exceeding 250°F (122°C).
- Improper use of sterilization equipment can result in burns from steam and exposure to ethylene oxide.
- The use of ethylene oxide requires aeration of sterilized items and exhaust ventilation for the waste gas.
- Detailed operating instructions should be posted at or near sterilization units.
- Autoclaves and other steam-pressured vessels should be inspected periodically, and records of the inspections should be maintained.
- Piping ethylene oxide through the hospital from a storage area may increase the potential for exposure to this hazard.
- If supply lines are not drained before tanks are changed, the gaseous mixture can spray maintenance workers before the pressure is released.
- Long supply lines from cylinders to sterilizers are also a potential source of exposure for many people and may make it difficult to locate and repair ruptures or leaks.
- By placing cylinders close to the sterilizer in a mechanical access room, the exposure and accident hazard can be contained and controlled.

- Although the mechanical access room is usually very warm and humid, these conditions can be controlled through adequate exhaust ventilation.
- Hospitals with sterilizers that use 100% ethylene oxide cartridges should store only a few cartridges in the department.
- Ethylene oxide cartridges should be kept in a cool dry place.
- Exhaust systems for ethylene oxide should be designed to prevent re-entry of the vapors into other areas of the building.
- Cuts, bruises, and puncture wounds from blades, needles, knives, and broken glass are among the most common accidents in central supply areas.
- Rules for gathering and disposing of sharps or other hazardous instruments should be reviewed regularly.
- Workers should handle items returned to central supply as if they contained sharp or hazardous instruments.
- Strains, sprains, and back injuries are common in central supply areas.
- Workers should be provided with appropriate carts, dollies, and other material handling aids and should be instructed in proper techniques for handling materials.
- Stepstools and ladders should be available and checked frequently for serviceability.
- Chairs, boxes, and other makeshift devices should not be used for climbing because they are a frequent cause of falls.
- Safety procedures for central supply gas systems can be found in NFPA 99, Chapter 4-3.1.8.

O. PHARMACEUTICAL SAFETY

Pharmacy Safety Requirements—Pharmacies must comply with Joint Commission, American Osteopathic Association, or Medicare accreditation/certification standards. Pharmacies obtain, distribute, and control all drugs and pharmaceutical agents within the organization. Other functions include directing special drug programs, providing services to satellite locations, and managing drug information systems.

Drug Quality—The American Society of Hospital Pharmacists (ASHP) sets guidelines on drug quality and specifications. Pharmacy procedures should require that all drugs and medications meet the standards of the U.S. Pharmacopeia–National Formulary (USP/NF). Drugs not included in the USP/NF should be approved by the FDA.

- Drugs must be obtained from known sources and meet identity, purity, and potency requirements.
- Drugs should comply with FDA current manufacturing practices.

Evaluating Drug Suppliers—According to the ASHP, the following technical factors should be considered when assessing sources of drugs and medications:

- Data on sterility and analytical controls
- Bio-availability and bio-equivalency information

- Information about raw materials and finished products
- Miscellaneous information on the quality of the drug or medication

Packaging and Protection—Finished drugs must be packaged to protect the product from light, moisture, temperature, air, and handling. Containers must be clean and never re-used. The ASHP publishes guidelines on environmental controls such as proper light, temperature, humidity, ventilation, and segregation.

Traditional Dispensing Systems—

- The traditional system requires nursing personnel to prepare individual doses from a supply sent by the pharmacy.
- The patient care professional is more involved in the process.
- There is an increased chance of medication errors.
- The system may be more effective in some patient care units, such as pediatrics and obstetrics.

Unit Dose Dispensing System—

- The unit dose system is now used by most facilities.
- The system provides a chest with compartments for each patient.
- The system requires more frequent deliveries to patient care areas but is more effective for a number of reasons:
    1. Reduces time required of nursing and pharmacy personnel
    2. Decreases the number of medication errors
    3. Allows better control and recording by the pharmacy
    4. Improves communication between the pharmacy and patient care units
    5. If an order cannot be read, the prescribing professional must be contacted for clarification
- Drugs should be labeled with the patient's full name and room number. Stock should be inspected regularly, and outdated drugs should be disposed of properly.
- Outdated drugs should be returned to the supplier for credit if possible.
- Cytotoxic drugs should be handled cautiously, according to relevant guidelines.

Pharmacy Hazards—Pharmacy workers are subject to slips and falls, back injuries, cuts from broken bottles and equipment, and exposure to chemicals (such as alcohols and solvents), dusts (such as talc and zinc oxide), and antineoplastic drugs. The following hazard control measures should be considered:

- Stepladders should be provided to help personnel reach items stored on high shelves.
- Spills should be cleaned up promptly.
- Broken bottles and unusable pharmaceutical products should be disposed of according to established procedures.
- Mixers, packaging and bottling equipment, and labeling machinery should be guarded properly.
- Adequate exhaust hoods should be provided where needed.

- Laminar airflow hoods should be checked frequently to determine if they are operating properly.
- Pharmacy personnel should be aware of the hazards associated with handling antineoplastic agents and be familiar with safety guidelines.
- Workers should be instructed in safe practices for lifting and carrying to prevent injuries.
- Thermometers, manometers, and other instruments that contain mercury should not be repaired in the pharmacy. Such equipment should either be repaired in an appropriate hospital shop or sent out for repair.
- Opening devices should be installed on the inside of walk-in vaults and refrigerators to prevent workers from being accidentally locked inside.
- The adverse effects of exposure to any medications that are packaged or dispensed in the pharmacy should be identified through medical surveillance.
- Workers should not be permitted to smoke or eat in pharmacy preparation areas because drug aerosols may be inhaled or drugs may be ingested.

Pharmacy Departmental Safety—Most healthcare organizations have a safety management program. The ASHP recommend that pharmacies develop and implement a safety management program.

- *Safety Committee Participation*—Pharmacy representation on the safety committee is important not only to the department but also to the effectiveness of the safety committee. Concern over safe handling of cytotoxic materials has increased the pharmacy's presence on the committee.

- *Safety Education and Training*—The director of the pharmacy must ensure that the department conducts an effective orientation and training program to address:
  1. The importance of practicing safety on the job
  2. The department's disaster planning and emergency response roles
  3. Hazards found in specific jobs or processes
  4. Organizational/departmental safety policies and procedures

Hazardous Materials Safety—The Joint Commission provides guidance on hazardous materials and waste management. Pharmacies not only comply with OSHA and Joint Commission requirements but also follow guidelines in the "ASHP Technical Assistance Bulletin on Handling Cytotoxic and Hazardous Drugs."

ASHP-Defined Hazardous Drugs—

- Can cause DNA damage or mutations
- Are carcinogenic or teratogenic
- Can be toxic at low doses
- Can adversely affect human organs
- Should be handled in a class II, contained, vertical-flow biological safety cabinet
- Radioactive materials should be handled in accordance with Nuclear Regulatory Commission (Title 10 CFR Part 20) guidelines

Hazard Communication Program—

- Identify areas where hazardous materials are used or stored.
- Develop a written HAZCOM plan outlining policies and procedures.
- Develop and maintain a current listing of all hazardous materials located in the department.
- Conduct a hazardous materials handling training session upon assignment of a worker to the department and when a new hazardous material is introduced into the workplace.
- Maintain a Material Safety Data Sheet file or binder in a convenient location.

Drug Storage—

- Drugs for external use should be stored separately from medications that can be taken internally.
- Respiratory care drugs and those used to prepare irrigation solutions should not be kept with other injectable drugs.
- Large quantities of acids or other hazardous materials should be stored close to floor level.
- Large or heavy drug containers should be stored on lower shelves.
- Hazardous storage areas should be identified with an appropriate caution or warning.
- Poisons should be stored in a secure area and clearly labeled.
- Drugs should never be stored in a refrigerator that contains food or drink.

Disaster Preparedness—

- Pharmacy personnel must be familiar with the organization's emergency planning program.
- Pharmacy personnel must be trained in the department's responsibilities in supporting the plan.
- The department should develop a plan for obtaining and distributing drugs during emergency situations.

Fire Safety—All pharmacy staff members should know:

- Fire identification and reporting procedures
- The classes and hazards of fire
- How to activate the fire alarm and notify others
- How to select and use the proper fire extinguisher
- Techniques for controlling smoke and fire
- Evacuation routes and egress responsibilities

SUMMARY

This chapter covered a wide variety of hazards found in patient care areas. Healthcare lifting and the resulting back injuries continue to be a major occupational hazard in many healthcare organizations. Special emphasis was placed on back injury prevention, including brief sections on lifting, transferring, and moving patients. Special emphasis was also placed on slip, trip, and fall prevention in patient care areas and nursing homes. The chapter covered a number of other nursing-related topics, including aerosol-delivered drug precautions, medication safety, incident reporting, and dealing with patient emotions. Information was also presented on Safe Medical Device Act reporting requirements, pharmacy safety, central supply hazards, home health nursing, and managing stress.

REVIEW QUESTIONS

1   Back injury prevention in the healthcare environment should focus on:
    a.   eliminating lifts wherever possible
    b.   providing mechanical lift devices to assist with patient moving tasks
    c.   implementing employee exercise and strengthening programs
    d.   all of the above

2   The most common occupational injury experienced by nursing personnel is:
    a.   needlestick
    b.   back strain
    c.   burn
    d.   slip or fall

3   Medication administration errors can be reduced by:
    a.   nursing personnel following established procedures
    b.   properly storing all drugs
    c.   personnel following good handwashing discipline
    d.   all of the above

4   Heating appliances should not exceed _____°F and should never be placed directly against the skin.
    a.   100
    b.   110
    c.   120
    d.   130

5   One of the greatest exposure hazards found in the central supply area is:
    a.   material handling
    b.   ethylene oxide
    c.   benzene
    d.   used sharps

6   Which of the following statements concerning hospital side rails is true?
    a.   most incidents involve elderly patients
    b.   since 1990 the FDA has received more than 100 reports of patient entrapment

   c.   the FDA recently released a "safety alert" concerning the use and operation of side rails

   d.   all of the above

7   A patient restraint device should:

   a.   be visually checked every 15 minutes

   b.   be physically checked every 30 minutes

   c.   not be used on a patient for more than 2 hours

   d.   a and b

Answers: 1–d   2–b   3–a   4–c   5–b   6–d   7–d

CHAPTER 10

HEALTHCARE SUPPORT AREA SAFETY

A. ENVIRONMENTAL SERVICES

The housekeeping or environmental services department provides services to virtually all patient and non-patient areas in a facility. The department plays an important role in keeping the facility clean, safe, and functional. The major tasks for environmental service workers include:

- Daily tasks such as mopping, dusting, and disinfecting
- Cleaning patient care areas including rooms
- Maintaining common areas, corridors, and offices
- Cleaning windows, extracting carpets, and cleaning vents
- Assist with moving furniture and equipment
- Terminal cleaning of patient rooms after discharge
- Moving trash and refuse to containers or pickup points
- Cleaning and disinfecting contaminated areas
- Pickup of hazardous and infectious waste materials
- Cleanup of chemical spills and releases

Cleaning for Health—The environmental services department plays a key role in controlling infection within a facility. Workers must be trained to "clean for health" first and then clean for appearance. As this approach to cleaning becomes more widely accepted, the important role environmental services personnel play in keeping healthcare facilities safe and healthful will become more appreciated. It can be described as removing contaminants from the environment. The levels of cleaning are as follows:

- *Sanitary*—Cleaning to the degree that general health is protected. A risk of disease may exist, but the risk is considered to be acceptable.
- *Disinfected*—95% of contaminants, pathogens, or pollutants are removed or killed.
- *Sterile*—Environment is 100% free of contamination.

Guidelines for Cleaning—

- The safety of patients, visitors, and staff must be emphasized during the entire cleaning process.
- The removal of contaminants should be a priority while keeping cleaning residues to a minimum.
- Humans should be protected from exposure to chemicals during the process.
- The cleaning process should be assessed in relation to the entire organization.
- Waste materials must be handled and disposed of properly.

Cleaning Effectiveness—The cleaning effectiveness and safety performance of the environmental services department impact the quality of patient care. A clean and safe facility impacts the morale and job performance of all healthcare workers.

- Environmental services workers are potentially exposed to all of the health and safety hazards found in the healthcare or hospital environment.
- Workers are protected under Occupational Safety and Health Administration (OSHA) regulations, including hazard communication (29 CFR 1910.1200) and bloodborne pathogens (29 CFR 1910.1030) standards.
- Workers are also prone to strain-related injuries and to slip, trip, and fall hazards.
- Workers must be thoroughly trained on chemical safety and how to use the information found in the Material Safety Data Sheets (MSDSs) in accordance with OSHA standards.
- Workers should receive regular training in how to clean for health and awareness training on the specific hazards in each department. Hazards encountered by workers include exposure to radiation, compressed gases, infectious materials, and hazardous chemicals.
- The environmental services supervisor plays an important role in keeping custodial workers safe. Workers need a proactive supervisor who ensures that all personnel receive the proper tools, equipment, cleaning supplies, and training.
- Environmental services must be integrated into everyday operations. Workers must understand the hazards of chemical interactions and the importance of proper storage. Storage areas must be kept clean and disinfected.

Preventing Accidents—Environmental services workers can help prevent accidents by keeping alert to identify and correct hazards. The following safety rules should be observed:

- Wipe up spills, leaks, and tracked-in water immediately.
- Keep oily substances off of floors.
- Replace worn tiles, fix loose floor moldings, and repair torn carpeting as soon as possible.
- Use care in mixing detergents, germicides, and cleansers.
- Never mix chemical solutions because a dangerous reaction could occur.
- Avoid careless placement of tools, supplies, and equipment.
- Post WET FLOOR signs and barriers when cleaning or refinishing floors.
- Ground electric cords while operating floor machines, vacuums, and other electrical appliances.

- Check electrical equipment for frayed wires, loose plugs and connections before using.
- Never run cleaning machines over electrical cords.
- Never use a product for other than intended purposes.

Chemical and Physical Hazards—Rules that should be followed when dealing with hazardous chemical and physical agents are as follows:

1. Know which substances cause sensitization reactions.
2. Measure and mix cleaning solutions correctly.
3. Use appropriate protective equipment.
4. Reduce the use of solvents in cleaning processes.
5. Follow safety rules for flammable materials.
6. Follow safety rules when using caustic substances.
7. Train personnel in the safe use of all disinfectants including:
   - Quaternary ammonia compounds
   - Phenols
   - Iodophors
   - Hypochlorite
8. Protect face when exposed to splash or splatter hazards.
9. Wear hearing protection in noisy environments.
10. Know the location of the HAZCOM plan and the MSDS file.

Eye and Face Protection—Workers may need eye protection such as goggles, safety glasses, or face shields when mixing or using chemical compounds and certain disinfectants. Eye protection devices should be considered optical instruments and be carefully selected, fitted, and used.

Hand Protection—Glove use determines glove material. For most light work, a canvas glove is both satisfactory and inexpensive. For handling rough or abrasive material, leather gloves reinforced with metal stitching will be required. Many plastic and plastic-coated gloves are available. They are designed to give protection for a variety of hazards. Other gloves have granules or rough materials incorporated in the plastic for better gripping quality.

Foot Protection—About a quarter of a million disabling occupational foot injuries take place each year. Many of these injuries could be avoided by wearing proper foot protection. Housekeeping employees should wear well-fitted, sturdy, low-heeled shoes. The soles should be made of skid-resistant material, and there should be some toe reinforcement to protect the wearer against injury from dropped objects.

Other Personal Protective Equipment (PPE)—In some cases, the use of other protective clothing and equipment might be required by housekeeping employees, such as aprons for torso protection, "bump" hats for head protection, or noise control devices for hearing protection.

Worker Training—The following topics should be included in the environmental services training program:

- Cleaning techniques for how to clean for health
- Infection control and disinfecting procedures
- Workplace safety principles and policies
- Universal precautions
- Infection control procedures
- Selection and use of PPE
- Storage of equipment and chemicals
- Waste management and handling requirements
- Safe lifting and material handling
- Ladder use and safety
- Electrical equipment safety
- Service cart operation and maintenance
- Slip, trip, and fall hazard identification
- Emergency action and injury reporting procedures

Cleaning and Polishing Agents—The proper cleaner in the specified amount should be used for each cleaning task. Improper usage can result in failure to disinfect or cause overexposure, resulting in dermatitis, slipping hazards, or deterioration of floors and furniture. Other precautions include:

- Never mix cleaning chemicals because some mixtures can create hazardous gases. Avoid mixing chlorine with substances such as vinegar, toilet bowl cleaners, or ammonia.
- Use chemicals from containers that are properly labeled.
- Never transfer cleaners to containers used for food or drink.
- Avoid skin contact when using caustics because even diluted solutions can cause skin burns.
- Never leave cleaning, disinfecting, or polishing agents unattended in areas where patients or visitors might mistakenly ingest or touch them.
- Follow all precautions contained on labels or in the MSDS when using toxic substances.
- Post the Poison Control Center number in highly visible locations.

Electrical Fixtures and Equipment—Electrical equipment should only be handled with dry hands and should always be disconnected before repair or maintenance. Key electrical safety precautions include:

- Disconnect electrical appliances by grasping the plug, not the cord.
- Clean electrical fixtures with a damp cloth after turning off and unplugging.
- Clear the area before changing bulbs or fluorescent tubes.
- Never change bulbs when people are below or nearby.
- Never force a fluorescent tube from its socket.

Cleaning Windows—

- When cleaning windows, always wear proper clothing.
- Follow the suggestions for safe ladder selection or use.

- Before cleaning, inspect safety belts and attachments and window belt hooks.
- Take down shades, curtains, drapes, and screens that are to be cleaned and stack them neatly in a place where they will not become a tripping hazard.
- Tie or hook any drapes to be left on the window so staff can get in and out safely.

Slip, Trip, and Fall Prevention—Many slipping and falling hazards result from improper housekeeping practices. Ways to reduce this problem are as follows:

- Immediately clean up all foreign materials on floors, such as liquids, foods, and dirt, as well as other obstructions such as trash and extension cords.
- Ensure that entrances, steps, and outside walks are free from ice.
- Keep entrances mopped up in wet weather to prevent slipping hazards.
- Use the type and amount of wax specified for each kind of floor. Too much wax, improper buffing, or polishing with an oiled mop creates a slipping hazard.
- Before mopping, rope off corridors and public rooms or put WET FLOOR markers at exits and near stairways.
- Mop or wax only one side of a corridor at a time so that there will always be a dry path for people to use.
- Keep equipment on the side being cleaned. When using electrical equipment, plug it in on the side where working. If this is not possible, use a CAUTION floor marker to call attention to the cord. Do not block doorways or elevator entrances with cleaning equipment.
- Tape vacuum cleaner hoses and extension cords with high visibility windings to make them more conspicuous.
- Place wet mops, brooms, electrical equipment, and supplies in their proper storage areas as soon as a job is completed.

Maintaining Equipment—Housekeeping personnel should be provided with well-made and properly maintained equipment. Safety features and cost should be considered when equipment is being purchased. Adequate storage facilities should be provided for housekeeping equipment and supplies. Closets should be organized to provide a designated space, shelf, hanger, or rack for each item used.

Inspection Requirements—Equipment should be inspected at least monthly. A system for employees to report defective items and conditions should be established. Hazards should be corrected and defective equipment immediately taken out of service and replaced.

Equipment Carts—Equipment carts should be carefully moved through corridors to avoid collision and tripping hazards.

- Reduce speed near stairways, corridor intersections, elevators, and down ramps.
- Pull a cart through swinging doors, rather than shoving it.
- Push a cart only when there is clear visibility.
- Never leave carts, equipment, or supplies in a location that creates a hazard.
- Immediately report a cart that needs to be repaired.
- Arrange items on a cart for safe and efficient use.

Handling Wastes—Housekeeping personnel who handle waste should follow all facility guidelines for handling waste materials. Personnel must wear PPE such as approved gloves and aprons. When handling potentially infectious materials, personnel must use universal precautions. Tips on handling waste include:

- Never reach into a wastebasket with hands. Empty the receptacle by tipping it into a collection container or by carefully removing the bag or lining.
- Sweep but never pick up broken glass by hand. Pick up the fine pieces with a damp cloth, paper towel, or cotton.
- Never store waste where food is being prepared or served.

B. LAUNDRY SAFETY

The laundry department is responsible for laundering, distributing, and storing all linen and other washable items in the facility. Laundry tasks can also include pressing, folding, and repairing. Laundry personnel must also accomplish special tasks such as inspection of surgical and sterile linen supplies. Some hospitals may use a shared facility located off-site.

General Considerations—

- Laundry personnel must be trained in how to process contaminated laundry items. Laundry personnel can be exposed to both biological and chemical hazards.
- Sharps and needles left in linens can be a real hazard during processing. Laundry personnel must follow universal precautions and take measures to protect themselves from exposure to bloodborne pathogens.
- Personnel are also exposed to chemical substances such as conditioners, softeners, and detergents. Many laundry systems now use an automatic dispensing system which reduces the frequency of chemical exposure.

Linen Handling—Exposure results from handling linen contaminated by drainage and body wastes. Exposure can be reduced by using the following precautions:

- Linen should be placed in plastic bags at collection locations.
- Color-coded bags should be used to alert laundry staff to potential hazards.
- Linen from isolation rooms should be double-bagged and washed separately.
- To reduce exposures, laundry should be emptied directly into washing machines.
- Handwashing and glove use are essential to basic infection control.

Safety Guidelines—The following points should be included in a health and safety program for hospital laundry workers:

- Keep floors dry; label wet floors.
- Provide non-skid mats or flooring in wet areas.
- Require workers to wear non-skid boots or shoes.
- Handle laundry carefully; watch for sharps and needles.
- Handle soiled linens gently to prevent contaminating the air.

- Use caution when handling linen from radiation/cytotoxic drug areas.
- Place dirty linens in color-coded and impervious bags at collection site.
- Use a barrier to separate soiled linen areas from the other laundry areas.
- The high temperature and excessive humidity in some laundry areas may be impossible to control with engineering devices alone, especially during the summer months.
- Administrative controls may be necessary, and persons working in excessively hot environments can be rotated to other jobs or shifts.
- Workers should be aware of the symptoms of heat stress and the need for water consumption and more frequent breaks.
- Workers who sort and wash contaminated linens should wear proper protective clothing and respirators.
- Laundry personnel should be instructed to wash their hands thoroughly before eating, drinking, and smoking; before and after using toilet facilities; and before going home.
- Workers who handle and sort soiled linen in the laundry department should be included in the hospital immunization program.
- The wrapping on steam lines should be adequately maintained to protect workers from burns.

C. FOOD SERVICE SAFETY

Healthcare food service operations prepare meals for both patients and employees. Many organizations also operate snack bars, coffee shops, and cafeterias. The food service operation in an acute care facility can employ up to 10% of the total workforce. The actual number of employees depends on size and type of facility, preparation equipment used, and the method of distributing prepared meals. Many food service departments also support special functions, meetings, and seminars. The nutrition function provides therapeutic meals, nutrition counseling, and educational sessions to support organizational and community programs. Many healthcare food service departments are actually two departments that often operate under a single manager or department head. One area is responsible for food preparation and delivery while the other function is concerned with nutrition, meal planning, and therapeutic dietary planning.

Types of Food Service Operations—

- *Food Service Operation*—The food service preparation function is normally managed by a professional with education or experience in hospitality, food, or hotel management.
- *Nutrition and Dietary Management*—The nutrition area is normally headed by a registered dietician. In many facilities, the registered dietician serves as the department head and supervises both the food preparation and nutrition planning functions.
- *Contract Food Service Management*—Many facilities now employ professional contract firms to provide senior management personnel or manage the entire food service operation.

Sanitation and Safety—Regardless of organization structure, the operation will require a number of workers, including cooks, food handlers, and sanitation workers, who are exposed to a number of safety hazards. Food service departments must meet inspection standards of the local health department and comply with a number of federal, state, and Joint Commission on Accreditation of Healthcare Organizations guidelines or standards.

Hiring Safe Workers—Management personnel can greatly affect the safety of their department by selecting and hiring safety-conscious workers. A strong orientation program can be important in providing employees with the information and motivation to work safely and avoid hazardous situations.

Food Service Orientation Topics—The following areas should be covered during initial orientation and periodically during in-service training sessions:

- Principles of personal hygiene and wearing proper clothing
- How to handle, prepare, and serve food properly
- Selection, use, and care of PPE
- Good housekeeping procedures
- Principles of sanitary food handling
- Safe food transportation and service methods
- Preventive maintenance and safe operation of equipment
- Food spoilage prevention methods
- Chemical safety and hazard communication procedures
- Emergency response and incident reporting procedures

Food Service Hazards—Most injuries occur in food service areas while workers are handling materials, processing food, or distributing meals. The following guidelines can reduce hazardous exposures:

- Service electrical components and equipment regularly.
- Train workers in correct material handling techniques.
- Properly guard all machinery and hot surfaces.
- Keep walking and working surfaces free of hazards.
- Require good work and housekeeping practices.
- Train workers about food service safety.

Food Service Equipment Safety—Meat saws, slicers, and grinders should be properly guarded. Workers should use tamps or push sticks to feed grinders and choppers. Wheels on food carts should be kept in good repair. Workers should be instructed to get help when moving a heavily loaded cart. Workers should observe the following rules:

- Push food carts instead of pulling them.
- Secure and properly store compressed gas tanks.
- Keep compressed tank gauges in good working order.
- Guard exposed drive belts, gears, chains, and sprockets.
- Secure dumbwaiters when not in use.

- Clearly label or mark all steam and water pipes.
- Train workers in the location and operation of shutoff valves.

Cutting and Mixing Equipment Hazards—Workers should observe the following safe practices when using cutting or mixing equipment:

- Check and secure guards before starting.
- Use non-metallic utensils to stir/test contents.
- Stir contents only when equipment is not in motion.
- Never remove food containers until all parts have stopped.
- Use a wooden stomper to feed food into a grinder.
- Wear wire-mesh gloves to clean machines with sharp blades.
- Disconnect all machines before cleaning or servicing.
- Restrict use of cutting equipment to authorized workers.

Burn Prevention Measures—All stoves, pots, and pans should be considered potential burn hazards. Cooking utensil handles should be turned away from the front of the stove. Workers should be required to use hand protection when handling hot utensils. The pan cover should be used to deflect steam when removing covers. Workers should stand to the side of the unit when lighting gas stoves and ovens. Scalds and burns can be prevented by following some basic safety rules:

- Wear well-fitting uniforms with long sleeves.
- Have enough dry potholders available.
- Pour hot foods away from the body.
- Turn handles of pans and pots in from the edge of the stove.
- Lift lids from the far side to prevent steam burns.
- Properly position cookers and equipment.
- Promptly treat minor burns or scalds with cold water or ice.
- Clean stove hoods and filters on a regular schedule.
- Clean grease from the stove hood flange on a regular basis.
- Never use a stove without hood filters in place.

Fire Extinguishing Equipment—Workers should be trained in the use of fire extinguishers and hood extinguishing systems.

- Educate workers on emergency and evacuation procedures.
- Train workers annually on fire extinguisher use.
- Extinguishers of the proper type and capacity should be available.
- Permanently mount all extinguishers and mark their locations.
- Keep areas around extinguishers free of clutter.
- Aim automatic sprinkler at the hazard location.
- Comply with local fire codes and Joint Commission and OSHA requirements.

Electrical Equipment Safety—

- Ground or double insulate toasters, blenders, hand mixers, fans, refrigerators, and radios.
- Check all personal items to ensure proper grounding for industrial application.

- Remove power before adjusting or cleaning equipment.
- Tag equipment being serviced or cleaned.
- Never plug in an electrical appliance or equipment with wet hands.
- Never stand in water while around electrical equipment.
- Disconnect power when servicing equipment.
- Follow prescribed lockout/tagout procedures.

Electrical Shock Prevention—Food service workers are exposed to electric shock hazards on a daily basis. The National Institute of Occupational Safety and Health (NIOSH) recommends the following ways to reduce the risk of electric shock:

- Install ground-fault circuit interrupter in all wet areas.
- Receptacle boxes should be of non-conductive materials.
- Identify outlets and fixtures for each circuit breaker or fuse.
- Never use breakers as on and off switches.
- Train all workers about electrical safety and hazards.
- Never touch a person experiencing electric shock.
- Post emergency procedures and numbers in all work areas.

Walking and Working Surfaces—Slip-resistant floor coverings should be standard in wet or greasy areas around sinks, dishwashers, and stoves.

- Immediately clean up spilled foods, liquids, and broken dishes.
- Identify all areas with appropriate signs until cleanup is complete.
- Require workers to wear shoes with slip-resistant soles.
- Repair or replace damaged floor mats promptly.
- Place appropriate mats in all wet areas.

Fall Prevention—Never use chairs, stools, or boxes as makeshift ladders. Safety stools or ladders should be provided for reaching high storage areas. Never block aisles, walkways, or exits with carts, boxes, or trash. The following can also help reduce the incidence of trips and falls:

- Repair broken/missing tiles, cracked boards, and damaged floors.
- Ensure the safe condition of all ramps, stairs, and steps.
- Do not place items in aisles or walkways, even temporarily.
- Treat or cover walkways with slip-resistant materials.
- Workers should wear shoes with slip-resistant soles.
- Immediately clean up all spills and breakages.
- Routinely inspect for hazards after each mealtime.
- Schedule thorough floor cleaning at least three times weekly.
- Provide proper lighting in all work areas, corridors, and stairways.
- Ensure that electric cords and wires do not cross traffic paths.
- Provide non-slip matting in clipper and dishwashing areas.

Strains—

- Instruct workers on back health and proper lifting techniques.
- Require workers to follow proper safe lifting techniques.

- Reduce the strain of handling bulky items by supplying workers with hand trucks, dollies, and other material handling equipment.
- Inspect all food service equipment on a regular basis and repair defective equipment immediately.
- Evaluate hazards and require employees to correctly select and use PPE.

Chemical/Physical Hazards—Food service workers are exposed to a variety of chemical and physical hazards. Although many of these hazards are covered elsewhere in this text, a summary of some of the common food service chemicals and associated hazards is provided below:

- Exercise caution when using ammonia and avoid skin contact.
- Flush affected skin areas or eyes promptly with water.
- Wear gloves and face/eye protection when working with ammonia.
- Use ammonia only in areas with good ventilation.
- Make sure stove hoods are working when using ammonia to clean grease.
- Chlorine mixed with ammonia creates a toxic chlorine gas. Exposure to chlorine can result in eye, nose, and throat irritations.
- Use protective clothing when working with caustic cleaning solutions.
- Workers in hot areas may need frequent water and rest periods.
- Regularly check microwaves for defective hinges, doors, and seals.

Cutting Hazards—Food preparation workers are exposed to a variety of cutting hazards. The following are some general safety rules to help prevent cutting accidents:

- Train workers in the safe handling and use of knives.
- Keep cutlery sharpened and in good condition.
- Use a cutting board or other firm surface.
- Always use cut strokes away from the body.
- Store knives, saws, and cleavers with blades not exposed.
- Use racks or carts to carry small amounts of breakable items.
- Drain the sink before removing broken tableware.
- Never pick up broken glass with bare hands.
- Brush broken glass into a dustpan with a brush or damp towel.
- Do not stack glass tumblers inside each other.
- Never drop silverware into tumblers or pitchers.
- Never use glassware for storing tacks, pins, or chemicals.
- Use hand guards to reduce cuts in case hands slip from the handle.
- Provide ample work space for workers doing cutting tasks.
- Provide slip-resistant standing surfaces in all cutting areas.
- Keep fatigue mats and wooden duck boards in good repair.
- Wipe knives with a towel, with cutting edge pointed away.
- Wash knives separately and never place them in soapy water.
- Discard food within a five-foot radius of glassware breakage.

Food Service Area Layout—The layout of food preparation and service areas can impact the safety of the entire operation. The relationship of worker tasks and the layout of work areas affects productivity and safety. Areas should be arranged so as to decrease handling time or distance. Unnecessary movement and unproductive arrangement increase chances for employee accidents or injuries.

Layout Suggestions—
- Store supplies where they are used most often.
- Reduce heavy lifting tasks and uncomfortable work heights.
- Workers should not stand for extended periods in one place.
- Use assembly or straight lines whenever possible.
- Food service areas should have sufficient space to properly store all foodstuffs and supplies.
- Clean all dishes and utensils effectively.
- Provide dining room space to serve guests efficiently.
- House water-heating equipment and other support systems.
- Store all cleaning and housekeeping supplies away from food.
- Allow for the orderly and sanitary handling/processing of food.
- Provide at least three feet between working areas.
- Provide a ceiling height of at least eight feet.

Dining Room Safety—Facilities with dining rooms or cafeterias should:

- Consider safety and ease of serving when arranging the dining area.
- Ensure that drapes and curtains are made of fire-resistant materials.
- Securely anchor pictures and wall coverings.
- Inspect chairs and tables regularly for defects.
- Arrange tables systematically to allow for dish and tray removal.
- Keep bus carts and racks in good condition.

Controlling Food Contamination—Healthcare facilities should follow all local department of health regulations concerning sanitation. An excellent guide to avoiding food contamination is the U.S. Public Health Service publication "Food Service Sanitation Manual."

General Precautions—
- Take precautions when handling potentially hazardous foods such as milk products, eggs, meat, and poultry. Fish, shellfish, and edible crustaceans need special attention.
- Maintain potentially hazardous foods at 40°F or lower unless being prepared or served.
- Reheating foods that have been at room temperature for some time will not protect diners from foodborne illnesses.

Foodborne Illnesses—Foodborne illnesses can be caused by a variety of bacteria, viruses, parasites, and chemicals.

1. *Common Types of Foodborne Bacteria—*
   - **Salmonella**—This bacteria is found in intestinal tracts of humans and animals and can cause salmonellosis. It can be transmitted to humans through

meat, fish, and eggs. The bacteria is destroyed by heat and certain chemical germicides.
- **Staphylococcus**—This common foodborne bacteria produces a toxin that causes food poisoning. The bacteria is not destroyed by normal cooking temperatures. It can be easily transmitted by unsanitary work habits.
- *E. coli*—This bacteria can cause serious illness and is common in undercooked ground beef. All ground meat must be cooked sufficiently to destroy this deadly bacteria.

Note—The contaminating agents of a foodborne illness usually cannot be detected through taste, odor, texture, or appearance.

2. *Prevention*—
   - Wash hands frequently and after smoking or using restroom.
   - Use antibacterial soap and wash for at least 30 seconds.
   - Do not use common towels for hand drying.
   - Cook/maintain hot foods at a temperature of 140°F or higher.
   - Heat poultry and stuffed meat dishes to 165°F.
   - Cook pork at 150°F or higher.
   - Minimum heating temperature for rare roast beef is 130°F.
   - Keep cold foods at 40 to 45°F or lower if possible.
   - Exclude ill workers from food handling and preparation.
   - Refrigerate food immediately.
   - Reheat leftovers rapidly to 165°F before serving.
   - Refrigerate potentially hazardous foods until used.
   - Eliminate all pests from food preparation and storage areas.

Preventing Cross-Contamination—Cross-contamination can be prevented by properly handling and storing containers, utensils, glassware, and dishes. The dishwashing cycle should include cleaning with warm soapy water and rinsing in clear water to remove detergent residue. Consider the following sanitizing methods:

1. Immerse in clean hot water (170°F) for 30 to 45 seconds.
2. Immerse in hypochlorite water solution (50 ppm) for at least one minute at a temperature of 75°F.
3. Immerse in quaternary ammonium compound solution and iodine germicides at 12.5 ppm.
4. Mechanical cleaning of dishes and utensils should adhere to the following water temperature requirements:
   - Wash cycle—140°F
   - Initial rinse—160°F
   - Final rinse—180°F

Emergency Food Service Operations—Healthcare food service and dietary departments must be responsive to the organizational disaster and emergency planning requirements. Emergency food-service planning should stress the delivery of safe food and water. Plans should cover the following:

1. Emergency supplies may need to be stored or made available from an off-campus site. Some of the ready-to-eat foods that should be available may include:
   • Boxed cereals, crackers, and cookies
   • Canned fruits, beans, and meats
   • Evaporated milk, bottled water, and canned juices
2. Emergency food and water supplies should be properly stored, inspected, and rotated on a regular basis.
3. Guidelines or procedures should be established to restore full meal capability as soon as possible.
4. The basic needs of the organization must be addressed in case there is no refrigeration, cooking capability, or adequate safe water supply.
5. Planning documents should guide food service personnel in the preparation and delivery of meals that do not require cooking or extensive preparation.
6. Emergency feeding plans should be approved by the organization's disaster committee and local health authorities.

Food Service Manager's Safety Responsibilities—

• Coordinate safety-related issues with the hazard control manager, safety director, and organizational safety committee.
• Maintain a written hazard communication plan and a current MSDS file.
• Ensure that all workers receive a safety orientation when assigned.
• Train workers in the organization's hazard communication procedures when initially assigned.
• Conduct HAZCOM retraining when a new hazardous material is introduced into the workplace or a worker changes job position.
• Ensure that all chemicals and hazardous materials are properly stored, used, and labeled.
• Conduct regular safety meetings and give employees an opportunity to attend in-service training and education sessions.
• Ensure that periodic safety self-inspections are conducted and all hazards are immediately corrected.

D. FACILITY SECURITY

The Joint Commission's security standards require facilities to have a security management plan. Hospitals must also accomplish a risk assessment of the local area and the campus. Facilities must also identify concerns and evaluate readiness in protecting workers, patients, and visitors from crime. Areas of concern include emergency rooms, mental institutions, and hospital psychiatric wards. Patient care staff and hospital security personnel should be trained to handle aggressive behavior of patients. Assaults by visitors are also a concern, especially in emergency rooms that handle violent gang-related incidents. Personal and property crimes are frequent because of the around-the-clock operation of healthcare facilities.

General Security Recommendations—

- Establish written security policies for patient restraint, weapons detection, prisoner restraint, and emergency response procedures.
- Maintain a working relationship with local law enforcement agencies.
- Establish a security surveillance system that monitors all areas and departments on a 24-hour basis.
- Provide security escort or shuttle service to parking areas, garages, and public transportation points.
- Ensure that lighting and security measures are adequate on all sidewalks, parking areas, and bus stops.
- Install emergency phones in parking lots, underground tunnels, and elevators.
- Limit access through doors by installing outside locking mechanisms, but do not hinder emergency egress.
- Install closed-circuit television system to help monitor high-risk areas.
- Provide separate emergency room facilities or take additional security measures with mentally disturbed patients.
- Provide physical barriers between reception personnel and patients/visitors.
- Install a buzzer at emergency room entrance.

Security Management Plan Requirements—The Joint Commission requires the documented management plan to:

- Establish security management program procedures
- Ensure security of patients, visitors, personnel, and property
- Establish identification policies for staff, visitors, and patients
- Control access where necessary to sensitive areas
- Manage traffic in emergency service areas
- Identify persons responsible for managing the security program
- Report and investigate all security-related incidents
- Conduct an annual evaluation of the effectiveness of the security program

Emergency Room Security—Emergency care departments are prime targets for violence. People in waiting and reception areas may sometimes be at risk, and many facilities have full-time security personnel located in the emergency room. Access to treatment areas must be controlled and sealed off so that emergency care procedures can be accomplished without fear or intimidation. Healthcare facilities should have a written policy to deal with firearms. Signs prohibiting firearms should be posted. Security officers should be trained in how to impound and store weapons. Medical staff members should be trained in the proper use of restraints. Emergency room policies must address:

- Diagnosis of dementia
- History of drug or alcohol abuse
- Inflexible treatment or milieu routines
- Risk factors considered in patient care

Security in Newborn Nurseries—Infant abduction is a great concern for facilities with obstetrical services. Most abductors are females who have no prior criminal record.

Nursery staff members should be alert for repeat visits. Information about floor layouts should be restricted. Nursery staff should never answer questions about nursery procedures or feeding times. Unusual behaviors should be reported to security personnel immediately.

Pharmacy Security—Controlling drugs is the primary security issue for the pharmacy. The pharmacy should have well-established procedures to ensure that all drugs are properly stored in locked containers. Security must also be maintained during delivery of medications to patients and to other departments.

Joint Commission Education Requirements—Security education programs must stress:

- Policies to minimize risks in security-sensitive areas
- Emergency procedures to follow during a security incident
- Organizational policies to follow in reporting security incidents that involve patients, visitors, personnel, or property

Security Incident Emergency Guidelines—Emergency procedures for security should address specific procedures to be followed:

- In the event of an incident or security failure
- During a civil disturbance
- When handling a situation involving VIPs or the media
- When additional staff is needed to control human/vehicle traffic

Parking Security—Healthcare facilities should provide managed parking areas if possible. At many hospitals, day workers park in remote areas and second-shift workers park in spaces closer to the facility. Lighting is one of the most important factors in parking area safety. Some hospitals have developed and implemented minimum lighting requirements for parking lots. All "dark spots" should be identified and additional lighting installed. Good lighting also reduces slip, trip, and fall risks. Healthcare facilities must also provide training to employees in parking area safety. The training should provide employees with some common-sense reminders such as:

- Look around and under the car when returning to it.
- Have keys and self-defense spray in hand.
- Check the back seat before unlocking the car.
- If working late, be sure to let someone know.
- Request an escort when working in a high-crime area after dark.
- Consider attending a self-defense training class.

Patient Security Guidelines—

- A plan should be developed to assure protection of patients, visitors, employees, hospital property, equipment, and patient valuables.
- Security personnel should be briefed when patients who pose a security risk are treated or admitted to the hospital.

- Security personnel should instruct the public relations office not to release information regarding at-risk patients.
- The business office should be instructed not to enter the name of an at-risk patient into the billing computer, and the medical records office should take precautions to protect privacy.
- All hospital departments should use patient care numbers for all transactions, reports, and tests.
- A program should be developed to deter and preclude individual or group acts ranging from personal theft to solicitation, strikes, or demonstrations.
- Security procedures should be established to clear lawyers, investigators, and law enforcement officers before granting access to a patient's room or family.
- Social and welfare workers should have proper clearance before being granted access to a patient.
- Written procedures should be developed to address patient security.
- Some patients are at risk due to gang involvement and some may be targets of domestic violence.
- When discharged, an at-risk patient's information must continue to be protected and an INFORMATION RESTRICTED label should be placed on his or her medical record.

Security Program Evaluation—Security programs vary depending on the size, location, and type of facility. Management should consider the following when evaluating security:

- Employee identification system
- Exit/entry control and internal traffic control
- Written plan for managing bomb threat and civil disturbances
- Coordinating plans and in-service training
- Management of prisoner-patients
- Illumination control of external building and grounds
- Surveillance system management by observation and alarm
- Control of packages and patient valuables
- Key control

Workplace Violence—OSHA recently published non-mandatory workplace violence control guidelines aimed primarily at healthcare facilities. Citations would still be issued under the General Duty Clause, Section 5(a)(1) of the OSH Act.

- *OSHA Workplace Violence Guidelines*—OSHA recently published its new "Guidelines for Preventing Workplace Violence for Health Care and Social Service Workers." The guidelines have been described as non-mandatory and merely "advisory in nature." The guidelines identify five core elements of a workplace violence prevention program: written procedures, worksite analysis, prevention and control measures, training and education, and recordkeeping/program evaluation.

   1. **Written Program**—The written program should:
      - Be communicated to all workers and staff members

- Establish a clear policy of zero tolerance for workplace violence (communicated to patients and visitors)
- Ensure no reprisals are taken against employees who report violence
- Encourage workers to report incidents immediately
- Outline a plan for maintaining a secure work environment
- Assign program responsibility to a team that has been properly trained
- Affirm management's commitment to employee safety
- Establish an organizational briefing to address safety issues

2. **Worksite Analysis**—Employers are advised to establish a threat assessment team to assess organizational vulnerability to workplace violence. The team should include personnel from senior management, operations, employee assistance, security, safety, legal, and human resources. Worksite analysis should include the following:
   - *Records Analysis/Tracking*—The threat assessment team should review medical, workers' compensation, and other records to pinpoint incidents of violence. The data should be analyzed and a baseline established to measure program effectiveness.
   - *Employee Screening Survey*—The assessment team should survey employees at least annually to get their opinions about the program and the potential for violence.
   - *Workplace Security Analysis*—The assessment team should periodically inspect the workplace to identify potentially hazardous locations/situations. The effectiveness of current security measures should be evaluated.

3. **Hazard Prevention/Control**—Organizations should develop and implement procedures to ensure security hazards are controlled.
   - Engineering controls to be considered include the use of alarm systems, cellular phones, metal detectors, closed-circuit video, enclosed nursing stations, and dual exits from patient care areas.
   - Administrative and work practice controls include requiring sign-in procedures for visitors, issuing visitor passes, ensuring adequate staffing in all patient care areas, and determining behavioral history of patients.
   - Post-incident response procedures for all victims of violence should include prompt medical treatment and psychological evaluation.

4. **Training and Education**—The guidelines identify appropriate training for all employees and address specialized training requirements for supervisors and security personnel. Training should:
   - Cover potential risks for assault and focus on prevention activities, including role playing, simulations, and drills
   - Address causes and techniques to recognize escalating behavior and warning signs, including such topics as risk factors that contribute to assaults and ways to diffuse volatile situations
   - Provide information on safe methods of restraint and escape
   - Cover ways workers can protect themselves and others from violence
   - Be conducted annually at all institutions and more frequently at large facilities

- Be given to newly hired and reassigned employees before they begin their job duties

5. **Recordkeeping**—The guidelines identify several categories of records that employers should maintain to help ensure the success of the program:
   - OSHA logs
   - Medical reports
   - Reports of aggressive behavior that does not result in injury
   - Information on patients with a past history of violence

6. **Program Evaluation**—Evaluation procedures should include establishing a uniform reporting system that requires regular review of reports and safety/security committee meetings. Facilities should also analyze trends, survey employees, conduct literature reviews, and evaluate worker experience with violent situations.

7. **Management Commitment**—The guidelines suggest in vague terms that an effective program must include "management commitment and worker involvement." Elements of management commitment include:
   - Equal commitment to worker safety and patient welfare
   - Appropriate allocation of authority and resources
   - Assigned responsibility for the various aspects of the program
   - A comprehensive treatment, counseling, and debriefing program for employees who experience or witness acts of violence

8. **Implementation Considerations**—Healthcare organizations should consider the implications before fully implementing all of the OSHA recommendations.
   - Some critics of the guidelines have expressed grave concerns about the staff survey included in the guidelines. The survey appears to have several questions that bear no obvious relationship to workplace violence issues. Some questions raise right of privacy issues.
   - Some questions could lead to future negligent supervision or retention claims against the employer if no action is taken on information received in the surveys.
   - Employers with unionized staff should take special care in implementing the guidelines to ensure they stay within the strictures of the National Labor Relations Act.

- *Prevention Tips*—Healthcare organizations can prevent workplace violence by training supervisors and management personnel how to:
  1. Recognize and deal with threats and violent behavior
  2. Obtain and use conflict resolution resources
  3. Train and educate workers about improving relationships
  4. Hire and fire employees in ways to reduce violent behavior
  5. Respond to emergency situations that involve potential violent behavior

- *Stages of Violent Behaviors*—
  1. **Primary Stage**—Refusal to cooperate with others, including supervisors, co-workers, and even customers.

2. **Intermediate Stage**—Arguments with others, refusal to follow company rules, damage to or theft of property, and threatening comments.
3. **Advanced Stage**—Frequent display of intense anger which can result in threats and physical contact. Guns or other weapons may be shown and even used.

- *Handling Tense Situations*—
  1. Keep a distance from workers who show signs of violent behavior.
  2. Try to stay calm and keep the lines of communication open.
  3. Use non-verbal communication to confront the worker; if speaking is required, do so in a firm but quiet tone.
  4. Never give an indication of fear or attempt to tell the worker the consequences of his or her action.
  5. Do not become emotional or make a move that suggests physical contact.

E. RADIATION SAFETY

The Atomic Energy Act of 1954, as amended, requires all persons using or procuring radioisotopes to be authorized to do so by an appropriate license. The only exception to this requirement involves quantities of 50 microcuries or less, depending on the nature of the material. Certain devices that contain radioisotopes, such as luminescent signs, are also exempt. The states listed below have entered into an agreement with the federal government to regulate radioisotopes under Section 274b of the Atomic Energy Act as amended in 73 Stat. 689.

Alabama	Kansas	North Carolina
Arizona	Kentucky	North Dakota
Arkansas	Louisiana	Oregon
California	Maryland	South Carolina
Colorado	Mississippi	Tennessee
Florida	Nebraska	Texas
Georgia	New Hampshire	Washington
Idaho	New York	

Nuclear Regulatory Commission (NRC)—The NRC was established under the Energy Reorganization Act of 1974. The NRC ensures that civilian uses of nuclear substances meet the requirements of safety, environmental, and national security laws. The NRC adopts and enforces standards for the department of nuclear medicine in healthcare facilities. Some states have agreements with the government to assume these regulatory responsibilities.

1. *NRC Licenses*—The NRC issues a five-year license to qualified healthcare organizations that follow prescribed safety precautions and standards.
   - Information that must be provided to the NRC includes identifying authorized users, designating a radiation safety officer, and identify the address/location where radioactive materials are used.
   - Any changes requires the organization to file for a licensure amendment.

- There are some situations for which healthcare facilities will need to obtain an NRC or state license.
- Diagnostic kits are subject to licensing under 10 CFR 31.11.
- Some vendors of radioactive materials are required by their license to verify that the facilities they supply also have a license before shipment of materials is made.

2. *Special Licenses*—Some radionuclide licenses apply to generators of infectious and medical wastes:
   - A general license is issued to physicians, clinical laboratories, and healthcare facilities under the requirements of 10 CFR 31.11.
   - A specific use license is required for physicians in private practice and healthcare institutions. Medical use pertains to human administration of radioactive substances or radiation.
   - A broad-scope license can be issued to facilities that provide patient care and conduct research using radioactive materials.

3. *Written Radiation Control Programs*—Each NRC licensee must have a written management program that covers the following:
   - Required participation in the program by all users, organizational administrators, and the radiation safety officer
   - Procedures to inform all workers about the types and amounts of materials used, including dosing information, safety precautions, recurring training, and continuing education
   - Information and guidelines to be used to keep doses at as low as reasonably achievable (ALARA) levels
   - Radiation safety officers must meet the training and education requirements of 35.900 of the NRC rules

4. *Radiation Safety Officer Responsibilities*—
   - Develop and implement written policies and procedures to cover the purchase, storage, use, and disposal of radioactive by-products.
   - Ensure all personnel are properly educated and trained.
   - Develop procedures to investigate all incidents, accidents, or other deviations from prescribed procedures.
   - Work with the radiation safety committee in overseeing the facility's radiation safety program.
   - Maintain documentation as required by the program and other applicable regulations.

5. *Radiation Safety Committee*—Each organization must establish a radiation safety committee. The committee must meet at least quarterly. Responsibilities include:
   - Approving/disapproving authorized users
   - Selecting or appointing the radiation safety officer
   - Approving minor changes in operating policies
   - Reviewing radiation dose records of all exposed personnel
   - Reviewing all radiation incident reports
   - Conducting an annual review of the entire radiation safety program

6. *Other NRC Requirements—*
   - Low-level radioactive waste is regulated by the NRC under the Low Level Radioactive Waste Policy Act of 1980. The NRC can delegate regulation under state-approved plans.
   - The NRC regulates roentgenogram sources under Title 21 CFR Parts 1000–1050. The NRC also regulates all radioactive isotope sources except radium under Title 10 CFR Parts 20 and 34.
   - The NRC does not regulate naturally occurring radioactive materials such as radium and radon. Disposal of all radioactive waste is regulated by the NRC, including low-level radioactive wastes, which are regulated under rules in 10 CFR Parts 19 and 20.
   - General rules for radioactive waste management can be found in 20 CFR Parts 301– 311, and personnel monitoring guidelines can be found in 10 CFR Part 20.202.

7. *NRC Performance-Based Standards—*Some parts of the NRC regulations outline detailed procedures and other parts state limits and leave the procedural details to the licensee. In the license application, the institution proposes its intended procedures and disposal methods to meet the performance standards.
   - Approval of the license is contingent on the NRC's approval of the proposal submitted with the application.
   - License conditions are often negotiable.
   - The NRC publishes a series of regulatory guides and model procedures for certain performance-based standards, including a model procedure for waste management.
   - The ultimate test of all radiation safety programs is the NRC philosophy that licensees should go beyond the regulatory standards to "make every reasonable effort to maintain radiation exposures, and releases of radioactive materials in effluent to unrestricted areas, as low as is reasonably achievable."
   - Occurrences of therapeutic or diagnostic misadministration must be reported. Therapy errors must be reported to the NRC regional office within 24 hours. The physician must also be informed. Diagnostic errors must be investigated and documented by the radiation safety officer. Some cases may require that a report be provided to the NRC within 15 days.

8. *Worker Safety—*NRC rules outline minimum safety requirements for workers and patients. Safety instructions and precautions are specified for three categories of therapeutic intervention:
   - Radiopharmaceutical
   - Sealed sources/implants
   - Teletherapy

9. *Worker Training—*Basic training requirements for workers should stress:
   - Visitor control procedures
   - Contamination controls
   - Waste management requirements
   - Emergency reporting and notification policies

10. *Radiation Terms—*
   - **Agreement state**—Any state with which the NRC has entered into an effective agreement under Section 274b of the Atomic Energy Act of 1954, as amended (73 Stat. 689).
   - **Airborne radioactive material**—Any radioactive material dispersed in the air in the form of a dust, fume, mist, aerosol, vapor, or gas.
   - **Alpha particle**—A positively charged particle emitted by certain radioactive materials. It is identical to the nucleus of the helium atom and consists of two neutrons and two protons bound together. It is the least penetrating type of radiation and may be stopped by a sheet of paper. Because of its low penetrating power, external exposure to alpha particles is not considered particularly dangerous. However, damage to internal body tissue by this particle is rather high and is considered quite harmful if the emitting substance enters the body.
   - **Atomic number (Z)**—The number of protons (positively charged particles) in the nucleus of an atom. Each chemical element has a characteristic atomic number. All isotopes of a given element have the same atomic number.
   - **Atomic weight**—The mass of an atom relative to other atoms based on the mass of the carbon-12 atom. The atomic weight of an atom is approximately equal to the total number of protons and neutrons in its nucleus.
   - **Background radiation**—The radiation in man's natural environment, including cosmic rays and radiation from naturally occurring radioactive elements.
   - **Beta particle**—A particle emitted from a nucleus during radioactive decay. It can be stopped by a sheet of metal or acrylic plastic, depending on the emitted energy level of a particular isotope. If it does reach the skin, beta radiation can cause burns, and beta emitters can be harmful to inside tissue if they enter the body.
   - **Contamination**—Deposition of radioactive material in any place where its presence may be harmful.
   - **Curie**—The basic unit used to describe the amount of radioactivity in a sample of material. It is the quantity of a radioactive substance which undergoes 3.7×10^{10} (37 billion) disintegrations per second (dps), which is approximately the rate of decay of one gram of radium. A commonly used submultiple of the curie is the microcurie (μCi), which is equal to 3.7×10^4 dps or 2.2×10^6 disintegrations per minute (dpm).
   - **Decay**—Disintegration of the nucleus of an unstable nuclide by spontaneous emission of charged particles or photons.
   - **Decontamination**—Removal of radioactive contaminants from a surface by cleaning and washing.
   - **Dose**—Quantity of radiation absorbed per unit of mass by the body or by any portion of the body. Units of dose measurement are the rad, the roentgen, and the rem.
   - **Film badge**—A package of photographic film worn like a badge by persons working with or around radioactive material to measure exposure to ionizing radiation; the absorbed dose can be calculated from the degree of film darkening caused by the irradiation.

- **Gamma radiation**—High-energy photon, short-wavelength, ionizing, electromagnetic radiation emitted by nuclei of radioactive atoms during radioactive decay.
- **Half-life**—The time required for half of the nucleus of one atom of a particular radioactive substance to disintegrate to another nuclear form. Half-lives may range from a few minutes to thousands of years.
- **Isotope**—One of two or more atoms with the same atomic number (the same chemical element) but with different atomic mass. The atomic mass determines the isotope and varies with the number of neutrons.
- **Leak test**—A test performed to detect leakage of a radiation source.
- **Monitor**—An instrument that measures the level of ionizing radiation in an area.
- **Rad**—A unit of absorbed dose of ionizing radiation equal to an energy of 100 ergs per gram of irradiated material.
- **Radioactivity**—The spontaneous decay or disintegration of an unstable atomic nucleus, usually accompanied by the emission of ionizing radiation.
- **Radioisotope**—A radioactive isotope of an element.
- **Rem (roentgen equivalent man)**—The unit of dose of any ionizing radiation that produces the same biological effect on human tissue as one roentgen of X-rays.
- **Roentgen**—The international unit of gamma or X-radiation; the amount of radiation that produces one electrostatic unit of charge of either sign in one cubic centimeter of dry air at 0°C and 760 mm Hg.
- **Thermoluminescent dosimeter (TLD) badge**—A badge that contains a thermoluminescent "chip" which is worn by persons working with or around radioactive materials. When the chip is heated in a special instrument, the light emitted is directly proportional to the quantity of ionizing radiation received by the badge and thus the wearer.
- **X-Radiation**—Short-wavelength, ionizing, electromagnetic radiation that normally results from the interaction of energetic electrons with a metal target, as in an X-ray tube. X-rays are similar to gamma rays, but the two are differentiated on the basis of origin.

Radiation Sources—Natural radiation results in the average person receiving about 125 mrem per year. NIOSH estimates that medical and dental irradiation adds another 60 mrem. Radiation exposure usually results from the scatter of the x-ray beam caused by reflection or deflection from the main beam. Another source of exposure for healthcare personnel is gamma rays emitted by patients undergoing nuclear medicine treatments. Patients with therapeutic implants emit both gamma and beta radiation.

- *Ionizing Sources of Radiation*—Ionizing radiation is used in healthcare diagnostic radiology, including X-rays, fluoroscopy, angiography, dental radiography, and computerized axial tomography (CAT scan).

- *Other Radiation Hazard Locations*—Radiation hazards also exist in areas where therapeutic radiology is used, in radiopharmaceutical laboratories, and in areas where radioactive materials are stored or discarded.

- *Portable X-Ray Equipment*—Emergency room and intensive care units often use portable X-ray equipment, which increases the chances for inadvertent exposure and inadequate monitoring.
    1. The amount of external radiation received depends on the amount present, duration of exposure, distance from the exposure source, and the effectiveness of protective barriers.
    2. According to NIOSH, radiation workers in hospitals receive an annual average dose that ranges from 260 to 540 mrem; dental worker exposure averages about 40 mrem.

Occupational Exposure to Radiation—Workplace exposure is usually localized, and acute radiation syndrome is rare. Exposures even at low levels could result in biological damage. Other variables such as age, gender, genetic makeup, tobacco use, overall health, and endocrine system status can modify the effects of ionizing radiation exposure.

1. *Protection*—The amount of protection needed for a particular gamma or X-ray radiation hazard depends on the energy present and the length of time in use. Methods for reducing exposures include:
    - Limiting the length of exposure
    - Increasing distance from source
    - Shielding the exposed individual with protective covering
    - Avoiding unnecessary exposures

2. *Reducing Radiation Exposure*—The radiation control or protection officer should direct an integrated program that stresses radiation safety throughout the facility. The radiation control officer must monitor patients and staff members to ensure radiation exposure limits are never exceeded. An effective control program will also monitor the flow of radioactive materials to, through, and from the facility. An effective radiation safety training program should emphasize safe handling and disposal methods. The education program should also cover information about equipment maintenance, personal monitoring guidelines, and documentation procedures. Exposure goals should be 0.05 rem annually.

3. *General Radiation Control Measures*—
    - Properly identify any areas that contain a radioactive source.
    - Restrict entry to authorized individuals.
    - Enclose all areas where radioactive materials are located.
    - Maintain control boundaries around all sources.
    - Locate X-ray machine controls to prevent unintentional activation.
    - Check X-ray machine before use.
    - Ensure that secondary radiation cones and filters are in place.
    - Keep X-ray doors closed when equipment is in use.
    - Permit only trained staff and patients in radiation areas.
    - Clearly identify patients with radioactive implants.
    - Ensure workers wear protective shielding clothing when necessary.
    - Check equipment annually for cracks in the lead.

TABLE 10.1   Exposure Standards for Ionizing Radiation

Type of Standard	Federal Radiation Council	National Council on Radiation Protection (NCRP #39)	NRC (10 CFR 20)	OSHA (20 CFR 1910.96)
Radiation worker[a]				
Whole body	5 rem/year, 3 rem/quarter, not to exceed the cumulative lifetime limit	5 rem/year, 3 rem/quarter, not to exceed the cumulative lifetime limit	5 rem/year, 3 rem/quarter, not to exceed the cumulative lifetime limit	3 rem/quarter
Cumulative lifetime limit	5(N–18) rem[b]	5(N–18) rem	5(N–18) rem	5(N–18) rem
General population				
Individual, whole body	0.5 rem/year	0.5 rem/year	0.5 rem/year	—

[a] Workers in the radiation department of other job categories potentially exposed to ionizing radiation.
[b] N–18 = age of worker minus 18 years.

Adapted from *Documentation of the Threshold Limit Values*. American Conference of Governmental Industrial Hygienists, Cincinnati, OH, 1986.

Ionizing Radiation—Ionizing radiation is produced naturally by the decay of radioactive materials or by the operation of X-ray devices. A radioactive element spontaneously changes to a lower energy state and in the process emits particles and gamma rays from the nucleus. The particles are referred to as alpha or beta particles. X-rays are produced when highly energized electrons strike the nuclei of the target material. The electrons are deflected from their path and then release energy in the form of electromagnetic radiation or X-rays. Table 10.1 provides information about exposure standards.

- *Alpha Particles*—Particles with energies of 4 to 8 million electron volts. Alpha particles will not penetrate the skin and are not considered an external hazard. If ingested or inhaled, they can cause serious health problems, including cancer. Radium implants (222 and 226) are examples of alpha particle emitters.
- *Beta Particles*—Interact much less readily with matter than alpha particles. They can travel a few centimeters into tissue. External and internal exposure is potentially hazardous. Examples are the isotopes carbon-14, iodine-131, cobalt-60, selenium-75, and chromium-51.
- *Protons*—Produced by high-energy accelerators and can produce tissue ionization. Protons have a path length longer than alpha particles.
- *X-Rays*—Have longer wavelengths, lower frequencies, and therefore lower energies than gamma rays. X-rays can be encountered during the use of microscopes and electronic tubes. Gamma emitters are cobalt-60, cesium-137, iridium-192, and radium-226.

Radiation Measurement—Measurements and definitions are given in SI units:

- *Curie (Ci)*—Unit of measure of the radioactivity of a substance. One curie is equivalent to 3.7×10^{10} disintegrations per second.
- *Absorbed Dose*—The amount of radiation that is absorbed by the body.
- *Exposure*—The amount of radiation to which the body is exposed.
- *Radioactive Half-Life*—The time required for the radioactivity of an isotope to decrease by 50%.
- *Roentgen*—Unit of measure for the quantity of radiation produced by gamma or X-rays. One roentgen (R) is equivalent to 2.58×10^{-4} coulomb per kilogram.
- *Rad (radiation absorbed dose)*—A measure of the absorbed dose of ionizing radiation. One rad = 100 erg per gram = 0.01 gray (Gy).
- *Rem (roentgen equivalent man)*—The dosage of any ionizing radiation that will cause biological injury to human tissue equal to the injury caused by one roentgen of X-ray or gamma-ray dosage. One rem is equivalent to 0.01 sievert (Sv).

Note—The different types of ionizing radiation vary in the number of ions produced while transversing matter.

OSHA Ionizing Radiation Standard (29 CFR 1910.96)—This standard was developed to protect workers not covered by the NRC guidelines published in 10 CFR 20. OSHA requires that each radiation area must be posted with a sign or signs bearing the caution symbol found in 29 CFR 1910.96. Symbols must use conventional radiation caution colors or magenta or purple on a yellow background.

1. *Airborne Radiation Warning Signs*—High radiation and airborne radioactivity areas must be posted with signs that bear the caution symbol and the following wording: CAUTION—HIGH RADIATION AREA or CAUTION—AIRBORNE RADIOACTIVITY AREA.

2. *High-Radiation Area Signs*—A sign or signs with the words CAUTION—RADIOACTIVE MATERIALS must be posted:
   - In each area or room where radioactive materials are used or stored in amounts exceeding ten times the amount specified in 10 CFR 20, Appendix C
   - On each container that has any radioactive material greater than the quantity specified in 10 CFR 20, Appendix C

3. *Recordkeeping*—Facilities should maintain the following records:
   - Personal radiation exposures
   - Radioisotope inventory
   - Receipt and disposition log
   - Radiation survey record and reports
   - NRC documentation required in 10 CFR 20.401

4. *Protective Equipment*—No part of the body should be directly exposed. Lead aprons, gloves, and goggles should be worn as necessary. A thyroid shield and lead glasses are recommended for operations with consistently elevated exposure, such as angioplasty procedures.

National Council on Radiation Protection (NCRP)—This group was created by Congress to collect, analyze, develop, and disseminate information and recommendations on radiation quantities, measurements, and units.

1. **NCRP Guidelines**—The NCRP publishes maximum permissible levels of external and internal radiation. The major sources of information are:
   - Handbook 22: Maximum Permissible Body Burdens and Maximum Permissible Concentrations of Radio-Nuclides in Air and Water for Occupational Exposure
   - Handbook 43: Review of Current State of Radiation Protection Philosophy

2. **Annual Permissible Dose**—The annual permissible whole body dose is 5 rem per year, with 3 rem permitted within a 13-week period. NCRP goals are primarily to prevent and reduce cataracts, erythema, and probability of cancer.

Food and Drug Administration (FDA)—The FDA, under the Federal Food, Drug, and Cosmetic Act, has the authority to regulate the manufacture and distribution of radiopharmaceuticals and medical devices that contain radioactive materials. The FDA also sets performance standards for X-ray and other radiation-emitting equipment that was manufactured after 1974. The FDA also issues recommendations for the use of X-ray machines and other radiation-emitting devices. Title 10 CFR Parts 20 and 34 contain NRC rules on isotope sources. FDA X-ray regulations are found in Title 21 CFR Parts 1000 and 1050.

Non-Ionizing Radiation—Electromagnetic radiation has different effects on humans depending on the wavelength and type of radiation involved. Low-frequency radiation such as that generated by broadcast radio and shortwave radio has generally been considered as not dangerous. Some new information suggests that exposure to electric power frequencies could have an adverse impact on human health. Other types of non-ionizing radiation include:

1. **Infrared (IR) Radiation**—All objects with temperatures above absolute zero emit IR radiation. Exposure can occur during the use of heating and warming equipment in food preparation areas. Exposure can also occur during procedures that involve lasers and thermography. The main hazards of IR radiation are skin burns, increased pigmentation, and enhanced dilation of capillary beds. Prolonged exposure can result in eye damage. Eye protection with correct filters should be worn while working in exposure areas.

2. **Microwave/Radiofrequency (RF) Radiation**—There are many exposure areas in healthcare facilities, the most obvious being microwave ovens used for heating food. Other sources include cancer therapy procedures, thawing organs for transplantation, sterilizing ampules, and enzyme activation in research animals. The greatest hazard associated with exposure to microwave radiation is thermal heating.

3. **Standards and Recommendations**—
   - The OSHA exposure standard is 10 mW/cm^2.
   - The American National Standards Institute (ANSI) and American Conference of Governmental Industrial Hygienists (ACGIH) have also established exposure guidelines.

- The FDA has set a limit of 5 mW/cm^2 for microwave ovens during normal use (21 CFR 1030.10).

4. **Exposure Control—**
   - Monitor diathermy equipment for leakage before each treatment.
   - NIOSH recommends that ovens be checked every three months.
   - Consider areas where levels exceed permissible levels as dangerous.
   - Areas should be identified and warning signs posted.
   - Access to dangerous areas should be controlled.

OSHA Non-Ionizing Radiation Standard (29 CFR 1910.97)—The warning symbol for radiofrequency hazards must consist of a red isosceles triangle above an inverted black isosceles triangle, separated and outlined by an aluminum-colored border. The upper triangle must have the following wording: WARNING—RADIOFREQUENCY RADIATION HAZARD.

Ultraviolet (UV) Radiation—UV radiation can be emitted from germicidal lamps, during some dermatology treatments, from nursery incubators, and even some hospital air filters. Overexposure can result in skin burns and serious eye damage. Long-term exposure can contribute to accelerated skin aging and increased risk of skin cancer. NIOSH recommendations for UV exposure range from 200 to 400 nanometers, depending on length of exposure. The ACGIH also published exposure recommendations in 1987.

Radiation Safety Committees—Some facilities have two committees with responsibilities for ensuring radiation safety is practiced.

1. **Medical Isotopes Committee—**
   - Ensures that all individuals who work with or in the vicinity of radioactive material have sufficient training and experience to enable them to perform their duties safely and in accordance with NRC regulations and the conditions of the license.
   - Ensures that all use of radioactive material is conducted in a safe manner and in accordance with NRC regulations and the conditions of the license.
   - Reports to the safety committee quarterly.
   - Is coordinated by a specialist in nuclear medicine, a person with special competency in radiation safety, and a representative of the institution's management.
   - Is encouraged to include expertise in diagnostic radiology, clinical pathology, therapeutic radiology, and radiopharmacy.

2. **Radiation Safety Committee—**
   - Reviews and grants permission for, or disapproves, the use of radioactive material within the institution from the standpoint of radiological health and safety of patients or working personnel and other factors that the committee may wish to establish for medical uses of by-product materials prior to submission of an application to the division of licensing action.

- Prescribes special conditions that will be required during a proposed use of radioactive material, such as requirements for bioassays, physical examinations of users, minimum level of training, and experience of users.
- Should address film badge return rates, exposures that exceed guidelines, information on safety education and training, quality improvement, and issues relating to radiation regulation.
- Receives review records and reports from the radiological safety officer or other individuals to whom responsibility for health safety practices in the institution is delegated.
- Recommends remedial action to correct safety infractions, formulates and reviews institutional training programs for the safe use of radioisotopes, and maintains written record of actions taken by the committee.
- Informs the division of any changes in committee membership. The committee is normally chaired by a specialist in nuclear medicine, a person with special competency in radiation safety, and a representative of the institution's management. The committee is encouraged to include expertise in diagnostic radiology, clinical pathology, therapeutic radiology, and radiopharmacy.

Joint Commission Standards—Healthcare facilities should have written policies and procedures that address safety issues in the radiology and/or nuclear medicine departments, including:

- Use of collimation in all X-rays performed
- Required gonadal shielding procedures
- Use of appropriate shielding for all workers
- Fail-safe determination of pregnancy before radiological examination
- Requirement that all personnel wear a film monitoring badge
- Check of techniques to be made before exposure

Radiation Control Procedures—

1. *Diagnostic Procedures*—Workers who handle patients receiving diagnostic radioactive materials should:
   - Be monitored by the radiation control officer to ensure that all radioactive materials including body wastes are handled properly
   - Use gloves during the collection and transfer of radioactive materials
   - Dispose of patient body wastes in the sanitary sewer
   - Dispose of other materials, such as syringes, as radiation waste according to established procedures
   - Exhaust radioactive gases expired by patients through a shield duct system

2. *Therapeutic Procedures*—Proper control depends on the class of radioactive procedures being used:
   - **Class A**—Procedures that require administration of radioactive materials by mouth
   - **Class B**—Procedures that inject radioactive materials into body cavities
   - **Class C**—Procedures that inject radioactive materials into tumors and such materials are left there permanently

- **Class D**—Procedures that deliver radiation at distances of up to a few centimeters (brachytherapy)

3. *Therapeutic Radiation Safety Procedures—*
   - Precaution tags should be attached to the bed, chart, and patient.
   - Workers should enter patient's room to perform normal duties only.
   - Patients should care for themselves as much as possible.
   - Visitors should be no closer than six feet from the patient.
   - Pregnant workers should not be assigned to routine care duties for radioactive patients.
   - Contaminated clothing and equipment should be disposed of in a proper manner.
   - Radiation control officer should establish limits on time that a worker can spend with a patient.

General Radiographic Procedures—

- Ensure that all walls and barriers are sufficiently protective.
- Ensure that equipment works properly and complies with applicable regulations and standards.
- Survey rooms and floors adjacent to designated radiation-restrictive areas.
- Permit only patients in unshielded areas when X-rays are generated.
- Technicians must remain in a booth or behind a shield barrier.
- Never allow a person to hold or restrain a patient undergoing diagnostic radiology unless absolutely necessary.
- Always provide protective gear.
- Operators using a portable X-ray machine should remain at least six feet from the patient.
- [a] Personnel involved in fluoroscopy and angiography should follow basic radiation protection measures.

Monitoring Devices—Monitoring devices include the film badge worn on the outer clothing to monitor for gamma, X-ray, and beta radiation. Thermoluminescence detectors have become more popular for gamma and beta radiation monitoring. The pocket dosimeter is a direct-reading portable device. A monitoring program should include analysis and recording of units. The program must also have procedures for informing workers of their documented measured exposure.

Radioactive Waste Management—The NCRP issues recommendations for dealing with radioactive wastes. Wastes can be solids, liquids, or gases. Solid wastes can include rags and papers from cleanup operations, solid chemicals, contaminated equipment, experimental animal carcasses, or human or experimental animal fecal matter.

1. *Disposal*—The major ways to manage waste products include dilution, containment, incineration, and return to supplier. The properties of the material must be considered in the method of disposal. Specific disposal methods vary according to the material involved and the licensing authority of the user. Half-life and relative biological hazard should also be taken into consideration. Disposal depends on the half-life of the radionuclide. Material with a short half-life should be disposed of by a commercial contractor.

2. **General Safety Procedures—**
   - Equipment should be cleaned with large amounts of water and the water treated as radioactive liquid waste.
   - Radioactive urine and fecal matter are normally disposed of through the sewer system.
   - Toilets should be flushed several times after each use.
   - In cases where patients receive large doses of radioactive iodine, urine is generally collected for 48 hours after administration. It is analyzed in the laboratory and then flushed with large quantities of water down the sewer system.
   - Other liquid wastes should be handled in the same manner as solid wastes.
   - Substances with short half-lives can be stored in sealed containers until the radioactivity decays. Materials with long half-lives should be disposed of by a licensed contractor.
   - Gaseous radioactive wastes should be vented to the outside, away from makeup air sources to prevent re-entry into the facility.

3. **Radiation Waste Plan—**Healthcare facilities should develop a plan to ensure that radioactive wastes are disposed of in accordance with government guidelines and regulations. The plan should cover procedures for waste that contains radioactive materials as defined by the NRC as being hazardous to humans, animals, and the environment.
   - The emergency plan must be implemented in response to a radiation accident or incident.
   - Personnel monitoring must be conducted to ensure exposure is within acceptable limits.
   - Radiation measurement instruments must be properly calibrated before use.
   - Workers handling and exposed to radiation wastes must be properly trained.
   - Each department or service area that generates/handles radioactive wastes must develop written procedures that cover handling, transportation, and disposal activities.

4. **Identifying Radioactive Materials—**Location of hazardous radioactive materials in most facilities would include the laboratory, nuclear medicine, and radiology. Common radioactive materials found in facilities include:
   - Sodium pertechnetate Tc-99m, used in nuclear medicine procedures
   - Thallium-201, used for cardiac scans
   - Xenon-133 gas, used for lung ventilation study
   - Sodium iodine I-131, used for thyroid uptake studies and therapeutic treatments

5. **Storage and Handling Procedures—**
   - Properly secure and store all waste materials.
   - Designate controlled areas.
   - Dispense or draw materials only behind a protective barrier.
   - Label refrigerators that contain stored materials.
   - Monitor all incoming shipments prior to opening.

- Notify the radiation control officer if a contaminated shipment is received.
- Follow spill, leak, and contamination procedures.
- Handlers should wear rubber gloves and other protection.

6. ***Emergency Procedures***—Facilities should develop procedures for minor and major spills. Response to a major spill could include the following:
   - Require all involved persons to vacate the room or area.
   - Right a spilled container if it can be done safely.
   - Wash and flush contaminated skin areas immediately.
   - Discard contaminated clothing at once.
   - Turn off all fans and vacate the room.
   - Notify the radiation control officer as soon as possible.
   - Ensure that exposed personnel are decontaminated at once.
   - Decontaminate the area as soon as possible.
   - Do not permit work to resume until the area is declared safe.
   - Complete a radiation incident report.

7. ***Decontamination Procedures***—
   - Remove all external and surface contamination with soap and warm water. Wash for three minutes and focus on the areas between the fingers and around the fingernails. Repeat as necessary to remove all radioactive materials.
   - Wash all wounds, needlesticks, and lacerations at once under running water.
   - Call a physician to treat radiation injuries.
   - Treat exposed persons as if contaminated with particulate radioactive material.
   - Isolate exposed persons at least six feet from others.
   - Place removed clothing, used towels, and washcloths in bags and label.
   - A total body shower is recommended if possible under the circumstances.
   - Special medical procedures apply in situations where a person has inhaled or ingested radioactive material.
   - Report all radiation exposure incidents to the radiation control officer.
   - The radiation control officer upon arrival will determine if a person is contaminated.
   - A person with no radiation hazard will receive appropriate medical care with radiation precautions.
   - If a person is determined to have detectable radiation contamination, clothing removal and cleansing operations will begin.
   - Attending emergency personnel should wear lead aprons.
   - No person involved in the incident should be permitted to return to work unless approved by the radiation control officer.
   - Monitor exposed personnel for the development of late radiation illness.

F. LASER SAFETY

Laser (Light Amplification by Stimulated Emission of Radiation)—A laser emits electromagnetic radiation in the visible spectrum. Laser use is increasing at a very fast pace in the healthcare environment. New laser surgery techniques are being developed almost daily. Lasers are used in radiology departments to help align patients for treatment.

Laser Hazards—

1. The main hazard of lasers is the potential for eye damage. The nature of eye damage depends of the type, power, and duration of the laser exposure. There are three types of eye injuries:
   - The most common type of injury results when a light beam heats the retina and causes loss of vision in a point of a person's field of vision.
   - A beam from a pulse laser can cause an explosion in the retina and result in severe damage.
   - A laser with enough energy can cause retina cell death. Any damage is permanent but not as severe as thermal or acoustic damage.
2. Lasers striking the skin can result in erythema, blistering, and charring. The extent of the damage depends on wavelength, power, and length of exposure. Lasers also use high voltage and should be considered potential electrical hazards. The performance of lasers is regulated by the FDA's Bureau of Radiological Health under 21 CFR 1040.

FDA Laser Classes—

- *Class 1*—Lasers capable of producing damaging radiation
- *Class 2*—Lasers that may be viewed under strict controls
- *Class 3*—Lasers that require controls to prevent direct viewing and eye damage
- *Class 4*—Lasers that must be controlled to prevent eye and skin damage

FDA Laser Incident Reporting Requirements—The FDA Center for Devices and Radiological Health requires equipment producers to report all incidents of death or injury. There appears to be a lack of medical injury data related to laser incidents. Many experts feel that the greatest injury threat occurs during the alignment of the laser beam. Safety glasses are designed for unplanned viewing, and most injuries occur when workers fail to protect their eyes.

Laser Regulations—

- Laser injury incidents fall under the Safe Medical Device Act reporting requirements.
- OSHA uses ANSI Z-136.1, Guidelines for Laser Safety and Hazard Assessment when enforcing workplace laser hazards under its general duty clause. OSHA also relies on CFR 1910.132, the face and eye protection standard.
- ACGIH has published recommendations to reduce occupational exposure.
- ANSI Standard Z-136.3, Safe Use of Lasers in Healthcare Facilities provides guidance for laser usage in medicine.
- NIOSH recommends that a laser safety officer be appointed in facilities where laser use warrants extra precautions.
- The Laser Institute of America also publishes information on safely using lasers.

Laser Safety Guidelines—

- Primary worker protection measures include using effective eye protection and properly shielding high-energy beams.
- ANSI Z-136.1 and Z-136.3 require healthcare facilities to appoint a laser control officer.
- Only trained and authorized technicians should move, operate, or repair laser equipment.
- Lasers should be attached to an individual transformer and an emergency power source and should have a safety interlock.
- An approved fire extinguisher should be immediately available.
- Laser use areas should be identified and warning signs posted.
- Personnel should prevent laser beams from coming into contact with combustible, flammable, and reflective materials.
- Personnel using or exposed to lasers should be in an eye health medical surveillance program.
- Smoke generated during laser usage should be properly removed and filtered before being evacuated from the building.
- Laser equipment should be properly maintained and included in a preventive maintenance program.

Note—NFPA 99-1993, Annex 2 contains additional safety information for laser use in healthcare facilities.

Selecting Protective Eyewear—Managers selecting eyewear should consider the following to be important in the process:

- Eyewear must correspond to the wavelength of the laser.
- The power of the laser always determines the optical density (OD) or protection level.
- The duration of the laser exposure must also be considered.

G. LABORATORY SAFETY

Laboratory Safety—Laboratories are major ancillary departments in a healthcare facility. Depending on the size and type of facility, laboratories can have several functions:

- *Pathology*—Processes and tests tissue removed during surgical procedures
- *Cytology*—Processes specimens to determine abnormalities in cell structure
- *Chemistry*—Analyzes body fluids to determine such things as glucose, protein, enzyme, and hormone levels
- *Serology*—Analyzes body fluids for antigens and antibodies
- *Hematology*—Analyzes blood to determine information relating to red cells, white cells, and platelets
- *Microscopy*—Analyzes urine and body fluids
- *Microbiology*—Analyzes specimens to determine causes of infection

Chemical and Fire Hazards—Laboratory workers are routinely exposed to hazardous chemicals such as acetone, carbon monoxide, formaldehyde, hydrogen sulfide, mercury, nitric acid, and xylene. Many exposures occur annually in laboratories, resulting in chemical-related illnesses such as dermatitis, eye irritation, and even fatal pulmonary edema.

- OSHA's approach to controlling occupational exposures to hazardous chemicals has been through the development of substance-specific standards such as the standard for benzene.
- Where substance-specific standards do not exist, published permissible exposure limits (PELs) catalogued in the Z Tables of 29 CFR 1910.1000 should be used.
- The use of such chemicals is generally limited to small quantities on a short-term basis. However, operations where chemicals are used change frequently, exposing workers to many different chemicals.
- Refer to NFPA 99 for information on minimizing hazards of fire and explosions.

OSHA Laboratory Standard (29 CFR 1910.1450)—The standard emphasizes the use of safe work practices and appropriate worker protection as required by the laboratory environment. This performance-oriented standard allows employers the flexibility to implement specific safe work practices to help reduce worker injuries and illnesses.

1. *Overview of the OSHA Standard—*
   - The standard covers workers in laboratories located in industrial, clinical, and school environments.
   - The standard covers all chemicals that meet the definition of a health hazard as defined in the hazard communication standard (29 CFR 1910.1200).
   - The standard does not specify work practices necessary to protect employees from potential hazards associated with chemical use but does require that physical hazards be addressed in the employer's training program.
   - The requirement for training on physical hazards, when coupled with current safety regulations, will improve employee safety and protection in laboratories.
   - The standard requires continued compliance with all published PELs and with the employer's written chemical hygiene plan.
   - The standard requires that special consideration be given to substances that are particularly hazardous, including some selected carcinogens.
   - The standard requires that consideration also be given to reproductive toxins and substances that have a high degree of acute toxicity.

2. *Key Provisions of the OSHA Standard—*
   - Employee exposure monitoring (under certain conditions)
   - Standard operating procedures for laboratory work with chemicals
   - Employee training that supersedes but emphasizes the requirements of the hazard communication standard
   - Medical consultation and examination
   - Hazard identification information such as MSDS and labeling requirements.
   - Chemical fume hood performance certification

3. ***Chemical Hygiene Plan Requirements*—**
   - The written chemical hygiene plan provides the basis for meeting the requirements of the standard.
   - The plan also allows employers flexibility in providing the type of worker protection appropriate for a specific workplace.
   - The plan developed by the employer specifies the training and information requirement of the standard.
   - The plan establishes appropriate work practices, standard operating procedures, methods of control, and measures for appropriate maintenance and use of protective equipment.
   - The plan also details procedures regarding medical examinations and special precautions for work with particularly hazardous substances.
   - Some workplaces may meet certain criteria of the standard by relying on their existing safety and health plans.
   - The plan must be reviewed annually and updated as required.
   - The written program must be available to employees and their designated representatives and to the assistant secretary of labor for occupational safety and health.
   - Employers must appoint a chemical hygiene officer and, if appropriate, establish a chemical hygiene committee to participate in developing and implementing the plan.
   - The chemical hygiene officer may have a number of responsibilities, including duties such as monitoring safety, assisting with facility upgrade projects, and advising administrators on ways to improve chemical hygiene practices.

4. ***Employee Training*—**Employers must establish a training and information program for employees exposed to hazardous chemicals in the workplace.
   - The worker program should be initiated at the time of initial assignment and prior to assignments involving new exposure situations.
   - This provision incorporates the training and information requirements of the hazard communication standard.
   - Discussion and training topics must include the following:
     a. The location of the facility hygiene plan and requirements of the OSHA laboratory standard
     b. PELs for regulated substances and recommended exposure limits (RELs) where no regulatory standard applies
     c. Signs and symptoms associated with exposures to hazardous chemicals and substances
     d. Location and availability of HAZCOM plan, the chemical list, and MSDSs
     e. Policy documents on safe handling, storage, and disposal of hazardous chemicals in the workplace
     f. The components of the chemical hygiene plan and how it is implemented in the workplace
     g. The hazards of the chemicals used in work areas and the protective measures employees can take
     h. Specific procedures put into effect by the employer to increase worker protection, including engineering controls, work practices, and PPE

    i.   Detection methods and observation guidelines, including monitoring procedures, visual appearances, and/or odors that workers can use to detect the presence of hazardous chemicals

5. ***Medical Examinations and Consultations***—The laboratory standard does not mandate medical surveillance for all laboratory workers.
   - The employer must provide workers an opportunity for medical attention, including follow-up examinations and treatment recommended by an examining physician, if an employee:
     a. Exhibits signs or experiences symptoms associated with exposure to a hazardous chemical
     b. Is routinely exposed above the action level or PEL for a regulated substance
   - A medical consultation must be offered to any employee potentially exposed through a spill, leak, or explosion of a hazardous chemical.
   - The employer must provide information about the hazardous chemical, conditions under which the exposure occurred, and a description of symptoms experienced by the worker.
   - The employer must obtain from the treating physician any written opinion requiring follow-up examinations or medical tests.
   - The employer must obtain information on any medical condition that might pose an increased risk and a statement that the employee was informed of the results of the medical examination/consultation.

6. ***Control Methods and Personal Protective Equipment***—Employers must develop criteria for determining and implementing control measures that will reduce exposure to hazardous chemicals. These measures include engineering controls, work practice controls, and PPE.
   - **Engineering Controls**—Engineering controls include general ventilation, fume hoods, glove boxes, and other exhaust systems.
   - **Work Practice Controls**—Work practice controls may cover items such as restricting eating and drinking areas, prohibiting mouth pipetting, and performing work a manner that minimizes exposure and maximizes the effectiveness of engineering controls.

   Note—OSHA policy dictates that engineering and work practice controls be used to reduce employee exposure below the PEL.

7. ***Personal Protective Equipment***—
   - Respiratory protection is to be used only as an interim measure or when engineering or work practice controls are not feasible. Use of respiratory equipment must comply with the requirements in Title 29 CFR 1910.134.
   - Other PPE used in laboratories when appropriate includes safety glasses, whole body coverings, and gloves.

8. ***Safeguards for Hazardous Substances***—
   - Employers must include information on additional protective measures for work that involves carcinogens, reproductive toxins, and acutely toxic substances.

- Specific consideration must be given to incorporating the following provisions:
  a. Establishment of a designated area with appropriate signs warning of the hazards associated with the substance
  b. Information on safe and proper use of a fume hood or equivalent containment device
  c. Procedures for decontaminating the designated area
  d. Procedures for safe removal of contaminated waste

9. *Hazard Identification—*
   - Employers must ensure that hazardous chemical container labels are not removed or defaced.
   - MSDSs must accompany incoming shipments of chemicals and be made available to employees.
   - The employer is not required to prepare an MSDS except in cases where a chemical is produced in the laboratory for another user outside the laboratory.

10. *Recordkeeping—*
    - Employers must establish and maintain employee records of exposure monitoring results, medical examinations, and consultations.
    - Records must be kept, transferred, and made available in accordance with 29 CFR 1910.20.
    - Exposure records and related data analyses must be kept for 30 years.
    - Medical records must be kept for the duration of employment plus 30 years.
    - Medical records for employees who have worked for less than one year need not be retained after employment. *Note*—The employer must provide these records to the employee upon termination of employment.

11. *Appendices*—Appendices to the laboratory standard provide non-mandatory guidelines and recommendations. Appendix A is extracted from the 1981 National Research Council publication *Prudent Practices for Handling Hazardous Chemicals*. Appendix B contains references intended to assist in developing a chemical hygiene plan.

Laboratory Safety—

1. *Supervisor Responsibilities*—Department heads and supervisors within the facility should:
   - Ensure that employees know, understand, and follow the chemical hygiene plan and related standard operating procedures.
   - Ensure that the proper PPE is available and in good condition and that employees are trained in its proper use.
   - Perform quarterly chemical hygiene and housekeeping inspections.
   - Perform semi-annual chemical inventories of all laboratories and storage areas.
   - Determine PPE for the procedures and chemicals in use in their area.

2. *Laboratory Worker Responsibilities—*
   - Plan and conduct each laboratory operation in accordance with the chemical hygiene plan.

- Maintain all work areas in good order.
- Correctly select and use required PPE.
- Report exposures, injuries, or problems to the supervisor or chemical hygiene officer.

3. **General Safety Guidelines—**
   - Never stand on chairs, lab stools, boxes, or drums to reach high shelves or the ceiling area. Use stepladders or stepstools specially designed for such purposes.
   - Wash hands and arms several times during the course of the day to remove bits of irritating chemicals, animal dander, or biohazards.
   - Maintain adequate ventilation at all times. Check hood drafts regularly, and direct questions about the proper functioning of the hood to the maintenance department or the chemical hygiene officer.
   - Static electricity develops when liquids that are poor conductors are transferred from one container to another.
   - Electrical charges also develop when compressed gases are released rapidly from a cylinder. These charges can jump air gaps and form sparks which may ignite flammable vapors or gases.
   - Ensure cylinders are properly grounded by connecting the container and receiver by a ground wire.
   - Electrical charges may also build up on personnel wearing shoes with rubber or plastic soles.
   - Some areas may be designated regulated areas and the doors to these rooms will usually be marked AUTHORIZED PERSONNEL ONLY, RESTRICTED AREA, or NO ADMITTANCE.
   - Employees should never enter regulated areas or rooms without first consulting the responsible supervisor. The supervisor will then describe what personal protective clothing or safety equipment is required before entering the area. *Note*—In some cases, the room may be equipped with a red light which will be lighted when it is unsafe to enter the room.
   - Report sluggish drains the to maintenance department immediately.
   - Never pour down the drain materials which will attack the pipe material or react with each other. Do not dispose of carcinogenic materials by pouring down the drain.
   - Accompany materials poured down the drain with a large volume of cold water.
   - Never pour flammable materials down the drain.
   - Refer to NFPA 45, Fire Protection Standard for Laboratories Using Chemicals for information on construction, ventilation, and fire protection requirements.

4. **Biohazards—**Dangers are potentially present in the transport and handling of biological materials. Toxicity may arise through bites, scratches, allergic sensitization, inhalation, ingestion, and cutaneous absorption. Depending upon the materials or animals used in the laboratory, various safety precautions must be taken. Naturally, these precautions vary, and the specific details of laboratory procedures should therefore be explained by the laboratory supervisor.

5. *Laboratory Animals—*
   - Use care when handling animals to avoid being bitten or scratched. Use proper restraining or protective devices whenever possible.
   - Wear protective gloves when dissecting or conducting necropsy.
   - Use first-aid procedures to treat animal bites and scratches and report all incidents immediately.
   - Immediately report allergic reactions to animals or to the drugs used in treating animals.
   - When using animals to study the progress of disease, it is the responsibility of the supervisor to explain methods of protection to all workers.
   - [a] Euthanize infected animals and thoroughly wrap before disposal.
   - Thoroughly disinfect living area of infected animals.
   - Render all animal carcasses non-infectious by autoclaving or incineration.

6. *Holding and Disposing of Microorganisms—*
   - It is the responsibility of the laboratory supervisor to caution all workers about the hazards involved and to emphasize the importance of using sterile techniques.
   - Before disposal, autoclave or disinfect by other means cultures containing microorganisms or viruses.
   - Place pipettes used in handling infectious disease agents in a disinfectant solution immediately after use and autoclave as soon as possible.
   - Autoclave other contaminated glassware before washing.

Working Safely with Lab Instruments and Containers—

1. *Glass Tubing—*
   - The accepted method for preventing cuts from broken glass rods or tubing is to wrap the glass in cloth while inserting it into a stopper.
   - When putting glass tubing into stoppers or rubber tubing, lubricate the end of the tubing with water or glycerine.
   - Hold the glass tubing within an inch or two of the point of insertion and always use a towel to protect the hands.
   - Never hold a stopper against the palm of the hand.
   - The ends of the glass tubing should be fire polished.
   - Never attempt to push or pull glass tubing or a thermometer from a cork or rubber stopper that has hardened. Cut the cork or rubber away.

2. *Cork Borers—*
   - Borers can cause severe lacerations and should be lubricated with water or glycerine to reduce the danger of slipping.
   - Cork borers should be kept sharp and twisted when cutting.
   - Never force the borer through the stopper; always hold the cork between the thumb and forefinger and never in the palm of the hand.
   - The cork should be cut half way through from one side and then the remainder of the way from the other side; otherwise, it should be rested on a piece of cardboard or a wooden block while boring completely through from one side.

3. *Glass Apparatus—*
- Never use cracked, badly chipped, or otherwise dangerous glassware.
- Never return any broken apparatus to the stockroom.
- Place all broken, cracked, or chipped glassware in a rigid container marked GLASS to protect other workers from cuts.
- Never place broken fragments of glass in a sink or regular trash receptacle.
- Avoid flasks that have small cracks, "stars," or large bubbles.
- Never use flat-bottom flasks under vacuum unless especially designed for vacuum use.
- Glass under pressure may explode or implode. Use shields and towels to protect against flying glass.
- Wipe dry the exterior of large battery jars and beakers by using gloves to remove any liquid or other slippery material.

4. *Safety During Extractions—*
- When performing extractions or shaking volatile liquids, use separatory funnels and other closed equipment and ensure that pressure is released frequently.
- Too much pressure can cause the stopper to be blown out or the funnel to explode.

5. *Suction Flasks—*
- Suction flasks will collapse violently if cracked or otherwise weakened.
- Ensure that Gooch crucibles cannot slip through holders into flasks.
- Never tap flasks vigorously when full suction is applied.
- Never use thin-walled flat-bottom flasks under vacuum.

6. *Supporting Large Containers—*
- Flasks that contain large amounts of material should not be supported solely on iron rings or portable ring stands.
- Support the greater part of the weight on a tripod with clamps and ring stand to hold the apparatus in position.
- Use wire gauze between the ring and the flask.

7. *Vacuum Distillations—*
- Be sure that the vacuum has been released from all parts of the apparatus before disconnecting it.
- Sudden changes in pressure may cause breakage of the glass or spattering of the contents of the flask.
- Allow substances distilled at high temperatures under vacuum to cool before admitting to the system.
- Use ground glass joints instead of rubber stoppers in vacuum systems.
- Use a safety shield during vacuum distillation, especially in all cases involving flasks or receivers larger than 500 ml capacity.
- Place safety shields in direct line between the apparatus and the operator.
- When size or shape does not allow the use of shields, protect by surrounding with a wire screen or coating with plastic.
- Open valves leading to the source of vacuum before applying heat to an apparatus under vacuum.

- Completely wrap flasks and jars with adhesive tape to prevent flying particles in the event of breakage.

8. *Pipettes—*
   - Pipetting by mouth is dangerous and unnecessary.
   - Use a pipette bulb, atomizer bulb, or vacuum line.

9. *Safety with Benches—*
   - Scrub benches periodically and wash frequently with a cleaning agent appropriate for the type of work being done.
   - Develop the habit of putting away everything not absolutely needed for the experiment being performed.

10. *Parr Bombs—*
    - Wear goggles during parr bomb analysis when adding peroxide and when opening the cup.
    - Run all fusions in an approved hood with closed doors.
    - Protect face and hands, because parr bombs operate under high pressure.
    - Keep water away from both equipment and peroxide except when using the electric ignition type.
    - When necessary to discard the contents of an unignited bomb, remove the capsule and immerse the bomb in running water.
    - Never throw the contents of an unignited bomb into the wastecan.
    - Be sure to use the proper amount of water during the drowning step and use a shield in front of the beaker.

11. *Desiccators—*
    - Desiccators subjected to vacuum must be constructed of Pyrex glass and enclosed by a wooden box, wire cage, or other barrier to protect against flying glass.
    - Test vacuum desiccators before use and never use under a greater vacuum than recommended by the manufacturer and never without a desiccator guard.
    - Use solid desiccating agents whenever possible. *Note—*Take special precautions when using sulfuric acid.

12. *Special Equipment—*
    - Special equipment such as nitrometers, high-pressure autoclaves, and hydrogenation equipment requires special safety precautions.
    - Ensure that workers understand the operation and potential dangers of any new piece of apparatus.
    - Supervisors or other qualified persons should explain and/or demonstrate the correct use of equipment. Consult the manufacturer's directions when available.

Laboratory Infection Control—Organisms can be inhaled, ingested, or inoculated through the skin. Many laboratory-acquired infections, especially common diseases, are not reported. Laboratory-acquired tuberculosis and hepatitis are significantly underreported. Nearly all sizeable blood banks and serology laboratories report at least one case of hepatitis. Other commonly recognized exposure incidents are spills, breakages resulting in sprays (aerosols) of infectious material, injuries with broken glass, sharps injuries, and aspiration during mouth pipetting. Research laboratories are the most hazardous because many lack the standardized handling procedures found in large commercial laboratories.

1. **Aerosols**—Aerosols are the suspected source of infection where there is no recognized causal event. Aerosols are airborne droplets of infectious material that may be generated by:
   - Blowing out pipettes
   - Mixing test tube contents
   - Opening lyophilized cultures
   - Centrifuging suspensions
   - Pouring liquids
   - Using automatic pipetters
   - Mixing fluid cultures by pipette
   - Harvesting or dropping infected eggs
   - Mixing with high-speed blenders
   - Using poorly made, open, or large wire loops
   - Spilling liquids

2. **Aerosols Infection**—Small aerosol particles dry almost instantly and remain suspended in the air for long periods. When inhaled, they penetrate deep into the lungs and may cause infection. Larger and heavier particles settle slowly on laboratory surfaces and workers' skin.

3. **Reducing Aerosols Infection**—
   - Use smooth agar, a glass rod, or a cool wire loop if necessary for spreading.
   - Drain pipettes instead of blowing them out.
   - Mix cultures in a tube mixer.
   - Use disinfectants on work surfaces during transfer of biogenic materials.
   - Wrap needles and bottle tops in alcohol-soaked pledgets when withdrawing needles from stoppered vaccine bottles.
   - Properly maintain equipment such as high-speed blenders.
   - Use sealed centrifuge buckets.
   - Carefully package specimens during transport and storage.

Storage and Disposal of Laboratory Waste—Laboratory personnel must use correct waste storage and disposal procedures when working with infectious materials and chemicals.

1. **Laboratory Waste Hazards**—
   - Mercury trapped in porous sinks that continues to vaporize
   - Improper use of perchloric acid, which can be an explosion hazard
   - Azides that combine with metals such as copper, ammonium, or lead in plumbing systems and form explosive combinations when dry
   - Organic solvents that continue to vaporize and contaminate laboratory air after vigorous flushing

2. **Combustible Wastes**—Bag combustible contaminated material in a paper bag and mark as MEDICAL WASTE. Fold and staple the bag and incinerate it.

3. *Non-Combustible Wastes*—After treatment, bag non-combustible material in a paper bag, then fold and staple the bag, and mark it for the landfill as TREATED WASTE.
   - Use caution when packaging non-combustible items that are rigid and sharp enough to puncture the disposal bag. Repackage the waste before removing it from the restricted area.
   - Mark NON-COMBUSTIBLE MATERIAL on the outside of the bag. Pipettes, blades, and needles should be discarded in a rigid, leak-proof container that resists puncture.

4. *Radioactive Waste*—Radioactive medical waste disposal must be handled by the radiation safety officer.

Controlling Chemical and Physical Agents—

1. *Hazardous Materials List*—Compile a list of the common agents used in each laboratory, including:
   - Organic compounds such as acetone, formaldehyde, xylene, and other solvents
   - Inorganic compounds
   - Physical hazards such as UV radiation and ultrasonic devices
   - Radioactive isotopes such as those of iodine and cesium

2. *Worker Exposure*—
   - Inform workers potentially exposed to hazardous substances about the hazards and symptoms of exposure and the effects of overexposure.
   - Monitor exposures to ensure that airborne concentrations of specific contaminants are at least below the allowable limits.
   - Collect biologic samples to monitor worker exposures to toxic substances. Examples include mercury in the blood, hippuric acid in the urine due to toluene exposure, and enzyme activity levels to detect liver damage.
   - Establish a procedure for the proper storage, handling, and disposal of all chemicals.
   - Establish a procedure to ensure that biological safety cabinets are decontaminated routinely and certified annually.

Chemical Hazards—

1. *Identify Chemicals*—Identify all chemical hazards and establish appropriate training, precautions, PPE, and controls. Make sure that workers are aware of hazardous mixtures such as bleach, chromic acid, certain organic compounds, oxidants, and flammable liquids such as ether.

2. *American Association of Anatomists Chemical Listing*—The American Association of Anatomists has listed and reviewed the following chemicals ordinarily used in medical laboratories:
   - Fixatives such as acrolein, formaldehyde, glutaraldehyde, osmium tetroxide, phenol, picric acid, and potassium dichromate
   - Solvents including acetone, benzene, carbon tetrachloride, chloroform, dioxane, ether, ethoxyethanol, glycerol, methanol, propylene oxide, pyridine, tetrahydrofuran, toluene, trichloroethylene, and xylene

- Embedding media and reagents such as benzoyl, peroxide, benzyldimethyl-amine, dibutyl phthalate, dichlorobenzoyl, acrylic, epoxy, nitrocellulose, polyester, and tridimethylaminomethyl phenol
- Metals and metal compounds including chromic acid, lead acetate, mercury, osmium tetroxide, potassium permanganate, silver nitrate, uranyl acetate, and vanadium
- Explosive agents such as ammonium persulfate, benzene, dioxane, ether, glycerol, methanol, nitrocellulose, perchloric acid, picric acid, silver nitrate, and tetrahydrofuran

Carcinogens/Mutagens/Teratogens—

1. **Carcinogens**—About 30 chemicals have been established as human carcinogens, but several hundred have been found to cause cancer in test animals. Laboratory workers should use engineering controls and safe work practices to reduce exposure because many of these suspected carcinogens have not yet been tested. Carcinogens to which laboratory workers may frequently be exposed include:
   - Benzidine
   - Carbon tetrachloride
   - Ethylene oxide
   - Benzene
   - Dioxane

2. **Reproductive Hazards**—Laboratory workers can be exposed to chemicals that may cause mutations, genetic changes, or congenital malformations in the developing fetus of a pregnant worker. Most reproductive hazards affect both men and women; however, the fetus is particularly at risk from exposure to ionizing radiation, drugs, and biologic agents.
   - Chemicals such as benzene, ethylene oxide, ethylene diaminetetraacetic acid (EDTA), diazo dyes, lead acetate, mercury, sodium arsenate, toluene, and xylene all pose reproductive risks.
   - Biologic agents such as mumps, rubella, varicella, hepatitis viruses, and human immunodeficiency virus also pose risks.
   - Known and suspected reproductive hazards include ionizing radiation; alpha-, beta-, and gamma-emitting radionuclides; and X-rays.
   - Drugs that pose a risk include actinomycin D, antineoplastics, mitomycin, quinine, and streptomycin.

Spill Cleanup—

- Establish a detailed procedure for dealing with chemical spills.
- Check floors and benches for accumulations of spilled mercury.
- Post names and telephone numbers of persons to be notified in emergency situations.
- Procedures for decontamination and disposal of chemical spills depend on the properties of the specific chemical. Liquids should be absorbed with vermiculite, sand, or paper towels. Solids should be absorbed with moistened paper towels. Clean the area in cases with a solvent.

- Clean up and properly dispose of all contaminated material as soon as possible.
- Clean up spills of toxic chemicals as quickly as possible. Consult the radiation safety policy for spills containing radioactive materials.

Ventilation Hoods—Ventilation hoods can be effective for capturing and containing contaminants. Design specifications for laboratory fume hoods can be found in *Industrial Ventilation—A Manual of Recommended Practice* (ACGIH, 1986).

- Ventilation rates should be measured and recorded for all hoods. Measurement readings should be kept near the hood for quick reference.
- The entire ventilation system should be monitored monthly to check its efficiency.
- Chemical fume hoods must meet the requirements in NFPA 45, Laboratory Ventilating Systems and Hood Requirements (NFPA 1983, Volume 3).

Biosafety Levels—NIOSH published biosafety guidelines for large research laboratories. A summary of the four levels is provided below. Table 10.2 provides a quick reference summary for each biosafety level.

1. *Level 1*—This level of safety is for work that involves agents of no known or of minimal hazard to laboratory personnel and the environment. The laboratory is not separated from the general traffic patterns in the building. Work is generally conducted on open bench tops. Special containment equipment is not required or generally used. Laboratory personnel should have specific training in the procedures conducted in the laboratory and should be supervised by a scientist with general training in microbiology or a related science. The following standards, special practices, safety equipment, and facilities apply to the Biosafety Level 1 designation:
   - When experiments are in progress, access to the laboratory is limited or restricted at the discretion of the laboratory director.
   - Work surfaces must be decontaminated once a day and after any spill of viable material.
   - Contaminated liquid or solid wastes must be decontaminated before disposal.
   - Mechanical pipetting devices are used but mouth pipetting is prohibited.
   - Eating, drinking, smoking, and applying cosmetics are not permitted in the work area.
   - Food can be stored in cabinets or refrigerators designated for this purpose only. Food storage cabinets or refrigerators should be located outside of the work area.
   - Workers should wash their hands after they handle viable materials, work with animals, and before leaving the laboratory.
   - Procedures must be performed carefully to minimize the creation of aerosols.
   - It is recommended that laboratory coats, gowns, or uniforms be worn to prevent contamination or soiling of street clothes.
   - Materials to be decontaminated at a site away from the laboratory should be placed in a durable leak-proof container before removal.
   - The facility must have an effective insect and rodent control program.

- Special containment equipment is generally not required for manipulations of agents assigned to Biosafety Level 1.

2. *Level 2*—This level is similar to Level 1 and is suitable for work that involves agents of moderate potential hazard to personnel and the environment. Laboratory personnel must be given specific training in handling pathogenic agents and directed by a competent scientist. Access to the laboratory is limited when work

TABLE 10.2   Summary of Biosafety Levels

Biosafety Level	Agents	Practices	Safety Equipment (Primary Barriers)	Facilities (Secondary Barriers)
1	Not known to cause disease in healthy adults	Standard microbiological practices	None required	Open bench top sink required
2	Associated with human disease; hazards are auto-inoculation, ingestion, and mucous membrane exposure	BSL 1 practices plus: • Limited access • Biohazard warning signs • Sharps precautions • Biosafety manual defining any necessary waste decontamination or medical surveillance policies	Primary barriers: Class I or II biological safety cabinets or other physical containment devices used for all manipulations of agents that cause splashes or aerosols of infectious materials; PPE: laboratory coats, gloves, face protection as needed	BSL 1 plus: Autoclave available
3	Indigenous or exotic agents with potential for aerosol transmission; disease may have serious or lethal consequences	BSL 2 practices plus: • Controlled access • Decontamination of all waste • Decontamination of lab clothing before laundering • Baseline serum	Primary barriers: Class I or II biological safety cabinets or other physical containment devices used for all manipulations of agents; PPE: protective lab clothing, gloves, respiratory protection as needed	BSL 2 plus: • Physical separation from access corridors • Self-closing, double-door access • Exhausted air not recirculated • Negative airflow into laboratory
4	Dangerous or exotic agents that pose high risk of life-threatening disease, aerosol-transmitted lab infections, or related agents with unknown risk of transmission	BSL 3 practices plus: • Clothing change before entering • Showering on exit • All material decontaminated on exit from facility	Primary barriers: All procedures conducted in class I, II, or III biological safety cabinets in combination with full-body air-supplied positive-pressure personnel suit	BSL 3 plus: • Separate building or isolated zone • Dedicated supply/exhaust, vacuum, and decontamination systems • Other requirements outlined in the text

Adapted from NIOSH BioSafety Handbook, 3rd edition. May 1993.

is being conducted. Certain procedures in which infectious aerosols are created must be conducted in biological safety cabinets or other physical containment equipment. The following standard and special practices, safety equipment, and facilities apply to agents assigned to Biosafety Level 2:

- Standard microbiological practices are the same as Level 1.
- Contaminated materials to be decontaminated off-site should be placed in durable leak-proof containers before being removed from the laboratory.
- The laboratory director limits access to the laboratory, especially to persons at increased risk of acquiring infection. The director should assess each circumstance and determine who may enter or work in the laboratory.
- The laboratory director establishes procedures to advise persons of potential hazards. The director establishes specific entry requirements and makes provisions for entry.
- A hazard warning light with the universal biohazard symbol should be posted on the access door to the laboratory work area. The warning sign should identify the infectious agent and the names and telephone numbers of responsible persons. It should also indicate the special requirement(s) for entering the laboratory.
- The facility must have an effective insect and rodent control program.
- Laboratory coats, gowns, smocks, or uniforms must be worn while in the laboratory.
- When leaving the laboratory for non-laboratory areas, protective clothing must be removed and left in the laboratory.
- Animals not involved in the work being performed are not permitted in the laboratory.
- Special care must be taken to avoid skin contamination with infectious materials.
- Gloves should be worn when handling infected animals and when skin contact with infectious materials is unavoidable.
- Wastes from laboratories and animal rooms must be appropriately decontaminated before disposal.
- Hypodermic needles and syringes must only be used for parenteral injection and aspiration of fluids from laboratory animals and diaphragm bottles.
- Only needle-locking syringes or disposable syringe-needle units can be used for the injection or aspiration of infectious fluids.
- Extreme caution should be used when handling needles and syringes to avoid autoinoculation and the generation of aerosols during use and disposal.
- Needles and syringes should be promptly placed in a puncture-resistant container and decontaminated before discard or re-use.
- Spills and accidents resulting in exposures to infectious materials must be immediately reported to the laboratory director.
- Medical evaluation, surveillance, and treatment must be provided as appropriate and written records maintained.
- As appropriate, baseline serum samples for laboratory and other at-risk personnel should be collected and stored. *Note*—Additional serum specimens may be collected periodically depending on the agents handled.

- A biosafety manual must be prepared and adopted.
- Workers must be advised of special hazards and required to follow all practices, procedures, and instructions.
- Extreme caution should be exercised when performing procedures with a high potential for creating infectious aerosols, such as centrifuging, grinding, blending, vigorous mixing, sonic disruption, and opening containers of infectious materials.
- When working with large volumes of infectious agents, the materials may be centrifuged in the open laboratory if sealed heads or centrifuge safety cups are used. *Note*—They must be opened only in a biological safety cabinet.

3. *Level 3*—This level applies to clinical, diagnostic, teaching, research, or production facilities that work with indigenous or exotic agents which may cause serious or potentially lethal disease through exposure by inhalation. Laboratory personnel must have specific training in handling pathogenic and potentially lethal agents. Laboratory personnel are supervised by a competent scientist who is experienced in working with these agents. All procedures involving infectious material must be conducted within biological safety cabinets or other physical containment devices. Personnel must wear appropriate personal protective clothing and devices. The laboratory must have special engineering and design features, including those required under Levels 1 and 2.
   - Work surfaces of biological safety cabinets and other containment equipment must be decontaminated when work with infectious materials is finished.
   - Plastic-baked paper toweling must be used on non-perforated work surfaces within biological safety cabinets.
   - Laboratory clothing that protects street clothes must be worn in the laboratory. Laboratory clothing is not to be worn outside the laboratory and must be decontaminated before being laundered.
   - Molded surgical masks or respirators must be worn in rooms that contain infected animals.
   - Animals and plants not related to the work being conducted are not permitted in the laboratory.
   - Vacuum lines must be protected with high-efficiency particulate air filters and liquid disinfectant traps.
   - Biological safety cabinets or other appropriate combinations of personal protective or physical containment devices must be used for all activities that involve the threat of aerosol exposure.

4. *Level 4*—This level is required when working with dangerous and exotic agents that pose a high individual risk of life-threatening disease. Members of the laboratory staff must be given specific and thorough training in handling extremely hazardous infectious agents and must understand the primary and secondary containment equipment and laboratory design characteristics. They are supervised by a competent scientist who is trained and experienced in working with these agents. Access to the laboratory is strictly controlled by the laboratory director. The facility must be either in a separate building or in a controlled area within the building completely isolated from other areas. A specific facility

operations manual must be prepared or adopted. All activities are confined to class III biological safety cabinets or class I or II biological safety cabinets used along with one-piece positive-pressure personnel suits ventilated by a life support system. The maximum containment laboratory has special engineering and design features to prevent microorganisms from being disseminated into the environment. The following special safety practices, equipment, and facilities apply to agents assigned to Biosafety Level 4:

- Biological materials removed from the class III cabinet or from the maximum containment laboratory in a viable or intact state must be transferred to a non-breakable, sealed primary container and then enclosed in a non-breakable, sealed secondary container.
- Material must be removed from the facility through a disinfectant dunk tank, fumigation chamber, or an airlock designed for this purpose.
- Materials must be kept in a viable or intact state to be removed from the maximum containment laboratory unless they have been autoclaved or decontaminated.
- Equipment or material that might be damaged by high temperatures or steam must be decontaminated by gaseous or vapor methods in an airlock or chamber designed for this purpose.
- Only persons required to support the program are authorized to enter.
- Persons at increased risk of acquiring infection are not allowed in the laboratory or animal rooms. The supervisor has the final responsibility for assessing each circumstance and determining who may enter or work in the laboratory.
- Access to the facility is limited by means of secured and locked doors.
- Accessibility is managed by the laboratory director, biohazard control officer, or other person responsible for the physical security of the facility.
- Before entering, persons must be advised of the potential biohazards and instructed as to appropriate safeguards for ensuring their safety.
- Authorized persons must comply with instructions and all other applicable entry and exit procedures.
- A log book signed by all personnel indicates the date and time of each entry and exit. Practical and effective protocols for emergency situations must be established.
- Personnel may enter and leave the facility only through the clothing change and shower rooms. Personnel must shower each time they leave the facility but can only use the airlocks during an emergency.
- Street clothing must be removed in the outer clothing change room. Complete laboratory clothing, including undergarments, pants, shirt, jumpsuit, shoes, and gloves, must be provided and used by all personnel.
- When leaving the laboratory and before proceeding into the shower area, personnel must remove their laboratory clothing and store it in a locker or hamper in the inner change room.
- When infectious materials are present, a hazard warning sign with the biohazard symbol must be posted on all access doors. The sign identifies the infectious agent and lists the name of the laboratory director or other responsible

person(s). The sign also indicates any special requirements for entering the area, such as the need for immunizations or respirators.

- Supplies and materials must be brought in through a double-door autoclave, fumigation chamber, or airlock. The access area must be decontaminated between each use.
- A system must be established for reporting laboratory accidents and exposures, including employee absenteeism. The system must also monitor medical surveillance of potential laboratory-associated illnesses.
- Written records must be prepared and maintained. The facility must also establish procedures to deal with quarantine, isolation, and medical care of personnel with potential or known laboratory-associated illnesses.
- Containment equipment:
  1. All procedures within a facility with agents assigned to Biosafety Level 4 are conducted in a class III biological safety cabinet or in class I or II biological safety cabinets used in conjunction with one-piece positive-pressure personnel suits ventilated by a life support system.
  2. Activities with viral agents requiring Biosafety Level 4 secondary containment capabilities for which highly effective vaccines are available can be conducted with class I or II biological safety cabinets within the facility without a one-piece positive-pressure personnel suit being used if the facility has been decontaminated. *Note*—Applies if no work in the facility is being conducted with other agents assigned to Biosafety Level 4 and all other standard and special practices are followed.

SUMMARY

This chapter addressed safety and hazard control activities of several healthcare support areas including environmental services, laboratories, and food service operations. The employees of these departments face a number of hazards during their work each day. Environmental services is also a front-line defense against a number of hazards that affect the entire facility, including disinfecting, waste handling, and maintaining corridors to reduce slip, trip, and fall hazards. A large part of the chapter addressed radiation control requirements and the safe use of lasers. Laboratory safety was covered, including the OSHA chemical hygiene plan requirements. The chapter looked at healthcare security, including the OSHA workplace violence guidelines.

REVIEW QUESTIONS

1   The bacteria that can cause serious illness and is most commonly associated with ground meat is:
    a.   *E. coli*
    b.   salmonella
    c.   staphylococcus
    d.   none of the above

2 The Joint Commission now requires healthcare facilities to:
   a. develop a security management plan
   b. conduct a risk assessment of all grounds and facilities
   c. install emergency phones in all parking areas
   d. a and b

3 The Nuclear Regulatory Commission issues licenses to healthcare facilities for a period of _____.
   a. 2 years
   b. 3 years
   c. 4 years
   d. 5 years

4 The quantity of radiation absorbed per unit of mass is a dose. Dose measurements are expressed in:
   a. curies
   b. microcuries
   c. radioisotopes
   d. rads, roentgens, and rems

5 Which of the following provides guidance for safe laser usage in medicine?
   a. 29 CFR 1910.132
   b. 21 CFR 1049
   c. ANSI Z-136.3
   d. FDA laser guidelines

6 The OSHA laboratory standard requires:
   a. employee exposure monitoring
   b. development of a chemical hygiene plan
   c. chemical fume hood performance certification
   d. all of the above

7 Chemical exposures in a laboratory are regulated by:
   a. OSHA Hazard Communication Standard (29 CFR 1910.12001)
   b. OSHA Laboratory Standard (29 CFR 1910.1450)
   c. the formal chemical hygiene plan
   d. NFPA 45, Fire Protection Standard for Laboratories Using Chemicals

Answers: 1–a 2–d 3–d 4–d 5–c 6–d 7–b

CHAPTER 11

HEALTHCARE SAFETY TERMS AND ABBREVIATIONS

absorption—The process by which a liquid penetrates the solid structure of an absorbent's fibers or particles.

accident type—The description and classification of a mishap.

ACGIH—American Conference of Governmental Industrial Hygienists. An organization that publishes standards on exposure to toxic and hazardous materials in the air of work environments.

action level—The amount of a material in air at which certain OSHA regulations to protect employees take effect. Exposure at or above the action level is termed occupational exposure.

acute effect—An adverse effect on humans or animals, with symptoms developing rapidly and quickly becoming a crisis resulting from a short-term exposure.

ADA—Americans with Disabilities Act of 1990. A law that protects access and employment rights of people with disabilities.

adaptation—Change in the structure of an organism which results in its adjustment to its surroundings.

adsorption—A process by which a liquid adheres to the surface of a material but does not penetrate the fibers of the material.

aerosol—Liquid droplets or solid particles dispersed in air (0.01 to 100 μm).

AHA—American Hospital Association.

AIHA—American Industrial Hygiene Association.

air exchange rate—The speed at which outside air replaces air inside a building or the number of times the ventilation system replaces air within a room or building.

ALARA—As low as reasonably achievable.

allergen—A substance or particle that causes an allergic reaction.

alpha particle—A small electrically charged particle of very high velocity thrown off by many radioactive materials. It consists of two neutrons and two protons and has a positive electrical charge.

ambient air—Outside or surrounding air.

anemometer—A rotating vane, swinging vane, or hot-wire device used to measure air velocity.

annual summary—The occupational injury and illness totals for the year as reflected on OSHA Form 200. Includes company name and address, certification signature, and date.

annual survey—A survey conducted each year by the Bureau of Labor Statistics to produce national data on occupational injury and illness rate. Information is provided by employers from injury and illness records.

anosmia—Reduced sensitivity to odor detection.

ANSI—American National Standards Institute. A private organization that provides the mechanism for creating voluntary standards through consensus. Over 8000 standards are approved and used widely in industry and commerce.

antidote—An agent that neutralizes or counteracts the effects of a poison.

antimicrobial—An agent that destroys microbial organisms.

approved—Method, procedure, equipment, or tool that has been determined to be satisfactory for a particular purpose.

aromatic—Term applied to a group of hydrocarbons characterized by the presence of the benzene nucleus.

ASHRAE—American Society of Heating, Refrigerating, and Air Conditioning Engineers.

ASHRM—American Society of Healthcare Risk Management.

asphyxiant—A chemical gas or vapor that can cause unconsciousness or death by suffocation. Simple asphyxiants such as nitrogen use up or displace oxygen in the air. Chemical asphyxiants such as carbon monoxide interfere with the body's ability to receive or use oxygen.

ASTM—American Society for Testing and Materials. A technical organization that develops standards on characteristics and performance of materials, products, systems, and services. It is the world's largest source of voluntary consensus standards.

ATSDR—Agency for Toxic Substances and Disease Registry. A federal agency in the Public Health Service charged with carrying out health-related responsibilities of CERCLA and RCRA.

attenuated vaccine—Vaccine that has been weakened but is still required to be controlled as infectious by some regulatory programs.

autoclave—Device used to sterilize medical instruments and equipment by using steam under pressure.

autoignition temperature—Temperature at which a material will self-ignite and maintain combustion without a fire source.

automatic sprinklers—System built in or added to a structure that automatically delivers water in case of fire.

> *deluge*—All heads are open and water is released from a main valve.

> *dry pipe*—Piping is under pressure, and when head opens, air is released and water flows into the system.

> *preaction*—The main water control valve is opened by an actuating device.

> *wet pipe*—A water-filled pipe with a fusible mechanism in the sprinkler head.

background radiation—Radiation that comes from a source other than the radioactive material to be measured. This radiation is primarily due to cosmic rays which constantly bombard the earth's surface from outer space.

bacteria—Microscopic living organism.

beta particle—A small electrically charged particle thrown off by many radioactive materials. It is identical to the electron. Beta particles emerge from radioactive material at high speeds.

biocide—A substance that can kill living organisms.

biodegradable—A substance with the ability to decompose or break down into natural components.

bioremediation—The management of microorganisms.

bloodborne pathogens—Pathogenic microorganisms that are present in human blood and that can infect persons exposed to blood or other materials that contain these pathogens.

boiling point—The temperature at which a liquid changes to a vapor. Expressed in degrees Fahrenheit at sea level pressure. Flammable materials with a low boiling point generally present special fire hazards.

bolt ring—Closing device used to secure a cover to the body of an open-head drum. This ring requires a bolt and nut to secure the closure.

bonding—The interconnecting of two objects (tanks, cylinders, etc.) with clamps and bare wire as a safety practice to equalize the electrical potential between the objects and help prevent static sparks that could ignite flammable materials. Dispensing/receiving a flammable liquid requires dissipating the static charge by bonding between containers.

breakthrough time—The time from initial chemical contact to detection.

building-related illness (BRI)—Diagnosable illnesses with identifiable symptoms that can be attributed to airborne building contaminants.

bung—A threaded closure located in the head or body of a drum.

Bureau of Explosives—Division of the Association of American Railroads that regulates shipping specifications for hazardous products.

burnback—The distance a flame will travel from the ignition source back to the aerosol container.

CAA—Clean Air Act, Public Law PL 91-604, 40 CFR 50-80. Under EPA jurisdiction, this is the regulatory vehicle that sets limits and monitors airborne pollution that may harm public health or natural resources. The EPA sets national ambient air quality standards; enforcement and issue of discharge permits are carried out by the states under implementation plans.

carbon dioxide—A colorless, odorless, and non-toxic gas that results from human activity and fuel combustion indoors. Elevated levels indicate inadequate ventilation.

carbon monoxide—A colorless, odorless, toxic gas generated by the combustion of common fuels in the presence of insufficient air or where combustion is incomplete.

carcinogen—A chemical substance or agent that may cause cancer in animals or humans.

carpal tunnel syndrome—A common affliction caused by compression of the median nerve in the carpal tunnel. Often associated with tingling, pain, or numbness in the thumb and first three fingers.

CAS number—A number assigned to identify a chemical substance. Chemical Abstracts Service indexes information which appears in

Chemical Abstracts, published by the American Chemical Society. The CAS number identifies a specific chemical. It is assigned sequentially but possesses no chemical significance. The CAS number provides a concise means of material identification.

catastrophic loss—A loss of huge and extraordinary proportion.

caustic—Material that is able to burn, corrode, dissolve, or eat away another substance.

CDC—Centers for Disease Control and Prevention. A federal agency charged with protecting public health through prevention, control, and management procedures.

ceiling—Maximum allowable exposure limit not to be exceeded for an airborne substance.

ceiling concentration—Maximum concentration of a toxic substance allowed at any time or during a specific sampling period.

ceiling limit—Normally expressed as threshold limit value (TLV) and permissible exposure limit (PEL). Ceiling limit is the maximum allowable concentration to which an employee may be exposed in a given time period.

ceiling value—The concentration that should not be exceeded during the working exposure. An employee's exposure should at no time exceed the ceiling value (OSHA). *See also* TLV.

CERCLA—Comprehensive Environment Response, Compensation and Liability Act of 1980; also Superfund. Federal law authorizing identification and remediation of abandoned hazardous waste sites.

Certified Hazard Control Manager(CHCM)—A person certified by the Board of Hazard Control Management as having experience, education, and professional competence in the areas of hazard control and safety management.

Certified Safety Professional (CSP)—A person certified by the Board of Certified Safety Professionals as having achieved professional competence and ethics in the field of safety engineering.

CFR—Code of Federal Regulations. A codification of rules published in the Federal Register by the executive departments and agencies of the federal government. The code is divided into 50 titles which represent the broad areas subject to federal regulation.

cfm—Cubic feet per minute.

characteristic waste—Hazardous waste that exhibits one of four characteristics: ignitability, explosivity, toxicity, or corrosivity.

chemical—An element, compound, or mixture of elements and/or compounds.

chemical disinfection—The use of formulated chemical solutions to treat and decontaminate infectious waste.

chemical family—A group of compounds with related chemical and physical properties, such as the ketone or aldehyde family.

chemical hygiene plan—A written plan that addresses job procedures, work equipment, protective clothing, and training necessary to protect employees from chemical and toxic hazards. Required by OSHA under its laboratory safety standard.

chemical name—Scientific designation of a chemical substance.

CHEMTREC—Chemical Transportation Emergency Center. An organization that provides immediate information for members on what to do in case of spills, leaks, fires, or exposures. The toll-free number is 800-424-9300.

chlorinated solvent—Organic solvent that contains chlorine atoms. Examples are methylene chloride or perchloroethylene.

chronic effect—An adverse effect on animals or humans. Symptoms develop slowly over a long period of time or recur frequently.

class A fire—Fire that involves wood, paper, cloth, trash, or other ordinary materials.

class B fire—Fire that involves gasoline, grease, oil, paint, or other flammable liquids.

class C fire—Fire that involves live electrical equipment.

class D fire—Fire that involves flammable metals.

CO_2—Carbon dioxide. A heavy, colorless, non-flammable, relatively non-toxic gas. Produced by the combustion and decomposition of organic substances and as a by-product of many chemical processes. Also used as a fire-fighting agent.

combustible—A term used to classify certain liquids that will burn on the basis of flash point. NFPA and DOT define combustible liquids as having a flash point of 100°F (38°C) or higher. Non-liquid substances such as wood and paper are classified as ordinary combustibles by NFPA. OSHA defines combustible liquids under the hazard communication standard as any liquid having a flash point at or above 100°F (38°C) but below 200°F (93.3°C).

concentration—The amount of a substance in a stated unit of mixture or solution.

conductivity—The property of a circuit that permits the flow of an electrical current.

consensus standard—A standard developed according to a consensus of agreement among several organizations or individuals.

corrosive—A substance that causes visible destruction or permanent change in human skin tissue at the site of contact. A liquid that has a severe corrosion rate on steel.

criteria—Standards against which performance can be measured.

CSHO—Compliance Safety and Health Officer. An OSHA representative whose primary job is to conduct workplace inspections.

CTD—Cumulative trauma disorder.

cubic feet per minute (cfm)—Measure of the volume of a substance flowing through air within a specified time period. Used to measure air exchanged in ventilation systems.

CWA—Clean Water Act, Public Law PH 95-500, 40 CFR, Parts 100–140 and 400–470. The CWA regulates the discharge of non-toxic and toxic pollutants into surface waters. Its ultimate goal is to eliminate all discharges into surface waters. The EPA sets guidelines; states issue permits (NPDES, Natural Pollutant Discharge Elimination System permit) specifying the types of control equipment and discharges for each facility.

damper—Control that varies airflow through an air inlet, outlet, or duct.

dB—Decibel. A unit to express the relative intensity of a sound on a scale from 0 (average least perceptible) to 130 (average pain level). Sound doubles every 10 decibels.

dBA—Decibel time-weighted average.

decomposition—The breakdown of a chemical or substance into different parts or simpler compounds. Decomposition can occur due to heat, chemical reaction, decay, etc.

defatting—The removal of natural oils from the skin by the use of a fat-dissolving solvent.

density—Mass of substance per unit volume. Usually compared to water, which has a density of 1.

dermatitis—An inflammation of the skin caused by defatting of the dermis.

desiccant—A chemical substance that absorbs moisture.

dielectric—A material that is an electrical insulator or in which an electric field can be sustained with a minimum dissipation of power.

disabling injury (ANSI Z16.1)—An injury that prevents a person from performing a regularly established job for 24 hours beyond the day of occurrence.

disinfectant—Chemical agent with the ability to kill more than 99% of microorganisms. Registered by the EPA for public health use. The EPA registers three types: limited, broad-spectrum, and hospital disinfectants.

dose–response—The relationship between the amount of a toxic or hazardous substance and the extent of illness or injury produced in humans.

DOT—U.S. Department of Transportation.

DOT ID number—Four-digit identification number which is preceded by UN or NA. The number identifies hazardous substances regulated during transport. DOT regulations can be found in 49 CFR Part 172.

drop test—A test required by DOT regulations for determination of the quality of a container or finished product.

dry bulb temperature—The temperature of air measured with a dry bulb thermometer in a psychrometer to measure relative humidity.

dusts—Solid particles generated by handling, crushing, grinding, rapid impact, detonation, and decrepitation of organic or inorganic materials such as rock, ore, metal, coal, wood, and grain. Dusts do not tend to flocculate, except under electrostatic forces; they do not diffuse in air but settle under the influence of gravity.

EDA—Emergency Declaration Area. Officially designated area for cleanup of hazardous waste or materials, as at an NPL hazardous waste site.

emulsion—A stable mixture of two or more liquids held in suspension by small percentages of substances called emulsifiers.

engineering controls—The preferred method of controlling employee exposures in the workplace. Engineering controls may be accomplished by modifying the workplace environment.

enzyme—Complex protein produced by living cells that starts up biochemical reactions.

EPA—U.S. Environmental Protection Agency. A federal agency with environmental protection regulatory and enforcement authority. Administers the Clean Air Act, Clean Water Act, Resource Conservation and Recovery Act, Toxic Substances Control Act, and other federal environmental laws.

ergonomics—A multi-disciplinary activity that deals with interactions between workers and their total working environment plus stresses related to such environmental elements as atmosphere, heat, light, and sound, as well as tools and equipment in the workplace.

etiologic agent—A viable microorganism or its toxin which can cause human disease.

evaporation rate—The rate at which a material is converted to a vapor at a given temperature and pressure when compared to the evaporation rate of a given substance.

exhaust ventilation—The removal of air from any space, usually by mechanical means. The flow of air between two points is due to the occurrence of a pressure difference between the two points. This pressure difference causes air to flow from the high-pressure zone to the low-pressure zone.

experience rating—Process of basing insurance or workers' compensation fund premiums on the insured's record (also called merit rating).

explosion class 1—Flammable gas/vapor.

explosion class 2—Combustible dust.

explosion class 3—Ignitable fibers.

explosion-proof—An electrical apparatus designed so that the explosion of flammable gas or vapor inside an enclosure will not ignite flammable gas or vapor outside.

explosive limit—The amount of vapor in the air which forms an explosive mixture.

exposure—Subjection to a hazardous chemical or biological substance through any route of entry (inhalation, ingestion, skin contact, absorption, etc.).

exposure level—The level or concentration of a physical or chemical hazard to which an individual is exposed.

exposure limit—Concentration of a substance under which it is believed that nearly all workers may be repeatedly exposed day after day without adverse effects. ACGIH limits are called TLV and OSHA limits are called PEL.

exudate—Material such as fluid, cells, or cellular debris which has escaped from blood vessels and has been in tissue material.

face velocity—Average air velocity into the exhaust system measured at the opening into the hood or booth.

FDA—U.S. Food and Drug Administration.

Federal Register—A publication that officially documents rules and regulations promulgated under law. It is published each day following a government working day and is the supplement to the Code of Federal Regulations.

FEMA—Federal Emergency Management Agency.

FIFRA—Federal Insecticide, Fungicide, and Rodenticide Act.

first aid—According to OSHA, any one-time treatment and subsequent observation of minor scratches, cuts, burns, and splinters that normally does not require medical care. It is also aid provided by qualified first responders until more qualified medical aid arrives.

first report—State-published workers' compensation form used to report work-related injuries and illnesses. The form may qualify as a substitute for OSHA Form 101.

first responder—The first personnel trained to arrive on the scene of a hazardous or emergency situation.

flame arrestor—A mesh or perforated metal insert within a flammable storage can which protects its contents from external flame or ignition.

flame extension—The distance a flame will travel from an aerosol container when exposed to an ignition source.

flammable—Flash point less than 100°F and vapor pressure not over 50 psia at 100°F. (Definition may vary by organization.)

flammable liquid—A liquid with a flash point below 100°F (37.8°C).

flash back—A phenomenon characterized by vapor ignition and flame traveling back to the vapor source.

flash point—The lowest temperature at which a flammable vapor–air mixture will ignite when an ignition source is introduced.

flocculation—The process to make solids in water increase in size by biological or chemical means so they may be separated from water.

FM—Factory Mutual. A national testing laboratory and approval service recognized by OSHA.

fomite—An object not harmful in itself, but one that can harbor pathogenic microorganisms.

fumes—Particulate matter consisting of the solid particles generated by condensation from the gaseous state, generally after violation from melted substances, and often accompanied by a chemical reaction, such as oxidation.

fungi—Organisms that lack chlorophyll and must receive food from decaying matter.

gamma rays—The most penetrating of all radiation. High-energy photons, especially as emitted by a nucleus in a transition between two energy levels.

gas—A state of matter in which a material has very low density and viscosity. Gases expand and contract greatly in response to changes in temperature and pressure.

gas/vapor sterilization—A waste treatment technique that uses gases or vaporized chemicals such as ethylene oxide and formaldehyde as sterilizing agents.

gauge—Thickness of the steel used to manufacture a drum. The lower the gauge, the thicker the material. Also used to measure glove thickness in inches.

generator—Any person, organization, or agency whose act or process produces medical waste or causes waste to become subject to regulation.

hazard classes—Nine descriptive terms established by the United Nations Committee of Experts to categorize hazardous chemical, physical, and biological materials. Categories are flammable liquids, explosives, gaseous oxidizers, radioactive materials, corrosives, flammable solids, poisons, infectious substances, and dangerous substances.

hazardous chemical—Any chemical that poses a physical or health hazard.

hazardous material—A substance or material which has been determined by the DOT to pose an unreasonable risk to health, safety, and property when transported in commerce (49 CFR 171.8).

hazardous waste—Under RCRA, any solid or combination of solid wastes which because of its physical, chemical, or infectious characteristics may pose a hazard when not managed properly.

HBV—Hepatitis B virus.

HCFA—Health Care Financing Administration.

HCS—Hazard communication standard. The OSHA standard cited in 29 CFR 1910.1200 requiring communication of risks from hazardous substances to workers in regulated facilities.

health hazard—A chemical for which there is statistically significant evidence that acute or chronic health effects may occur in exposed individuals.

hearing conservation—Preventing or minimizing noise-induced deafness through the use of hearing protection devices, engineering methods, annual audiometric tests, and employee training.

HIV—Human immunodeficiency virus.

HMAC—Hazardous Materials Advisory Council. National organization that represents the hazardous materials industry. Devoted to safety in transportation and handling of hazardous materials.

HMIG—Hazardous Materials Identification Guide.

HMR—Hazardous Materials Regulations. Regulations administered and enforced by DOT covering the transportation of hazardous materials by air, highway, rail, water, and intermodal means.

HSWA—Hazardous and Solid Waste Amendments. The 1984 amendments to the RCRA establishing a timetable for more stringent regulation of RCRA-covered activities.

HVAC—Heating, ventilating, and air conditioning system.

hypersensitivity diseases—Diseases characterized by allergic responses to animal antigens. Often associated with indoor air quality conditions such as asthma and rhinitis.

I.D./O.D.—Inside/outside diameter of a container.

IDLH—Immediately Dangerous to Life or Health. An exposure to a concentration of a hazardous substance in the air which can threaten life or cause irreversible health effects.

ignitable—A solid, liquid, or compressed gas that has a flash point less than 140°F.

ignition temperature—Lowest temperature at which a substance can catch fire and continue to burn.

illumination—The amount of light a surface receives per unit area, expressed in lumens per square foot or footcandles.

ILSM—Interim life safety measures.

incidence rate—The number of injuries or lost workdays related to a common exposure base of 100 full-time employees working 40 hours per week for 50 weeks. The formula allows employers to compute rates for industry comparisons, trend analysis, or comparisons with previous years. The rate is calculated as:

$$\frac{\text{Number of injuries} \times 200,000}{\text{Total hours worked by all employees for the year}}$$

incompatible—The term used to indicate that one material cannot be mixed with another without the possibility of a dangerous reaction.

indicator—A measurement used to evaluate program effectiveness within an organization.

indoor air quality (IAQ)—The study, evaluation, and control of indoor air quality related to temperature, humidity, and airborne contaminants.

infectious—Capable of invading a susceptible host, replicating, and causing an altered host reaction. Commonly referred to as a disease.

infectious waste—Waste containing pathogens that can cause an infectious disease in humans.

ingestion—Taking a substance into the body through the mouth.

inhalation—The breathing of an airborne substance into the body. May be in the form of a gas, vapor, fume, mist, or dust.

inhibitor—A substance that is added to another substance to prevent or slow down an unwanted reaction or change.

innocuous—Harmless.

irradiation sterilization—The use of ionizing radiation for the treatment of infectious waste.

isotonic—Having the same osmotic pressure as the fluid phase of a cell or tissue.

JCAHO—Joint Commission on Accreditation of Healthcare Organizations.

job hazard analysis—The breaking down of methods, tasks, or procedures into components to determine hazards.

joule—Unit of energy used to describe a single pulsed output of a laser. It is equal to 1 watt-second or 0.239 calories.

kilogram—About 2.2 pounds (1 ounce = 28 grams).

lab pack—Generally refers to any small container of hazardous waste in an overpacked drum. Not restricted to laboratory wastes.

laser—Light amplification by stimulated emission of radiation.

LCD—Liquid crystal display. A constantly operating display that consists of segments of a liquid crystal whose reflectivity varies according to the voltage applied to them.

LED—Light-emitting diode. A semiconductor diode that converts electric energy efficiently into spontaneous and non-coherent electromagnetic radiation at visible and near-infrared wavelengths.

LEL—Lower explosive limit. The lowest concentration of a substance that will produce a fire or flash when an ignition source is present. It is expressed as a percent of vapor or gas in the air by volume.

LEPC—Local Emergency Planning Committee. A group defined under SARA as responsible for developing emergency plans.

local exhaust ventilation—A ventilation system that captures/removes contaminants at the point produced before they escape into the work area.

loss ratio—A fraction calculated by dividing losses by the amount of premiums.

lost workdays—The number of workdays an employee is away from work beyond the day of injury or onset of illness. OSHA also considers time of restricted work activity due to injury or illness as lost workdays.

LP (liquefied petroleum) gas—A gas usually comprised of propane and some butane created as a by-product of petroleum refining.

lumbar—The section of the lower vertebral column immediately above the sacrum. Located in the small of the back and consists of five large lumbar vertebrae. It is a highly stressed area in work situations and in supporting the body structure.

measure—A term used in the quality field for the collection of quantifiable data and information about performance, production, and goal accomplishment.

medical waste—Any solid waste generated in the diagnosis, treatment, or immunization of humans or animals.

melting point—The temperature at which a solid substance changes to a liquid.

mg/m³—Milligrams per cubic meter. Unit used to measure air concentration of dust, gas, mist, and fume.

microbe—Minute organism, including bacteria, protozoa, and fungi, which is capable of causing disease.

mil—One mil equals 1/1000 of an inch. Used in reference to glove thickness.

mist—Suspended liquid droplets generated by condensation from the gaseous to the liquid state. Breaking up a liquid into a dispersed state such as by splashing, foaming, or atomizing. Mist is formed when a finely divided liquid is suspended in air. Mist particles measure between 40 and 500 microns.

mixture—Any combination of two or more chemicals where the combination is not, in whole or in part, the result of a chemical reaction.

mppcf—Million particles per cubic foot.

MSDS—Material Safety Data Sheet. A document that contains descriptive information on hazardous chemicals under OSHA's hazard communication standard. Data sheets also provide precautionary information, safe handling procedures, and emergency first-aid procedures.

MSHA—Mine Safety and Health Administration. A federal agency that regulates the health and safety of the mining industry.

mutagen—A substance or agent capable of changing the genetic material of a living cell.

narcosis—Stupor or unconsciousness caused by exposure to a chemical.

natural gas—A combustible gas composed largely of methane and other hydrocarbons obtained from natural earth fissures.

NEC—National Electrical Code, NFPA 70.

necrosis—Death of plant or animal cells.

negative pressure—A condition caused when less air is supplied to a space than is exhausted

from the space. The air pressure in the space is less than that in surrounding areas.

negligence—Failure to do what reasonable and prudent persons would do under similar or existing circumstances.

NFPA—National Fire Protection Association. An organization that publishes standards relating to fire and hazardous material protection. NFPA is well known for its Life Safety Code. Many NFPA standards are used in fire codes, and others are promulgated in standards by governmental agencies.

NFPA hazard rating—A classification system that uses a four-color diamond to communicate health, flammability, reactivity, and specific hazard information for a chemical substance. A numbering system that rates hazards from 0 (lowest) to 4 (highest).

NIOSH—National Institute of Occupational Safety and Health. This agency does research in the fields of occupational safety and health and makes recommendations to OSHA.

nitrogen oxide—Compound produced by combustion.

NOAA—National Oceanic and Atmospheric Administration. A scientific support organization that serves regulatory agencies charged with enforcing environmental laws that affect oceans and the atmosphere.

NPL—National Priority List. The official list of hazardous waste sites to be addressed by CERCLA.

NRC—Nuclear Regulatory Commission.

NRR—Noise reduction rating.

NTP—Normal temperature and pressure (70°F and 14.696 psia).

occupational exposure limit—Maximum allowable amount of a toxic material in workroom air. Established to protect exposure during a lifetime of work.

occupational illness—An illness caused by environmental exposure during employment.

occupational injury—An injury such as a cut, fracture, sprain, amputation, etc. which results from an on-the-job accident or from a single exposure in the workplace.

occurrence—An incident classified as major or minor which results from apparent or foreseen causal factors.

odor threshold—The minimum concentration of a substance at which most people can detect and identify its characteristic odor.

oleophilic—Having an affinity for, attracting, adsorbing, or absorbing oil.

opacity—The amount of light obscured by particulate matter in air.

optical density (OD)—A logarithmic expression of the attenuation afforded by a filter.

organic—Designation of any chemical compound that contains carbon. Also used to describe substances derived from living organisms.

OSHA—Occupational Safety and Health Administration. Oversees and regulates workplace health and safety.

OSHA log and summary—The OSHA recordkeeping form (OSHA 200) used to list illnesses and injuries.

outcome—A result reached due to performance (or non-performance) of a task, job, or process.

oxidant—An oxygen-containing substance that reacts chemically to produce a new substance.

oxidizer—A material that may cause the ignition of a combustible material without the aid of an external ignition source.

ozone—A reactive oxidant that contains three atoms of oxygen.

PAPR—Powered air-purifying respirator.

parenteral—Exposure that occurs through a break in the skin.

Part B permit—The second, narrative section submitted by generators in the RCRA permitting process. It covers in detail the procedures followed at a facility to protect human health and the environment.

particulates—Fine solid or liquid particles found in air and other emissions.

PCB—Polychlorinated biphenyl. A pathogenic and teratogenic industrial compound used as a heat transfer agent. PCBs accumulate in human or animal tissue.

PEL—Permissible exposure limit. The OSHA limit for employee exposure to chemicals. Found primarily in 29 CFR 1910.1000.

permeation rate—An invisible process by which a hazardous chemical moves through a protective material. Measured in $mg/m^2/sec$.

pH—A measure of how acidic or caustic a substance is on a scale from 1 to 14 (1 = very acidic, 7 = neutral, 14 = very caustic).

physical hazard—A chemical validated as being or having one of the following characteristics: combustible liquid, compressed gas, explosive, flammable, organic peroxide, oxidizing qualities, pyrophoric, unstable, or water reactive.

poison A—A poisonous gas or liquid of such toxicity that when mixed with air a very small amount is dangerous to life.

poison B—A solid or liquid substance that is known to be so toxic to humans as to afford a hazard to health during transportation.

polymerization—A chemical reaction in which one or more small molecules combine to form larger molecules. Hazardous polymerization is the reaction that takes place at a rate at which large amounts of energy are released.

potential—That which is currently latent or unrealized.

PPE—Personal protective equipment. Devices such as respirators, gloves, and hearing protectors worn by workers to protect against hazards in the environment.

ppm—parts per million. Unit for measuring the concentration of a gas or vapor in contaminated air. Also used to indicate the concentration of a particular substance in a liquid or solid.

prefilter—A filter used in conjunction with a cartridge on an air-purifying respirator.

process—A method of interrelating steps, events, and mechanisms to accomplish an action or goal.

psi—Pounds per square inch.

psia—Pounds per square inch absolute. The absolute thermodynamic pressure exerted upon an area of one square inch.

PVC—Polyvinyl chloride. A member of the family of vinyl resins.

pyrophoric—A chemical that will ignite spontaneously in air at a temperature of 130°F or below.

qualitative analysis—The analysis of a gas, liquid, or solid sample to identify the elements or compounds the sample is composed of.

quantitative analysis—The analysis of a gas, liquid, or solid to determine the precise percentage composition of elements or compounds.

quaternary ammonium—Chemical substance used to disinfect or sanitize by rupturing the cell walls of microorganisms.

radionuclide—A nuclide with the capability of spontaneously emitting radiation.

RCRA—Resource Conservation and Recovery Act. An act that regulates waste materials (including hazardous wastes) from generation through final disposal. Called the "cradle-to-grave" regulation.

reactivity—The susceptibility of a substance to undergo chemical reaction and change which could result in an explosion or fire. Reactive materials may also produce corrosive or toxic emissions.

recordkeeping system—According to OSHA, a nationwide system of reporting and recording occupational injuries and illnesses. Provisions of the standard are found in 29 CFR 1904.

regulated material—A substance or material that is subject to regulations set forth by a federal agency, such as the EPA or DOT.

REL—Relative exposure limit.

relative humidity—The ratio of the quantity of water vapor present in air to the quantity that would saturate the air at any specific temperature.

relief valve—A valve designed to release excess pressure within a system without damaging the system.

retainer ring or cap—A plastic ring that holds a cartridge or filter on a respirator mask or holds a prefilter on a cartridge.

risk—The probability of injury, illness, disease, loss, or death under specific circumstances.

RMI—Repetitive motion injury.

RMIS—Risk management information system.

safety belt—A belt worn to prevent falls when working in high places; a belt used to secure passengers in vehicles or airplanes. Back safety belts provide support for planned lifting tasks.

safety can—An approved container of not more than five-gallon capacity with a spring-closing lid and a spout cover designed to safely relieve internal pressure when exposed to fire.

safety hat (hard hat)—Rigid headgear worn to protect a worker from head injuries, flying particles, and electric shock.

sanitize—Destroy microorganisms on a surface to a safe level.

sanitizer—One of three groups of EPA-registered microbials. To be a registered sanitizer, test results must show that organisms are reduced by at least 99.9%.

SARA—Superfund Amendment and Re-Authorization Act. Federal law that re-authorizes and expands the jurisdiction of CERCLA.

SARA Title III—Part of SARA which mandates public disclosure of chemical information and development of emergency response plans.

SCBA—Self-contained breathing apparatus. Designed for entry into and escape from atmospheres Immediately Dangerous to Life or Health (IDLH) or oxygen deficient.

scfm—Standard cubic feet per minute.

sensitizer—A substance that may cause no reaction in a person during initial exposure, but will cause an allergic response upon further exposure.

serious injury—An injury classification that includes disabling work injuries and injuries in the following categories: eye injuries, fractures, hospitalization for observation, loss of consciousness, and any other injury that requires medical treatment by a physician.

sharp—An object that can penetrate the skin, such as needles, scalpels, and lancets.

sick building syndrome (SBS)—A term that describes a situation where building occupants experience acute health or comfort effects that appear to be linked to time spent in the building. No specific illness or cause can be determined.

sludge—A solid material that collects as the result of air or water treatment processes.

SMDA—Safe Medical Device Act of 1990.

solubility—The percentage of a material (by weight) that will dissolve in water at a specified temperature.

solution—A mixture in which the components lose their identities and are uniformly dispersed.

solvent—A substance that dissolves or disperses another substance.

specific gravity—The weight of a material compared to the weight of an equal volume of water.

spectrum—A range of frequencies within which radiation has some specified characteristic, such as audio-frequency spectrum, ultraviolet spectrum, and radio spectrum.

Standard Industrial Classification (SIC)—A classification developed by the Office of Management and Budget. Used to assign each establishment an industry code which is determined by the product manufactured or service provided. The system uses a two-, three-, or four-digit code according to information available.

standard procedure—A written instruction which establishes what action is required, who is to act, and when the action is to take place.

staphylococcus—Any of various spherical parasitic bacteria which occur in grapelike clusters and cause infections, such as septicemia.

static pressure—The potential pressure exerted in all directions by a fluid at rest. When combined with velocity pressure, it gives total pressure.

steam sterilization—A treatment method for infectious waste using saturated steam within a pressurized vessel (autoclave).

STEL—Short-term exposure limit. Maximum concentration for a continuous 15-minute exposure period. Maximum of 4 such periods per day, 60 minutes minimum between exposure periods, and the daily TLV–TWA must not be exceeded.

sterilize—To destroy all living microorganisms on a surface or in water.

sterilizer—One of three groups of antimicrobials registered by the EPA. A sterilizer removes all forms of bacteria, fungi, and viruses and their spores. A sporicide is considered to be a sterilizer.

streptococcus—Any of various rounded, disease-causing bacteria that occur in pairs or chains.

substandard condition—Any physical state that deviates from what is acceptable, normal, or correct and is a potential hazard.

supplied air—Breathable air supplied to a worker's mask/hood from a source outside the contaminated area.

supported gloves—Gloves that are constructed of a coated fabric.

survey—A comprehensive study or assessment of a facility, workplace, or activity for insurance or loss control purposes.

systematic—Striving toward goal accomplishment in a planned manner using predetermined steps or procedures.

teratogen—A substance or agent which when a pregnant female is exposed to can cause malformations in the fetus.

TLD—Thermoluminescent dosimeter.

TLV—Threshold limit value. The airborne concentration of a hazardous/toxic substance to which workers may be repeatedly exposed day after day without adverse effect. TLV values are published yearly by the ACGIH. *Note*— TLV is a registered trademark of the ACGIH.

toxicity—The potential of a substance to have a harmful effect and a description of the effect and the conditions or concentration under which the effect takes place.

toxic substance—Any substance that can cause acute or chronic injury or illness to the human body.

toxin—A substance that is poisonous to varying degrees.

TSCA—Toxic Substances Control Act.

TWA—Time-weighted average. Usually a personal eight-hour average exposure concentration to an airborne chemical hazard. Expressed in ppm or mg/m^3.

UEL—Upper explosive limit. The highest concentration of a substance that will burn or explode when an ignition source is present. Expressed in percent of vapor or gas in the air by volume.

UFC—Uniform Fire Code. Regulations consistent with nationally recognized good practice for safeguarding life and property from the hazards of fire and explosion that arise from the storage, handling, and use of hazardous substances, materials, and devices.

UL—Underwriters Laboratories. An independent non-profit organization that operates laboratories for the investigation of devices and materials in respect to hazards which affect life and property.

Ultraviolet—Wavelengths of the electromagnetic spectrum which are shorter than those of visible light and longer than X-rays. Wavelength 10^{-5} to 10^{-6} cm.

universal precautions (standard precautions)—A method of infection control in which all human blood and certain other materials are treated as infectious for bloodborne pathogens.

unstable—A chemical which when in the pure state will vigorously polymerize, decompose, condense, or become self-reactive under conditions of shock, pressure, or temperature.

unsupported glove—Unlined glove without any type of fabric lining.

USC—United States Code.

vapor—The gaseous form of a substance that is normally in the solid or liquid state at room temperature and pressure.

vapor density—The weight of a vapor or gas compared to the weight of an equal volume of air. An expression of density of a vapor or gas. Materials lighter than air have a vapor density less than 1.

vapor pressure—The pressure exerted by a saturated vapor above its own liquid in a closed container. Vapor pressure is usually expressed as pounds per square inch but on MSDSs is in millimeters of mercury (mm HG) at 68°F. The lower the boiling point of a substance, the higher its vapor pressure.

vector—An organism that carries disease, such as insects or rodents.

viscosity—A relative measure of how slowly a substance pours or flows.

visible radiation—Wavelengths of the electromagnetic spectrum between 10^{-4} and 10^{-5} cm.

volatile—The tendency or ability of a liquid to vaporize. Liquids such as alcohol or gasoline are volatile because they have a tendency to evaporate quickly.

volatile organic compounds (VOC)—Compounds that evaporate from many housekeeping, maintenance, and building products made from organic chemicals. VOCs can cause eye, nose, and throat irritation. Some can also cause headaches, dizziness, visual problems, and memory impairment. Highly toxic VOCs can cause cancer in animals.

water-reactive—A chemical that reacts with water to release a gas that either is flammable or presents a health hazard.

wavelength—The distance in the line of advance of a wave from any point to a like point on the next wave. Usually measured in angstroms, microns, micrometers, or nanometers.

xenobiotic—A man-made substance, such as plastic, found in the environment.

X-ray—Highly penetrating radiation. Unlike gamma rays, X-rays come not from the nucleus of atoms but from the surrounding electrons.

CHAPTER 12

APPENDICES

APPENDIX A: SAMPLE HAZARD COMMUNICATION PROGRAM

The purpose of this Hazard Communication Program is to ensure the protection of employees at this establishment by providing information and training on the hazards associated with hazardous chemicals and substances imported into or produced within this facility. This Hazard Communication Program has been prepared in accordance with the OSHA Hazard Communication Standard, 29 CFR 1910.1200. As required by this standard, training will be provided upon initial assignment to duties and repeated or augmented whenever new hazards are introduced into the workplace.

Employees are encouraged to review this program very carefully and direct all questions concerning its content to the facility safety coordinator. Management wishes to encourage all employees to share their thoughts, ideas, and concerns regarding this program. This program is administered by the safety coordinator, who has full authority to make necessary decisions concerning the implementation of this program for the improved safety and health of our employees. Copies of this written program may be obtained by contacting your supervisor or at the following location(s): _____
_____ .

If any of the information in this program is not clear to you, contact your immediate supervisor for additional information.

I. Hazard Communication Program

Hazard Determination—A number of materials used in our daily activities, if not handled properly, could be hazardous to safety and health.

This Hazard Communication Program has been developed so that you will be able to recognize those materials considered to be hazardous under certain conditions and learn how to handle such substances safely. The hazard of a substance in our facility is evaluated based on the information supplied by the manufacturer's Material Safety Data Sheet (MSDS).

The person(s) responsible for determining which materials are considered hazardous within this facility is _____ .

If you have any questions regarding the hazards posed by any substance brought into or created within this facility, direct them to your immediate supervisor.

This facility does not utilize any other criteria and/or sources of information to determine chemical hazards within the facility.

II. Chemical Inventory List

Each department head will be responsible for the development of a chemical inventory list which will identify all of the hazardous chemicals located within the work area. The safety coordinator will be responsible for updating the list. This list is designed to link the label information on chemicals in this facility to the correct MSDS.

If you have difficulty locating a chemical in this facility on this chemical inventory list, or if the MSDS for a listed chemical is missing, notify your supervisor for assistance or contact the person(s) listed above. If you have a question concerning the use of a particular chemical or substance, consult the MSDS. If your question is not clearly answered, consult your supervisor before using the chemical.

Direct questions regarding our chemical inventory list and how to locate MSDS information to your immediate supervisor.

III. Material Safety Data Sheets

MSDSs provide you with specific information on the chemicals you use and are located behind the chemical inventory list in each MSDS binder. Each department head is responsible for obtaining the MSDS for each hazardous substance in the facility and all new chemicals ordered. NO PRODUCT WILL BE USED BY AN EMPLOYEE UNTIL THE APPROPRIATE MSDS IS AVAILABLE FOR REVIEW. The safety coordinator is responsible for reviewing MSDSs to verify that each data item has been addressed in some manner by the MSDS preparer and for contacting the manufacturer if information on the MSDS is not satisfactory or if an MSDS was not provided with the original shipment of a chemical.

MSDSs will be available for review by all employees during each work shift. To review MSDSs, contact your immediate supervisor. MSDSs will be updated on an annual basis.

Other questions relating to the location, maintenance, use, and availability of MSDSs should be addressed to the safety coordinator.

IV. Labels and Other Forms of Warning

Each department head will ensure that all hazardous chemicals are properly labeled and updated as necessary. All labels on containers of hazardous chemicals will have, as a minimum, the following information:

A. The identity of the product or chemical. This may be a common name or a chemical name; whichever is used will be referenced on the chemical inventory list.

B. A direct hazard warning which specifies the organs that may be affected by exposure to this chemical.

Labels provided by the manufacturer should also have the name and address of the manufacturer or responsible party to contact in the event of an emergency.

If a container is not labeled or if a label fails to meet the requirements stated above, it must be set aside until a proper label may be affixed. If you find a container of hazardous or unknown chemical in this facility which has no label, notify your supervisor immediately. If you have difficulty locating a chemical on the inventory list by using the identity on the label, notify the safety coordinator for assistance. DO NOT USE A CHEMICAL WITHOUT KNOWING EXACTLY WHAT YOU ARE WORKING WITH AND UNDERSTANDING ITS HAZARDS!

All secondary containers are to be labeled with the information listed above. Containers which have their contents labeled by use of placards may be removed from that location without affixing a proper hazard label first. If you transfer chemicals from a labeled container into a portable container for IMMEDIATE use, then no label is required for that portable container.

V. Hazardous Non-Routine Tasks

Tasks performed by employees in this facility which are regarded as non-routine are listed below, along with any hazards associated with the procedure.

Task	*Hazardous Exposure*
_____	_____
_____	_____
_____	_____

Prior to beginning work on these tasks, employees will be briefed on the hazards involved in the procedure, hazardous chemicals to which they may be exposed, and the precautions to take to reduce or avoid exposure. After training, do not begin work unless your immediate supervisor is aware of your activity.

VI. Chemicals in Unlabeled Pipes

When work is to be performed involving an unlabeled pipe, DO NOT proceed until you have learned what is carried by the pipe and the procedures to be used in closing and opening valves, working on or near the pipe, etc.

VII. Employee Training and Information

Before starting work, each new employee will receive training which includes the following information:

- The requirements of the HAZCOM standard
- Operations in the work area where hazardous materials are present
- The location and availability of the written Hazard Communication Program

- Methods and observations that may be used to detect the presence or release of a hazardous chemical in the work area
- The physical and health hazards of chemicals in your work area
- Measures employees can take to protect themselves from these hazards, including specific procedures we have implemented to protect employees, including specific safe work practices, engineering controls, emergency procedures, and personal protective equipment to be used
- The details of our Hazard Communication Program, which includes an explanation of the labeling system, MSDSs, and how employees can obtain and use the appropriate hazard information
- Procedures to follow in the event of an emergency spill or release of a hazardous substance

Before any new hazardous substance is introduced into the work area, employees will receive training and information which are appropriate to the hazard.

After receiving formal training, employees will be required to sign a statement acknowledging their attendance at the program and their understanding of its contents.

The designated trainer for this facility will be: _____.

Other individuals within this facility who are trained or are knowledgeable in the hazards in your work area and who will be available to answer questions from employees are as follows: _____

_____.

The manner of training, training sites, scheduling, and other information relative to our training program will be published by the in-service education coordinator.

VIII. Information to Contractors

The maintenance supervisor will provide the following information to contractors for distribution to their employees:

- The hazardous materials to which they may be exposed while working at this facility
- Protective measures the contractor's employees must take in order to avoid the risk of exposure
- The labeling system in use at this facility
- The location of MSDSs at this facility

Contractors at this facility will be required to report to the maintenance supervisor the steps they have taken to ensure the safety of their employees while on our premises.

Contractors who bring chemicals to this facility will be required to provide us with appropriate hazard information on these substances, including the labels and precautionary measures to take for proper handling and storage.

Additional Information—If you are in need of any additional information relative to this Hazard Communication Program, the standard, or any of the hazardous substances in your work areas, you should notify: _____

_____.

Hazardous Materials List

Chemical Name   *Location/Department*

_____   _____

_____   _____

_____   _____

_____   _____

_____   _____

_____   _____

_____   _____

_____   _____

_____   _____

_____   _____

_____   _____

_____   _____

_____   _____

_____   _____

_____   _____

_____   _____

_____   _____

_____   _____

_____   _____

_____   _____

_____   _____

_____   _____

_____   _____

APPENDIX B: BLOODBORNE PATHOGENS EXPOSURE CONTROL PLAN (29 CFR 1910.1030)

I. Employee Protection Policy

In the interest of preventing accidental exposure to bloodborne pathogens and other infectious materials, the management of this facility has established a policy of employee protection. Direct questions or comments to _____.
This plan will be subject to review and revision as needed. Annual review of the plan will be scheduled for _____.

II. Definitions

All references to *occupational exposure* will mean reasonably anticipated skin, eye, mucous membrane, or parenteral contact with blood or other potentially infectious materials that may result from an employee's duties. *Regulated waste* means liquid or semi-liquid blood or other potentially infectious materials, contaminated items that would release blood or other potentially infectious materials in a liquid or semi-liquid state if compressed, items that are caked with dried blood or other potentially infectious materials and are capable of releasing these materials during handling, contaminated sharps, and pathological and microbiological wastes containing blood or other potentially infectious materials. For more definitions, consult Paragraph (B) of the OSHA bloodborne pathogens standard.

III. Recognizing Hazards

_____ has determined that the listed employees may be exposed to hazards of bloodborne pathogens while performing certain jobs or tasks in this facility. These employees are covered under the provisions of OSHA's bloodborne pathogens standard and our Exposure Control Plan.

Employee	Job Title	Procedure	Location
_____	_____	_____	_____
_____	_____	_____	_____
_____	_____	_____	_____
_____	_____	_____	_____
_____	_____	_____	_____
_____	_____	_____	_____

IV. Exposure Control Procedures

Universal (standard) precautions, as recommended or defined by the CDC and/or OSHA, will be observed in order to prevent contact with blood and other potentially infectious materials, unless they interfere with the proper delivery of healthcare or would create a significant risk to the personal safety of the worker.

Engineering Controls—Wherever possible, engineering controls will be utilized to re-duce potential exposure. Listed below are all controls in this facility:

Control	*Location*	*Installation Date*	*Maintenance Due Date*
_____	_____	_____	_____
_____	_____	_____	_____
_____	_____	_____	_____
_____	_____	_____	_____

_____ will be responsible for inspection and maintenance of these controls. Records of frequency of inspection and repairs will be maintained.

V. Required Work Practices (General)

- Employees must wash their hands immediately or as soon as possible after removal of gloves or other personal protective equipment (PPE) and after hand contact with blood or other potentially infectious materials.
- All PPE must be removed immediately upon leaving the work area or as soon as possible if overtly contaminated and placed in an appropriately designated area or container for storage, washing, decontamination, or disposal.
- Used needles and other sharps may not be sheared, bent, broken, recapped, or resheathed by hand. Used needles may not be removed from disposable syringes. Recapping is permitted only if no other alternative is feasible and must be done using an approved mechanical device or one-handed technique.
- Eating, drinking, smoking, applying cosmetics or lip balm, and handling contact lenses are prohibited in work areas where there is a potential for occupational exposure.
- Food and drink should not be stored in refrigerators, freezers, or cabinets where blood or other potentially infectious materials are stored or in areas of possible contamination.
- All procedures involving blood or other potentially infectious materials will be done in a manner which minimizes splashing, spraying, and aerosolization of these substances.
- Mouth pipetting/suctioning is prohibited.
- If conditions are such that handwashing facilities are not available, antiseptic hand cleaners are to be used. Because this is an interim measure, employees are to wash hands at the first available opportunity.
- The following hygienic work practices will also apply:

_____

_____

_____

VI. Personal Protective Equipment

Where there is potential for occupational exposure, employees will be provided with and required to use PPE including but not limited to gloves, aprons, gowns, lab coats, head and foot coverings, and eye protectors (i.e., goggles, glasses with side shields, face shields). This equipment will be provided at no cost to employees. When necessary, hypoallergenic, powderless, or other alternative gloving will be provided to employees who are allergic to types normally provided. Supplies may be obtained at the following locations: _____

_____ .

- Disposable gloves may not be decontaminated or washed for re-use.
- Prior to leaving the work area, PPE (including lab coats) must be removed and properly disposed of or placed into designated storage or laundry areas. Employees are not permitted to carry any type of PPE home for cleaning or other use.
- PPE will be considered appropriate only if it does not permit blood or other potentially infectious materials to pass through or contact the employee's clothing, skin, mouth, or mucous membranes.
- Listed below are types of PPE available for employees' use and circumstances under which they must be used:

Item	*Procedure*
_____	_____
_____	_____
_____	_____

VII. Decontamination of Personal Protective Equipment

Decontamination of PPE will be performed in the following manner:

_____

_____

_____

_____

VIII. Housekeeping

- Work surfaces must be decontaminated with an appropriate disinfectant after completion of a procedure, when surfaces are overtly contaminated, immediately after any spill of blood or other potentially infectious materials, and/or at the end of the work shift.
- Protective coverings such as plastic wrap, aluminum foil, or imperviously backed absorbent paper may be used to cover equipment and environmental surfaces.
- Equipment which may become contaminated with blood or other potentially infec-

tious materials will be checked routinely and prior to servicing or shipping and must be decontaminated as necessary.

- All bins, pails, cans, and similar receptacles intended for re-use which have a potential for becoming contaminated with blood or other potentially infectious materials must be inspected, cleaned, and disinfected immediately or as soon as possible upon visible contamination.
- Broken glassware which may be contaminated should not be picked up directly with the hands. It should be cleaned up using mechanical means such as a brush and dust pan, tongs, or forceps.
- Specimens of blood or other potentially infectious materials should be placed into a closable, leak-proof container labeled or color-coded according to OSHA requirements prior to being stored or transported.
- Re-usable items contaminated with blood or other potentially infectious materials must be decontaminated prior to washing and/or reprocessing.

IX. Worksite Maintenance

_____ is responsible for ensuring the worksite is maintained in a clean and sanitary condition. Facilities will be cleaned and disinfected with an appropriate agent according to the following schedule:

Location	_Equipment_	_Cleaner/Disinfectant_	_Frequency_
_____	_____	_____	_____
_____	_____	_____	_____
_____	_____	_____	_____

X. Waste Disposal

All infectious waste destined for disposal must be placed in closable, leak-proof containers or bags that are colored-coded or labeled as herein described to ensure that waste is properly disposed of and the following rules are observed.

- If outside contamination of the container or bag is likely to occur, then a second leak-proof container or bag which is closable and labeled or color-coded will be placed over the outside of the first and closed to prevent leakage during handling, storage, and transport.
- Immediately after use, sharps must be disposed of in closable, puncture-resistant, disposable containers which are leak-proof on the sides and bottom and are labeled or color-coded per OSHA specifications.
- These containers will be easily accessible to personnel and located in the immediate area of use.
- These containers will be replaced routinely and not allowed to overfill.
- Employees must not have to insert hands into the container in order to dispose of a sharp.
- When moving containers of sharps from the area of use, they must be closed immediately prior to removal or transport.

- Re-usable containers may not be opened, emptied, or cleaned manually or in any other manner which would pose the risk of percutaneous injury.
- Disposal of contaminated PPE will be provided at no cost to employees.
- In accordance with other applicable federal, state, and local regulations concerning medical waste, the following disposal procedures will be observed:

_____

_____

XI. Laundry

- Laundry which has been contaminated with blood or other potentially infectious materials or may contain contaminated sharps will be handled as little as possible and with a minimum of agitation.
- Contaminated laundry must be bagged at the location where it was used and should not be sorted or rinsed in patient care areas.
- Contaminated laundry should be placed and transported in bags that are labeled or color coded as described herein.
- Employees responsible for handling potentially contaminated laundry are required to wear protective gloves and other appropriate PPE to prevent occupational exposure during handling or sorting.
- Laundering of PPE is to be provided by the employer at no cost to employees.
- If laundry is shipped off-site to a second facility which does not utilize universal precautions in its handling of all laundry, bags or containers with appropriate labeling and/or color coding will be used to communicate the hazards associated with this material.
- Persons responsible for ensuring the proper handling, storage, shipping, or cleaning of contaminated laundry are: _____

_____.

- Additional requirements pertaining to the handling of laundry are as follows:

_____

_____

XII. Communication of Hazards to Workers

- Signs with the biohazard symbol will be posted at the entrance to waste storage areas.
- Warning labels must be affixed to containers of infectious waste.
- Labels will bear the legend described in the OSHA standard for bloodborne disease prevention. They will be fluorescent orange or orange-red or predominantly so, with lettering or symbols in a contrasting color.
- All labels will be an integral part of the container or will be affixed as close as safely possible to the container.
- Red bags or red containers may be substituted for labels on containers of infectious waste.

- The person responsible for ensuring that containers of biohazardous waste are properly labeled is: _____ .

XIII. Information and Training

- All workers with occupational exposure will participate in exposure control training prior to their initial assignment and at least annually thereafter. This training will be free of charge to employees and scheduled during working hours.
- The person responsible for coordinating the program is: _____

  _____ .

- At the end of each training session, employees will acknowledge their participation in the program by signing a form provided by the company.

XIV. Scope of Training

Employees will receive training and information in the following areas:

- A copy of the standard and an explanation of its contents
- A general explanation of the epidemiology and symptoms of bloodborne diseases
- Information on the modes of transmission of bloodborne pathogens
- An explanation of the Exposure Control Plan and its location
- An explanation of the methods for recognizing tasks and procedures that may involve exposure to blood or other potentially infectious materials
- An explanation of the use and limitations of practices that will prevent or reduce exposure, including appropriate engineering controls, work practices, and PPE
- Information on PPE, which will address types available, proper use, location, removal, handling, decontamination, and/or disposal
- An explanation of the basis for selection of PPE
- Information on the hepatitis B vaccine, including information on its efficacy, safety, and the benefits of being vaccinated
- Information on the appropriate actions to take and persons to contact in the event of an emergency
- Procedures to follow if an exposure incident occurs, including the method for reporting the incident
- Information on the medical follow-up that will be made available and on medical counseling provided to exposed individuals
- An explanation of signs, labels, and/or color coding
- A question-and-answer session with the trainer

XV. Medical Surveillance

Employees who may be exposed to potentially infectious materials within this company will be offered, at no cost, a vaccination for hepatitis B, unless the employee has had a previous vaccination or antibody testing reveals the employee to be immune. If an employee declines the vaccination, he/she must sign a waiver form. Should an employee be exposed to a potentially infectious material, post-exposure evaluations will be provided.

XVI. Medical Evaluation and Follow-Up

The exposed employee will be provided a confidential medical evaluation and follow-up to include:

- Documentation of the route(s) of exposure, hepatitis B virus and HIV antibody status of the source patient(s) (if known), and the circumstances under which the exposure occurred.
- If the source patient can be determined and permission is obtained, collection and testing of the source patient's blood to determine the presence of HIV or hepatitis B infection.
- Collection of blood from the exposed employee as soon as possible after the exposure incident for determination of HIV/hepatitis B virus status. Actual antibody or antigen testing of the blood or serum sample may be done at that time or at a later date, if the employee so requests. Samples will be preserved for at least 90 days.
- Follow-up of the exposed employee, including antibody or antigen testing, counseling, illness reporting, and safe and effective post-exposure prophylaxis, according to standard recommendations for medical practices.

XVII. Information Provided to Attending Physician

- A copy of the OSHA bloodborne pathogens standard and its appendices
- A description of the affected employee's duties as they relate to the employee's occupational exposure
- Results of the source individual's blood testing
- All of the employee's medical records, including vaccination records relevant to the treatment of the employee

XVIII. Attending Physician Written Opinion

- The physician's recommended limitations on the employee's ability to receive the hepatitis B vaccination
- A statement that the employee has been informed of the results of the medical evaluation and that the employee has been told about any medical conditions resulting from exposure to blood or other potentially infectious materials which require further evaluation or treatment
- Specific findings or diagnoses which are related to the employee's ability to receive the hepatitis B vaccination; any other findings and diagnoses will remain confidential

XIX. Written Opinion Notification

For each evaluation under this section, the company will obtain and provide the employee with a copy of the attending physician's written opinion within 15 days of completion of the evaluation.

XX. Recordkeeping

- Medical records will be kept for the length of the workers' employment plus 30 years.
- Training records will be kept for three years.

Bloodborne Pathogens Training

I attended company-provided training on bloodborne pathogens on _____.
Topics covered in this training included:

- The OSHA standard and explanation of its contents
- A general explanation of the epidemiology and symptoms of bloodborne diseases
- Modes of transmission of bloodborne pathogens
- An explanation of the Exposure Control Plan
- An explanation of the appropriate methods for recognizing tasks and procedures that may involve exposure to blood or other potentially infectious materials
- An explanation of the use and limitations of practices that will prevent or reduce exposure, including engineering controls, work practices, and PPE
- Information on types, proper use, location, removal, handling, decontamination, and/or disposal of PPE
- The basis for selecting PPE
- Information on the hepatitis B vaccine, including its efficacy, safety, and the benefits of being vaccinated
- Information on how to respond to emergencies
- Procedures to follow if an exposure incident occurs, including the method of reporting the incident
- The medical follow-up that will be made available, as well as information on medical counseling provided to exposed individuals
- An explanation of signs, labels, and/or color coding
- A question-and-answer session with the trainer

_____   _____
          Trainer Initials   *Employee Signature*

Definitions

AIDS—Acquired immunodeficiency syndrome, a disease that results from HIV.

Antigen—A substance that causes antibody formation.

Blood—The OSHA standard refers to human blood, human blood components, and products made with human blood.

Bloodborne pathogen—A pathogenic microorganism that is present in human blood and can cause disease in humans.

Contaminated—The presence or the reasonably anticipated presence of blood or other potentially infectious materials on an item or surface.

Contaminated laundry—Laundry that has been soiled with blood or other potentially infectious materials or may contain sharps.

Contaminated sharp—Any contaminated object that can penetrate the skin including but not limited to needles, scalpels, broken capillary tubes, and exposed ends of dental wires.

Decontamination—The use of physical or chemical means to remove, inactivate, or destroy bloodborne pathogens on a surface or item to the point where they are no longer capable of transmitting infectious particles and the surface or item is rendered safe for handling, use, or disposal.

Engineering controls—Physical controls (e.g., sharps disposal containers, self-sheathing needles, etc.) that isolate or remove the hazards of bloodborne pathogens from the workplace.

Exposure incident—A specific eye, mouth, or other mucous membrane, non-intact skin, or parenteral contact with blood or other potentially infectious materials that results from the performance of an employee's duties.

HBV—Hepatitis B virus. One of the viruses that cause illness directly affecting the liver. It is a bloodborne pathogen.

Hepatitis—A disease that causes swelling, soreness, and loss of normal function of the liver. Symptoms include weakness, fatigue, anorexia, nausea, abdominal pain, fever, and headache. Jaundice is a symptom that may develop later.

Immune—Resistant to infectious disease.

Immunization—A process or procedure by which resistance to infectious disease is produced in a person.

Mucous membrane—Any one of the four types of thin sheets of tissue that cover or line various parts of the body. An example is the skin lining the nose and mouth.

Mucus—The clear secretion of the mucous membrane.

Non-intact skin—Skin that has a break in the surface. It includes but is not limited to abrasions, cuts, hangnails, paper cuts, and burns.

Occupational exposure—Reasonably anticipated skin, eye, mucous membrane, or parenteral contact with blood or other potentially infectious materials that may result from the performance of an employee's duties.

Parenteral—Piercing mucous membranes or skin barrier through such events as needlesticks, human bites, cuts, and abrasions.

Pathogen—Any virus, microorganism, or other substance that is capable of causing disease.

Percutaneous—Performed through the skin, as in draining fluid from an abscess using a needle.

Personal protective equipment—Specialized clothing or equipment worn by an employee for protection against a hazard.

Source individual—Any individual, living or dead, whose blood or other potentially infectious material may be a source of occupational exposure to an employee.

Sterilize—The use of a physical or chemical procedure to destroy all microbial life.

T_4—A cell in the immune system that acts as a sensor to activate the immune system.

Universal (standard) precautions—A comprehensive approach to infection control that treats all human blood and certain human body fluids as if known to be infectious for HIV, hepatitis B virus, and other bloodborne pathogens.

Vaccine—A suspension of inactive or killed microorganisms administered orally or injected to induce active immunity to infectious disease.

Sample Bloodborne Pathogens Test

1. Training in bloodborne pathogens must be provided:
   a. once a month
   b. twice a year
   c. annually
   d. only when hired

2. Bloodborne pathogens include:
   a. HIV, hepatitis B virus, and syphilis
   b. cancer and pneumonia
   c. diabetes and heart disease

3. Infection with hepatitis B may include:
   a. increased energy and normal liver function
   b. weakness, fatigue, and loss of appetite
   c. a spotted rash on the palms of the hands and soles of the feet

4. Infection with HIV is detected by:
   a. how a person looks
   b. a special test done only at the Centers for Disease Control
   c. tests for antibodies in blood and urine

5. An engineering control is:
   a. an attempt to design safety into tools and the workspace
   b. the practice of using special equipment
   c. a process of controlling people infected with hepatitis B virus

6. Personal protective equipment:
   a. is designed to scare people
   b. is not necessary to use
   c. includes things such as masks, gowns, and face shields

7. Universal precautions apply to:
   a. blood, any body fluid with blood, and semen
   b. sputum, urine, and tears
   c. only people known to be infected with syphilis

8. Hepatitis B virus immunization requires the worker to:
   a. purchase the vaccine
   b. receive the vaccine for hepatitis B if the employer insists
   c. receive three doses over six months if the employee elects to be vaccinated

9. The hepatitis B vaccine:
   a. is less than 20% effective
   b. is 96% effective after three doses
   c. no one has ever tested the effectiveness of the vaccine

10. Handwashing:
    a. is only necessary before meals
    b. with antiseptic towels is never approved
    c. is the most effective means of preventing disease

APPENDIX C: CHEMICAL HYGIENE PLAN

Facility Name

The general intent of the Chemical Hygiene Plan for _____ (*insert name of organization*) is:

1. To protect laboratory employees from health hazards associated with the use of hazardous chemicals in our laboratory
2. To assure that our laboratory employees are not exposed to substances in excess of the permissible exposure limits as defined by OSHA in 29 CFR 1910, Sub-part Z

The plan will be available to all employees for review and a copy will be located in the following areas: _____

_____.

This plan will be reviewed annually by _____ (*insert name or position*) and updated as necessary.

_____ (*insert name*) is designated as the chemical hygiene officer. (See Sections VI and VII for details.)

I. Standard Operating Procedures

Standard operating procedures are general procedures of laboratory operation relevant to safety and health when using chemicals. Section E of Appendix A of 29 CFR 1910.1450 lists the following considerations:

A. Accidents and spills
B. Avoidance of routine exposure
C. Choice of chemicals
D. Eating, drinking, smoking, etc.
E. Equipment and glassware
F. Exiting
G. Horseplay
H. Mouth suction
I. Personal apparel
J. Personal housekeeping
K. Personal protection
L. Planning
M. Unattended operations
N. Use of hood
O. Vigilance
P. Waste disposal and storage
Q. Working alone

Section F, Appendix A of 1901.1450 includes additional safety recommendations for:

A. Corrosive agents
B. Electrically powered laboratory apparatus
C. Fires and explosions
D. Low-temperature procedures
E. Pressurized and vacuum operations
F. Compressed gases
G. Chemical storage

Attached to this plan in Appendix _____ are the standard operating procedures in place at _____ (*insert name of organization*) for the safe handling of chemicals in our laboratory. (Often this will be your laboratory safety manual which is already in place. If you have the following programs, they can also be referenced.)

The written portion of the *Laser Safety Program* is located _____ .
The written portion of the *Radiation Safety Program* is located _____ .
The written portion of the *Biological Safety Program* is located _____ .

II. Criteria for Use of Control Measures to Reduce Employee Exposure to Hazardous Chemicals

A. The following operations must be performed in *laboratory fume hoods*:

_____

_____

B. The following operations must be performed in *biological safety cabinets*:

_____

_____

C. The following operations must be performed in *glove boxes*:

_____

_____

D. Respirators must be used in accordance with the respiratory protection policy of _____ (*insert name of organization*) and with the OSHA respirator standard (29 CFR 1910.134). This policy and associated documentation are filed _____ (*insert location*) for employee review.

E. Appropriate protective apparel compatible with the required degree of protection for substances handled must be used. _____ (*insert name of position*) will advise employees on the use of gloves, gowns, eye protection, barrier creams, etc., Permeability charts are available _____ _____ (*insert location*).

F. Employees will be instructed on the location and use of eyewash stations and safety showers. _____ (*insert name of position*) is responsible for this instruction.

G. Employees will be trained _____ (*insert how often, e.g., annually*) in the use of fire extinguishers and other fire protection systems.

III. Maintenance of Fume Hoods and Other Protective Equipment

Fume hoods will be inspected every _____ months by _____ _____ (*insert name of position*). Adequacy of face velocity will be determined by _____ (*insert method*). Reports of hood inspections are filed _____ (*insert location*) for employee review.

(Repeat the above for each additional major category of protective equipment, such as biological safety cabinets, ventilation of storage cabinets, interlocks on high-voltage equipment, safety showers, eyewash stations, etc., indicating how often they are inspected, by whom, what is measured, and where the inspection records and checklists are filed.)

IV. Employee Information and Training

A. Each employee covered by the laboratory standard will be provided with information and training so that he/she is apprised of the hazards of chemicals present in his/her work area. This training will be given at the time of initial assignment and prior to new assignments that involve different exposure situations. Refresher training will be given _____ (*insert how often*).

B. The training/information session will include:
    1. The contents of 1910.1450 and its appendices, available to employees at _____ (*insert location*)
    2. The availability and location of the written Chemical Hygiene Plan
    3. Information on OSHA permissible exposure limits where they exist and other recommended exposure limits
    4. Signs and symptoms associated with exposure to hazardous chemicals in laboratories
    5. Location of reference materials, including all MSDSs received, on the safe handling of chemicals in laboratories
    6. Methods to detect the presence or release of chemicals (i.e., monitoring, odor thresholds, etc.)
    7. The physical and health hazards of chemicals in laboratory work areas
    8. Measures to protect employees from these hazards, including:
        a. Standard operating procedures
        b. Work practices
        c. Emergency procedures
        d. Personal protective equipment
        e. Details of the Chemical Hygiene Plan

C. _____ (*insert name of position*) is responsible for conducting the training sessions, which will consist of _____ _____ (*insert training methods, e.g., videos, slides, lectures, etc.*). An outline of the training program is provided in Appendix _____ .

D. Each employee must sign a form documenting that he/she has received training. (Note that a signed form does not necessarily mean that the person has understood and retained the training provided. An enforcement officer would determine understanding and retention based on employee interviews and employee knowledge.)

E. _____ (*insert name of position*) is responsible for developing standard operating procedures. _____ (*insert name of position*) is responsible for the portion of the training on standard operating procedures.

V. Prior Approval for Specific Laboratory Operations

Certain laboratory procedures which present a serious chemical hazard require prior approval by _____ (*insert name of position*) before work can begin. For this facility, these procedures include:

A. Work with select carcinogens
B. Work with reproductive hazards
C. Work with neurotoxins
D. Work with acutely hazardous chemicals (Consider the eight physical hazards as well as the health hazards in this determination.)

These chemicals include (*insert a list of the acutely hazardous chemicals, e.g., cyanide*):

_____ _____

_____ _____

_____ _____

_____ _____

(If the laboratory does not utilize these classes of chemicals, then include a sentence which states "Our laboratory does not at this time use any chemicals which are sufficiently hazardous to require prior approval before they are used.")

VI. Medical Consultation and Examination

_____ (*insert name of organization*) will provide to affected employees medical attention including follow-up examinations which _____ _____ (*insert name of clinic or physician*) determines are necessary under the following circumstances:

A. Whenever an employee develops signs and symptoms associated with a hazardous chemical to which he/she may have been exposed, the employee will be provided an opportunity to receive appropriate medical examination. The employee must contact the chemical hygiene officer to initiate the medical program.

B. Where exposure monitoring reveals an exposure level routinely above the OSHA action level (in the absence of an action level, exposure above the OSHA permissible exposure level for OSHA-regulated substances for which there are medical monitoring and medical surveillance requirements), medical surveillance will be established for that employee.

Currently our laboratory uses:

1. _____ (*e.g., benzene*)

2. _____ (*e.g., formaldehyde*)

3. _____ (*e.g., list other substances covered*)

All of these substances have a separate OSHA standard with medical surveillance requirements. (If none of these substances is used, indicate that no substances for which OSHA has medical monitoring requirements are being used.)

C. Whenever an event takes place in the work area, such as a spill, leak, explosion, or other occurrence resulting in the likelihood of a hazardous exposure, the affected employee, laboratory, or custodian will be provided an opportunity for a medical consultation. This consultation is for the purpose of determining the need for a medical examination.

D. All medical examinations and consultations are provided by _____ _____ (*insert physician's name*) or at _____ _____ (*insert clinic/hospital name*). All aspects of these examinations are provided by a licensed physician or supervised by a licensed physician. These examinations are provided without cost to the employee, without loss of pay, and at a reasonable time and place.

E. The _____ (*insert name of position, e.g., chemical hygiene officer*) will provide the following information to the physician:
    1. Identity of the hazardous chemical to which the employee may have been exposed
    2. A description of the conditions of the exposure, including exposure date if available
    3. A description of signs and symptoms of exposure that the employee is experiencing (if any)

F. The written opinion that the company receives from the physician will include:
    1. Recommendations for future medical follow-up
    2. Results of examination and associated tests
    3. Any medical condition revealed which may place the employee at increased risk as the result of a chemical exposure
    4. A statement that the employee has been informed by the physician of the results of the examination/consultation and told of any medical conditions that may require additional examination or treatment

G. The material returned to _____ (*insert name of organization*) by the physician will not include specific findings and diagnoses which are unrelated to occupational exposure.

VII. Responsibilities Under the Chemical Hygiene Plan

_____ (*insert name of position or individual*) is designated as the chemical hygiene officer for _____ (*insert name of organization*). (The qualifications of this individual are important. This person should have a background in both chemistry and safety.)

A chemical hygiene committee will be formed. The membership list and minutes of the committee meetings are filed _____ (*insert location*) for employee review.

(You may wish at this point to follow the categories in Appendix A of the Lab Standard 1910.1450 and assign some chemical hygiene duties to all staff. The categories used in this appendix are:

Chief executive officer
Department supervisor
Chemical hygiene officer
Laboratory supervisor
Project director
Laboratory worker

You may wish to designate your existing committee or a subgroup of that committee as your chemical hygiene committee.)

VIII. Additional Protection for Work with Select Carcinogens, Reproductive Toxins, and Chemicals with High Acute Toxicity

When any of these chemicals are used, the following provisions must be employed where appropriate:

1. Establishment of a designated area
2. Use of containment devices such as fume hoods or glove boxes
3. Procedures for safe removal of contaminated waste
4. Decontamination procedures

Appendix A of the standard has detailed programs for working with these chemicals. If you are using them, refer to Appendix A as a guide for your detailed procedures.

Note that according to the standard, a *select carcinogen* means any substance which meets one of the following criteria: (1) it is regulated by OSHA as a carcinogen; or (2) it is listed under the category "known to be carcinogens" in the Annual Report on Carcinogens published by the National Toxicology Program (NTP) (latest edition); or (3) it is listed under Group 1 ("carcinogenic to humans") in the *International Agency for Research on Cancer (IARC) Monographs* (latest edition); or (4) it is listed in either Group 2A or 2B by the IARC or under the category "reasonably anticipated to be carcinogens" by the NTP.

Appendix _____ to this plan includes the special procedures used in this laboratory for these chemicals.

IX. Emergency Response

Two additional OSHA standards interface with the Chemical Hygiene Plan: 1910.38 Employee Emergency Plans and Fire Prevention Plans and 1910.120(p) and (q) Hazardous Waste Operations and Emergency Response (developed in response to SARA Title III). (Review these two standards and develop appropriate emergency procedures for your facility if your facility is covered by one of these standards.) Appendix _____ is our organization's Emergency Action Plan under 1910.38. Appendix _____ is our organization's Emergency Response Plan under 1910.120.

APPENDIX D: PERSONAL PROTECTIVE EQUIPMENT

I. General Requirements

The facility will be evaluated to determine if hazards are present, or likely to be present, which necessitate the use of personal protective equipment (PPE).

A. If such hazards are present, or likely to be present, the facility will select and have each affected employee use the types of PPE that will protect against the identified hazards. PPE must properly fit each affected employee, and the employer must verify the hazard assessment in writing.

B. Damaged or defective PPE will not be used.

C. The facility will provide training to each employee required to use PPE. Training will include when PPE is necessary; what PPE is necessary; how to wear PPE; and the proper care, maintenance, useful life, and disposal of PPE.

II. Eye and Face Protection

Employees must use appropriate eye or face protection when exposed to eye or face hazards from flying particles, molten metal, liquid chemicals, acids or caustic liquids, chemical gases or vapors, or potentially injurious light radiation. Protective eye and face devices purchased after July 5, 1994 must comply with ANSI Z87.1-1989 or be demonstrated to be equally effective. Devices purchased before that date must comply with ANSI Z87.1-1968 or be equally effective.

III. Head Protection

Employees must wear protective helmets when working in areas where there is a potential for injury to the head from falling objects. Protective helmets designed to reduce electric shock hazard must be worn by each affected employee when near exposed electric conductors which could contact the head. Protective helmets purchased after July 5, 1994 must comply with ANSI Z89-1986 or be equally effective. Helmets purchased before that date must comply with ANSI Z89.1-1969 or be equally effective.

IV. Foot Protection

Employees must wear protective footwear when working in areas where there is a danger of foot injury due to falling or rolling objects or objects piercing the sole, and where employees' feet are exposed to electric hazards. Protective footwear purchased after July 5, 1994 must comply with ANSI Z41-1991 or be equally effective. Protective footwear purchased before that date must comply with ANSI Z41.1-1967 or be equally effective.

V. Hand Protection

Employees must use appropriate hand protection when hands are exposed to hazards such as those from skin absorption of harmful substances, severe cuts or lacerations, severe abrasions, punctures, chemical burns, thermal burns, and harmful temperature extremes. Employees should base the selection of appropriate hand protection on evaluation of the performance characteristics of the hand protection relative to the tasks to be performed, conditions present, duration of use, and the hazards and potential hazards identified.

VI. Substitute Controls

PPE drastically cuts the chance of injury and reduces the severity of injury if an accident does occur. PPE should not be used as a substitute for other more effective hazard controls. PPE should be used together with engineering and substitution controls to provide worker safety and health on the job.

Personal Protective Equipment Survey

A survey of this department to determine hazards requiring PPE was conducted on _____. The survey revealed the following PPE requirements:

Department	Hazard	PPE
_____	_____	_____
_____	_____	_____
_____	_____	_____
_____	_____	_____
_____	_____	_____
_____	_____	_____
_____	_____	_____
_____	_____	_____
_____	_____	_____

_____

Certification Signature/Title

COMMENTS: _____

_____

_____

_____

_____

_____

_____

APPENDIX E: HEALTHCARE SAFETY ORIENTATION/TRAINING RECORD

Employee Name: _____ Hire Date: _____

Safe Work Agreement—I understand that this facility is committed to providing a safe and healthful work environment. I acknowledge my personal responsibility to follow all safety rules, practice job safety, and report unsafe practices/conditions to my supervisor. I also understand that if I refuse or fail to use safety appliances provided by the company or willfully or intentionally disobey company safety rules, I forfeit any rights to recover workers' compensation benefits under Alabama Code 25-5-51.

_____   _____
Signature/Date   *Witness Signature*

Orientation Topics	*Employee Initials*	*Trainer Initials*	*Date*
Safety policies and rules			
Substance abuse screening procedures			
Accident/injury reporting policies			
Workers' compensation procedures			
Emergency action/disaster training			
Hazard communication training			
Bloodborne pathogens/TB			
Fire procedures			
Lifting/transfer/body mechanics			

Safety Training Topics	*Initials*	*Date*	*Initials*	*Date*	*Initials*	*Date*
Back and lifting training						
Personal protective equipment						
Hazard communication (refresher training)						
Bloodborne pathogens (annual training)						
Fire extinguisher (annual training)						
Respirator use and training						
Hazardous waste operations						
Formaldehyde (annual)						

Other Training	Initials	Date

COMMENTS:

APPENDIX F: TRADE ASSOCIATION INDOOR AIR QUALITY GUIDELINES

The following associations have developed guidelines of care that may have a direct or indirect impact on indoor air quality. The EPA and NIOSH do not endorse these standards.

Air Conditioning Contractors of America (ACCA)

Technical Reference Bulletin Series—Indoor air quality is one of the topics covered in this series of technical bulletins on heating, ventilation, and air conditioning (HVAC). Bulletins can be filed in the *ACCA Technical Reference Notebook*. The Air Side Design tab of the notebook includes bulletins devoted to indoor air quality control.

Air-Conditioning and Refrigeration Institute (ARI)

Air Conditioning and Refrigeration Equipment General Maintenance Guidelines for Improving the Indoor Environment (1991)—General maintenance requirements for heating, ventilation, air conditioning, and refrigeration (HVACR) equipment. Specific equipment/component maintenance is given for the following: air cleaning systems; ducts; registers/diffusers and air terminals; dampers/economizers; drain pans; air handlers; humidifiers; package terminal units; and evaporator, condenser, hydronic, and economizer coils. The guidelines do not supersede any maintenance instructions provided by the manufacturer. In addition, the institute has issued an *Indoor Air Quality Briefing Paper* that addresses the interactions between HVACR equipment and the quality of indoor air.

Associated Air Balance Council (AABC)

National Standards for Testing and Balancing Heating, Ventilation, and Air Conditioning Systems (1989)—Establishes a minimum set of field testing and balancing standards and provides comprehensive and current data on testing and balancing HVAC systems. Chapters receiving special attention include Cooling Tower Performance Tests, Sound Measurements, Vibration Measurements, Fume Hoods, and AABC General Specifications. The book contains a complete index to the technical data provided.

National Environmental Balancing Bureau (NEBB)

Procedural Standards for Testing, Adjusting, and Balancing of Environmental Systems (1991)—A "how-to" set of procedural standards that provide systematic methods for testing, adjusting, and balancing (TAB) of HVAC systems. Includes sections on TAB instruments and calibration, report forms, sample specifications, and engineering tables and charts. A valuable innovation is the "Systems Ready to Balance" start-up checklist to help organize jobs systematically. Other features include additional engineering data, condensed duct design tables/charts, hydronic design tables/charts, and pertinent HVAC equations in U.S. and metric units.

National Pest Control Association (NPCA)

Good Practice Statements—Periodically updated and officially approved and adopted by the association's board of directors, these "Good Practice Statements" are designed as guidelines for performing various services rather than standards of operation. In addition, the association produces a self-study series for technicians that covers five areas of pest control, management manuals, an encyclopedia of structural pest control, a number of specific subject-matter technical reference manuals, and a pamphlet series.

Sheet Metal and Air Conditioning Contractor's National Association (SMACNA)

HVAC Duct Construction Standards—Metals and Flexible (1985)—Primarily for commercial and institutional work, this set of construction standards is a collection of material from earlier editions of SMACNA's low-pressure, high-pressure, flexible duct, and duct liner standards. In addition, SMACNA has published a manual entitled *Indoor Air Quality* which contains basic information on many aspects of indoor air quality and guidance on conducting building evaluations and indoor air quality audits. Other related SMACNA publications include *HVAC Duct Systems Inspection Guide, HVAC Systems—Testing, Adjusting and Balancing,* and *HVAC Air Duct Leakage Test Manual.*

APPENDIX G: OSHA LABORATORY TRAINING REQUIREMENTS

I. Occupational Exposure to Hazardous Chemicals in Laboratories Standard (29 CFR 1910.1450)

A. Content of the standard and appendices
B. Location and explanation of the Chemical Hygiene Plan
C. Location of reference materials and Material Safety Data Sheets
D. Details of access to medical consultation and management system

II. Physical Hazards

A. Combustible liquids
B. Compressed gas
C. Explosive
D. Flammable
E. Organic peroxide
F. Pyrophoric
G. Unstable (reactive)
H. Water reactive

III. Health Hazards

A. Local
    1. Irritants
    2. Corrosives
B. Systemic
    1. Toxics
        a. Acute/chronic
        b. Nervous system effects
        c. Respiratory system effects
        d. Reproductive system effects
    2. Sensitizers
    3. Carcinogens

IV. Route of Exposure

A. Inhalation
B. Skin absorption
C. Ingestion

V. Amount of Absorption

A. Gases/vapors
B. Particulates
    1. Dust

    2. Mist
    3. Fume

VI. Dose

A. Work practices
B. Personal hygiene
C. Weight
D. Personal protective equipment
E. Environmental controls

VII. Duration of Exposure

VIII. Exposure Limits Including PEL

A. Definition
B. Established by
    1. Chemical similarity
    2. Animal studies
    3. Human studies

IX. Air Sampling

A. Required by OSHA
B. Employee reports of illness
C. Confined workspace
D. Other

X. Response

A. Age
B. Gender
C. Body size
D. Health status
E. Personal habits
F. Other exposures

XI. Employee Concerns

A. Symptoms limited/many causes
B. Documentation
C. Referral
D. Refusal to work

XII. Organizational Standard Operating Procedures

APPENDIX H: SAMPLE SUBSTANCE ABUSE POLICY

I. Policy Statement

This facility is committed to operating a drug- and alcohol-free workplace to ensure the safety and well-being of our employees, visitors, and residents. The facility will not tolerate the abuse of alcohol in the workplace. The facility will not tolerate the use of illegal drugs by its employees at any time.

The facility prohibits employees from unlawfully manufacturing, distributing, possessing, or using controlled substances. It is a condition of employment that employees refrain from using illegal drugs while on facility time or premises. In addition, employees are prohibited from working under the influence of any substance that may have an adverse effect on their behavior or ability to perform their job. **Violation of this policy will result in disciplinary action up to and including termination of employment.**

II. Rules and Regulations

A. The distribution, sale, purchase, use, or possession of intoxicants, non-prescribed narcotics, hallucinogenic drugs, marijuana, or other non-prescribed controlled substances while on facility property or during work hours is prohibited.

B. The distribution, sale, purchase, use, or possession of equipment, products, and materials which are used, intended for use, or designed for use with non-prescribed controlled substances while on facility property or during work hours is prohibited.

C. Reporting to or being at work with a measurable quantity of intoxicants, non-prescribed narcotics, hallucinogenic drugs, marijuana, or other non-prescribed controlled substances in blood or urine is prohibited. The term *measurable quantity* refers to cutoff levels specified by 49 CFR Part 40 (40.29 Laboratory Analysis Procedures).

D. Reporting to or being at work under the influence or with a measurable quantity of alcohol in blood or urine is prohibited. The term *measurable quantity* is an amount equal to one-half the standard used by the state of Alabama for determination of intoxication in DUI cases.

E. Reporting to or being at work with a measurable quantity of prescribed or over-the-counter narcotics or drugs in blood or urine or use of prescribed or over-the-counter narcotics or drugs where, in the opinion of the facility, such use prevents the employee from performing the duties of his/her job or poses a risk to the safety of the employee, other persons, or property is prohibited. Any employee taking a prescribed or over-the-counter narcotic or drug must advise his/her supervisor of its use. The employee may remain on his/her job or may be required to take a leave of absence or other appropriate action as determined by management.

F. Adherence to the facility's policy on drugs and alcohol is a condition of continued employment for all employees. All employees will be required to sign the attached acknowledgment form and to consent to this policy.

G. All employees **must** notify the facility of any criminal drug statute conviction for a violation arising out of conduct in the workplace within five days of such conviction. Failure to do so may result in disciplinary action up to and including termination.

H. Managers and supervisory employees are responsible for enforcing facility policies. The possession, distribution, or use of illegal drugs or unauthorized controlled substances, whether on or off duty, impairs their ability to enforce these policies and may result in disciplinary action up to and including termination.

I. An employee who refuses to submit immediately upon request to a search of his/her person or property or to a blood test, urinalysis, breathalyzer test, or other diagnostic test or who otherwise is in violation of this policy is subject to disciplinary action up to and including immediate dismissal. An employee who refuses to submit to or cooperate with a blood or urine test after an accident forfeits his/her right to recover workers' compensation benefits under Alabama Code § 25-5-51.

III. Testing for Prohibited Substances

Under the facility Drug and Alcohol Abuse Policy, an employee may be requested to undergo a urinalysis, blood test, or other diagnostic test. This facility reserves the right to test under the following circumstances:

A. *Pre-Employment*—As a condition of the employment application.

B. *Post-Accident Drug Screen*—Standard procedure for an employee who has a work-related accident includes testing to determine the presence of alcohol, intoxicants, and/or illegal drugs or substance. A post-accident drug screen will be required for all employees involved in a work-related accident which required medical treatment by a physician. An employee who tests positive for any of these substances will be immediately discharged and the employee's claim for workers' compensation benefits may be denied.

C. *Reasonable Suspicion*—When there is reason to believe in the opinion of the facility that an employee has reported to work or is on facility property with a measurable quantity of intoxicants, drugs, or narcotics in blood or urine.

D. *Random*—The facility management reserves the right to perform unannounced random substance abuse testing in a non-discriminatory manner.

Note—Drug testing procedures adopted by the U.S. Department of Transportation in 49 CFR Part 40 will be utilized.

IV. Search

When there is reason to believe in the opinion of facility management personnel that an employee is under the influence of intoxicants, drugs, or narcotics or is in possession of any intoxicants, drugs, narcotics, equipment, products, or materials which are used, intended for use, or designed for use with non-prescribed controlled substances, the facility may request that the employee submit to a search by facility representatives of his/her person and/or property (including offices, lockers, desks, cabinets, closets, and vehicles brought onto facility premises).

V. Drug-Free Awareness Education Program

The facility will conduct an education and training program to inform employees about the dangers of drug and alcohol abuse, the indicators of drug and alcohol abuse, the facility's policy of maintaining a drug-free workplace, the availability of community drug counseling and rehabilitation resources, and the penalties that may be imposed for violation of the policy.

Supervisory personnel will receive additional training on the conduct, behavior, and indicators of drug and alcohol abuse.

VI. Employee Assistance Programs

As part of this policy, the facility encourages any employee with a substance abuse problem to seek help. Any employee who acknowledges a drug or alcohol problem and requests leave for rehabilitation will be granted leave consistent with the facility's leave policy as long as the request is made **prior** to any request by the facility for a drug or alcohol screening.

Below is a partial list of public out-patient treatment facilities available throughout the state to give guidance and assistance to employees with substance abuse problems. Please be advised that this list is provided as a service by this facility and that this facility accepts no responsibility or liability as to the cost of the actual treatment. Any treatment must be paid for by the employee.

A. National Institute on Drug Abuse (NIDA)
   5600 Fishers Lane
   Rockville, MD 20857
   (301) 443-6245
B. National Council on Alcoholism and Drug Dependency
   (800) NCA-CALL
C. National Institute on Drug Abuse Hot Line
   (800) 662-HELP

VII. Records and Confidentiality

All information, interviews, reports, statements, memoranda, and drug test results, written or otherwise, received by the facility as a result of its drug testing program are confidential communications. Unless required by law, this facility will not release such information to third persons without the written consent of the applicant or employee. However, the facility may use such information as evidence, if relevant, in proceedings relating to a discharged employee's entitlement to unemployment compensation benefits and workers' compensation benefits or in other legal proceedings involving the discharged employee. The facility reserves the right to disclose such information to its attorneys or other persons defending the facility where such information is relevant to its claim or defense in a civil or administrative matter and to supervisors on a need-to-know basis. Release of such information under any other circumstances shall be solely pursuant to a written consent form signed voluntarily by the person tested, unless such release is compelled by a court or a hearing officer.

VIII. Interpretations

Facility property covered by this policy includes property of any nature owned, controlled, or used by the facility, including parking lots, offices, desks, lockers, and vehicles.

Nothing in this policy alters the fact that employees are employed for an indefinite period and that either the employee or the facility may terminate such employment with or without cause at any time for any reason. Neither this policy nor any related policies, practices, or guidelines are employment contracts or parts of any employment contract. Due to the nature of the facility's operations and the possible need to accommodate individual situations, the provisions of this policy or of any related policies, practices, or guidelines may not apply to every employee in every situation. This facility reserves the right to rescind, modify, or deviate from this or any other policy, practice, or guideline as it considers necessary in its sole discretion, either in individual or facility-wide situations with or without notice.

Drug and Alcohol Abuse Policy

Employee Acknowledgment

This facility's intention in initiating this policy is to help provide for the maximum safety and well-being of employees, residents, and visitors. Your assistance and cooperation in the achievement of this goal are vitally important!

I hereby acknowledge that:

- ☐ I have received a copy of the Drug and Alcohol Abuse Policy.

- ☐ I have read the Drug and Alcohol Abuse Policy.

- ☐ I have had read to me a copy of the Drug and Alcohol Abuse Policy.

- ☐ I understand the Drug and Alcohol Abuse Policy.

I understand that there may be situations where I will be required to take a drug test. I also understand that if I refuse to take the test or if the test proves positive, I will be discharged. I further understand that I will be required to submit to a drug test should I be involved in an on-the-job accident and that my failure or refusal to submit to or cooperate with a drug test following such an incident may result in the forfeiture of my rights to recover workers' compensation benefits. I understand and accept this policy as a condition of my employment.

_____

Employee Signature

_____

Date Read and Signed

_____

Witness Signature

APPENDIX I: SAMPLE LOCKOUT/TAGOUT POLICY

I. Basic Requirements

OSHA places four basic requirements upon employers whose workers are engaged in service and/or maintenance functions:

    A. Procedures for lockout/tagout
    B. Training of workers
    C. Accountability of engaged workers
    D. Administrative controls

II. Definitions

The definitions provided by the standard may require further clarification. Therefore, the following terms used during service and maintenance operations are related to the definitions provided in *Webster's Ninth New Collegiate Dictionary.*

    A. *Maintenance*—The act of maintaining; the state of being maintained
    B. *Maintain*—To keep in an existing state (as of repair, efficiency, or validity); preserve from failure or decline (maintaining machinery)
    C. *Servicing*—To repair or provide maintenance for
    D. *Refurbish*—To brighten or freshen up; to renovate
    E. *Renovate*—To restore to a former better state (as by cleaning, repairing, or rebuilding)
    F. *Modify*—To make basic or fundamental changes in, often to give a new orientation to or to serve a new end

III. General Policies

The purpose of the lockout procedure is to render inoperative electrical systems, pumps, pipelines, valves, and any other systems that could be energized while associates are working.

The lockout system is administered by the maintenance department. Locks and tags should be issued by the maintenance department.

All energy sources will be locked out and a DO NOT OPERATE tag affixed to the equipment or system, indicating who installed the lock and the reason the system was locked out. An example would be: electrical systems as they are energized will be locked out until they are released for service. Any time repairs or modifications are made to electrical systems, either temporary or permanent, they will be locked out. Locks will be applied to the main disconnect switch whenever possible.

IV. Enforcement of Lockout Procedures

Our organization is required by OSHA to enforce the use of lockout procedures. Any employee who intentionally fails to follow lockout procedures will face disciplinary action. Violation of lockout procedures will be handled in the same manner as any other violation of safety rules and may include reprimand, probation, suspension, or dismissal.

The purpose of lockout inspection is to ensure that procedures are being used properly and result in the safe control of hazardous energy.

V. OSHA Requirements

OSHA rules state that equipment must be locked out during equipment servicing and maintenance whenever employees are exposed to injury from unintentional machine movement or startup. According to OSHA's regulations, servicing or maintenance procedures that require lockout include:

> Workplace activities such as constructing, installing, setting up, adjusting, inspecting, modifying, and maintaining and/or servicing machines or equipment. These activities include lubrication, cleaning, or unjamming of machines or equipment and making adjustments or tool changes, *where the employee may be exposed to the unexpected energization or startup of the equipment or release of hazardous energy.*

VI. Cord and Plug Connected Equipment

OSHA gives the facility two options to protect employees working on cord and plug connected equipment from injury due to accidental startup or movement:

A. The first option is to lockout this equipment. Lockable covers that fit over a plug are available from several different sources.

B. The second option does not require lockout. It does, however, require that the employee maintain control over the cord and plug. This means that the plug is in the employee's possession or within arm's reach and in the employee's line of sight.

C. Use the following procedures to protect yourself from unexpected machine startup or movement when servicing cord and plug connected equipment:

1. Unplug equipment from its electric socket.

2. Keep the plug in your possession at all times during equipment servicing or keep the plug within arm's reach and in your line of sight at all times during equipment servicing.

3. Lockout or control any other sources of hazardous energy that could cause injury during servicing.

VII. Outside Contractors

If employees of outside contractors perform servicing or maintenance that requires lockout, the maintenance department will:

A. Inform the outside contractor of your company's lockout procedures

B. Obtain a copy or description of the outside contractor's lockout procedures

C. Ensure that your employees understand and comply with the restrictions and prohibitions of the outside contractor's lockout procedures

Outside contractor lockout checklist:

☐ Company lockout procedures provided to outside contractor

☐ Information on outside contractor lockout procedures obtained

☐ Company employees informed of outside contractor lockout procedure restrictions and prohibitions

VIII. Training of Authorized Employees

Authorized employees are the people who perform machine maintenance and servicing that requires lockout. They are the only individuals who will lockout equipment. Authorized employees will learn when lockout is and is not required. They will receive training covering:

- Recognition of hazardous energy sources
- Types and amounts of hazardous energy in the workplace
- Methods, devices, and procedures used to control hazardous energy on company equipment, including cord and plug connected equipment if it is to be locked out
- Procedures for removing lockouts and returning equipment to operation
- Transfer of lockout responsibilities
- Group lockout procedures

Note—All training and retraining must be documented and certified.

IX. Employee Retraining Requirements

According to OSHA, authorized and affected employees will receive retraining in the proper application of lockout procedures when there is a change in:

- Job assignment(s) that exposes an authorized or affected employee to new hazards or lockout procedures
- Machines, equipment, or processes that present a new hazard or require modified lockout procedures
- Lockout procedures for a piece or type of equipment

X. Other Employees

OSHA's requirements for other employees are quite simple. Other employees are those whose work may require them to be in areas where lockout is used from time to time. These employees will be made aware of lockout procedures and instructed that they should never attempt to operate a machine that is locked out. Further training is not required unless your company wishes to provide other employees more detailed lockout information.

XI. Lockout/Tagout Procedures

A. *General*—This procedure is to be used to ensure that machines or equipment are isolated from all potentially hazardous energy and locked out (or tagged out) before employees perform any servicing or maintenance activities where the unexpected energization, startup, or release of stored energy could cause injury.

B. ***Compliance with This Program***—All employees are required to comply with the restrictions and limitations imposed upon them during the use of lockout. Authorized employees are required to perform the lockout in accordance with this procedure. All employees, upon observing a machine or piece of equipment which is locked out to perform servicing or maintenance, should not attempt to start, energize, or use that machine or equipment. Employees attempting to remove, bypass, or otherwise remove the lockout (or tagout) will be subject to immediate disciplinary action up to and including termination.

C. ***Sequence of Lockout***—
   - Notify all affected employees that servicing or maintenance is required on a machine or piece of equipment and that the machine or equipment must be shut down and locked out to perform the maintenance or servicing.
   - The authorized employee must refer to the company procedure to identify the type and magnitude of the energy that the machine or equipment utilizes, understand the hazards of this energy, and know the methods to control the energy.
   - If the machine or equipment is operating, shut it down by the normal stopping procedure (depress stop button, open switch, close valve, etc.).
   - De-activate the energy-isolating device(s) so that the machine or equipment is isolated from the energy source.
   - Lockout the energy-isolating device(s) with assigned individual lock(s). Locks must be obtained from the director of maintenance. Do not attempt a lockout/ tagout without making the director or assistant director aware.
   - Stored or residual energy (such as that in capacitors, springs, elevated machine members, flywheels, hydraulic systems and air, gas, steam, or water pressure, etc.) must be dissipated or restrained by methods such as grounding, repositioning, blocking, bleeding down, etc.
   - Ensure that the equipment is disconnected from the energy source(s) by first checking that no personnel are exposed and then verify the isolation of the equipment by operating the push button or other normal operating control(s) or by testing to make certain that the equipment will not operate. *Caution*—Return operating control(s) to neutral or off position after verifying the isolation of the equipment.
   - The machine or equipment is now locked out.

D. ***Restoring Equipment to Service***—When the servicing or maintenance is completed and the machine or equipment is ready to return to normal operating condition, the following steps must be taken:
   - Check the machine or equipment and the immediate area around the machine and/or equipment to ensure that non-essential items have been removed and that the machine or equipment components are operationally intact.
   - Check the work area to ensure that all employees have been safely positioned or removed from the area.
   - Verify that the controls are in neutral.
   - Remove the lockout device(s) and re-energize the machine or equipment. *Note*—

The removal of some forms of blocking may require re-energization of the machine before safe removal.

E. List of personnel authorized to lockout/tagout equipment or energy:

Name   *Position*

_____   _____

_____   _____

_____   _____

Lockout/Tagout Checklist

	Yes	No	N/A	Comments
1. Does the facility have a written program that outlines procedures for lockout/tagout?				
2. Does the program address authorized maintenance personnel, affected employees who work with the equipment, and other workers who have access?				
3. Does the lockout/tagout program cover all energy types including electrical, mechanical, pneumatic, chemical, thermal, and residual energy found in elevated machine parts?				
4. Are employees properly trained in all aspects of the program, including recognizing hazardous energy sources?				
5. Does the program provide the sequence of each lockout/tagout procedure?				
6. Does the plan provide information on procedures involving more than one person?				
7. Are periodic inspections made to ensure that OSHA standards are being met?				
8. Does the company provide authorized employees with the necessary locks and tags?				
9. Are lockouts performed only by an authorized employee?				
10. Do all locks contain the name of the using employee?				
11. Are locks made of substantial material to preclude removal without using excessive force?				
12. Do employees follow a checklist when restoring energy? (highly recommended)				
13. When outside employers perform maintenance, are procedures established for exchanging information about lockout/tagout procedures?				
14. When a group lockout is necessary, does a single authorized individual have primary safety responsibility for the group?				

APPENDIX J: SAMPLE SAFETY POLICY STATEMENT WITH RULES

This organization is committed to operating in a manner that promotes safety, health, and efficiency while providing quality patient care. Our intention is to provide a place of employment free of recognized hazards. This commitment to providing a safe workplace requires that safety be fully considered in all phases of operation. It is the duty and responsibility of management and of every employee of the facility to give their full support to this program.

Our goal is to ensure that our most valuable assets, employees and patients, are protected from unsafe acts and conditions which might cause illness, injury, or financial loss to themselves, fellow workers, or others. This will be accomplished by:

- Operating under the belief that accidents can be prevented
- Making safety a condition of employment
- Developing and implementing written safety procedures
- Training employees in safe practices, methods, and procedures
- Providing appropriate safety equipment and requiring its proper use at all times
- Striving to eliminate all accidents by providing knowledge, skill, and direction on an ongoing basis
- Providing timely information to our employees on good health and safety
- Requiring management personnel to hold all employees accountable for creating a safe and productive work environment
- Not condoning any unsafe acts by employees

Employee Safety Rules

These rules should be followed. They are for the safety of patients, employees, and visitors.

1. Any injury, no matter how slight, must be reported at once and appropriate treatment administered.
2. Proper shoes and clothing must be worn by all employees.
3. Machinery or equipment will not be operated, adjusted, or repaired unless authorized.
4. Equipment must be checked before using, and defects should be reported immediately.
5. Good housekeeping is the responsibility of each employee, and all work areas will be kept clean.
6. Using, possessing, or being under the influence of alcohol or drugs on facility property is dangerous and prohibited.
7. When lifting materials or transferring/lifting patients, proper procedures will be followed. DO NOT STRAIN. Use proper mechanics.
8. Employees are to be aware of fire extinguisher locations and proper use.
9. All ladders, regardless of their height (with the exception of stepladders), must have non-skid feet.

10. All power sources are to be locked out prior to electrical equipment maintenance.
11. Any cords with exposed wires must be reported immediately.
12. Universal precautions will be followed at all times.
13. All spills will be cleaned up immediately by the employee who first sees the condition. Never leave any spill unattended.
14. Push (do not pull) all rolling items such as carts and chairs from the end or back. Avoid having your hands where they can strike door frames or other objects.
15. Floors will be mopped on only one side at time. WET FLOOR signs must be posted and removed immediately when floor is dry.
16. Employees will not stand on any object other than an approved stepstool or ladder designed for that purpose.
17. Keep guards on power equipment such as saws, food choppers, grinders, and slicers in place at all times.
18. Never leave bed cranks, cabinet drawers, or doors in a position where they will create a hazard.
19. Electrical cords will not be left across hallways, stairs, or doorways.
20. All electrical cords will be maintained in good condition. If a cord is frayed, a plug loose, or the grounding pin on a plug is broken, do not use it. Report it immediately to your supervisor.
21. The use of any extension cords should not be permitted in patient rooms.
22. Injection needles will be disposed of in the proper container and not in the regular trash.
23. Razors and other sharps will be disposed of in sharps containers.
24. Report any condition or practice that might cause an injury or damage equipment.
25. Failure to perform your job in a safe and efficient manner can cause severe injury to yourself, your fellow employees, and the patients of this facility. Violation of safety regulations will result in discipline up to and including suspension or discharge.

APPENDIX K: SAMPLE OSHA HAZCOM TRAINING PLAN

Suggested Teaching Time—One-half to one hour (expand as necessary to cover all hazards)

Teaching Methods—Video presentation and informal discussion

References—29 CFR Part 1910.1200 OSHA HAZCOM Standard, Facility Hazard Communication Plan

Visual Aid—HAZCOM video or chemical/safety video

Learning Objectives—At the conclusion of the session, each worker will understand, know, and comprehend the following HAZCOM principles and information:

1. The purpose and requirements of the OSHA standard
2. The location of the written HAZCOM program
3. The methods and observations necessary to detect hazardous materials in the workplace
4. The labeling system used by the employer, location of the Material Safety Data Sheet (MSDS) file, and definitions of the categories of hazards
5. The operations in the workplace with hazards present
6. The procedures to follow in the event of a spill or release of hazardous materials
7. The principles of safe handling to include proper use of personal protective equipment, safety rules, engineering controls, and specific emergency responsibilities
8. The specific hazardous materials in the employee's work area

Note—Training for Objectives 7 and 8 should be conducted in the employee's work area by the responsible supervisor if feasible.

Recommended Training Sequence

1. Show an appropriate video. A video is an excellent way to introduce the topic and create interest. At the conclusion, review steps for safe handling and emphasize personal responsibility.
2. Present the objectives by discussing the main points which cover the areas required by the OSHA standard. Allow employees to ask questions or make comments during the session.

   EMPHASIZE: Objectives 2, 4, and 5, since these refer to the local program.

3. Objectives 7 and 8 are best trained in the work area by a knowledgeable supervisor. However, if the number of workplace hazards is small, these objectives can be covered during the general session.
4. If Objectives 7 and 8 are not covered, explain to the workers that they will be covered in a later session.
5. Summarize important points of the lesson or areas that were questioned during the session. Use a written or oral quiz of ten questions to reinforce learning.
6. Ensure that all workers sign and initial the training documentation report before leaving the area.

7. If training is done in two sessions, ensure that the second session is completed within 30 days of the first.

8. _____

_____

_____

Teaching Outline

I. Introduction

This training session begins with a short video that shows the steps for working safely with hazardous materials. As you watch the video, try to think of hazardous materials in your work area or department. After the video, we will review the steps and discuss the OSHA HAZCOM standard and the local HAZCOM plan.

A. Review the video by asking the following questions:
   - **Step #1: Why are warning signs important?**
   - **Step #2: What information should be on hazard labels?**—Chemical identification and warning precautions to take when working with the substance and protective equipment
   - **Step #3: What information does the MSDS contain?**—Identification, hazardous ingredients, physical/chemical characteristics, fire/explosion information, reactivity data, health hazards, and precautions for safe handling
   - **Step #4: What safety equipment is needed for handling hazardous materials?**
   - **Step #5: Why is it important to follow safety procedures?**
   - **Step #6: What are symptoms of overexposure to hazardous materials?**— Dizziness, nausea, irritation (eyes, nose, throat), skin rash, nervousness, agitation, sluggishness
   - **Step #7: Why is hygiene a priority?**
   - **Step #8: What should you do in an emergency?**
   - **Step #9: Who should you contact with questions about hazardous materials?**—Supervisor, safety coordinator
   - **Step #10: Why is personal responsibility so important?**
B. The video showed us key elements of safety when working with hazardous materials. OSHA published the HAZCOM standard in 1987 to require employers to communicate hazard information to workers.

II. Overview of 29 CFR Part 1910.1200 (Objective 1)

A. The OSHA standard requires workers to be informed about workplace hazards to which they may be exposed under normal working conditions or in a foreseeable emergency situation.
B. The standard covers all types of hazards including solids, gases, and liquids.
C. Employers must publish a written HAZCOM plan which describes the following:
   1. Labeling systems used in the facility

2. Information on location and availability of MSDSs
3. Hazard evaluation procedures
4. Training procedures

III. Location of the Written HAZCOM Program (Objective 2)

A. The written HAZCOM plan for this facility is located _____
_____ .

B. Workers have access to all hazard information by contacting their immediate supervisor.

IV. Methods Used to Observe/Detect Hazards in the Work Area (Objective 3)

A. Detection methods
   1. Observation
   2. Presence of exposure symptoms (rash, headache, dizziness, etc.)
   3. Olfactory indications
   4. Employee monitoring
   5. Hygiene surveys
B. This facility relies on the information provided by the supplier on the MSDS to evaluate the hazardous potential of all substances.
C. Safety personnel will also monitor operations and work areas to determine and evaluate environmental exposure risks.

V. Labeling, MSDS, and Health Hazard Information (Objective 4)

A. Labeling system
   1. Explanation of labeling system used: _____
   _____ .
   2. Each container is checked to ensure the following information is on the label:
      a. Chemical identity (must match MSDS)
      b. Appropriate warnings in words or pictures
      c. Description of protective clothing/equipment in words, pictures, or symbols
      d. Special handling or storage information
      e. Manufacturer's name/address/phone number
      f. Numbers to call for emergency assistance/information
B. The MSDS files(s) are located: _____
_____ .

Employees may request hazard information from MSDS files by contacting their immediate supervisor. The MSDS contains information describing health and physical hazard evaluations. Appearance/length vary, but all should have the following:
1. Chemical description
2. Conditions that could increase the hazard
3. Manufacturer/supplier

4. Safe handling procedures
5. Protection required for safe handling
6. Why the substance is hazardous
7. What to do if overexposed
8. How a person can be exposed
9. What to do in case of spill or release of the substance

C. *Section 1*—Contains general information about the chemical and who produces it.
D. *Section 2*—Contains information about hazardous ingredients, normally those above the threshold set by OSHA. This section also has information about exposure limits.
E. *Section 3*—Has information about physical data such as boiling point, vapor density, volatility, specific gravity, color, odor, etc.
F. *Section 4*—Describes fire and explosion data, including flash point, flammable limits, unusual hazards, and fire-fighting procedures.
G. *Section 5*—Describes health hazard data, including effects of overexposure, primary routes of entry into the body, and emergency procedures.
H. *Section 6*—Contains toxicology information obtained from controlled studies.
I. *Section 7*—Describes reactivity and indicates the stability of the substance under normal foreseeable conditions. Also indicates conditions to avoid and incompatibility with other materials.
J. *Section 8*—Covers the steps to take in an accidental release or spill. Describes waste disposal methods and the best procedures to control the spill.
K. *Section 9*—Contains special protection information, including ventilation requirements, respiratory protection, gloves, eye protection, and any other equipment/clothing required to work safely with the substance.
L. *Section 10*—Details storage and handling information to assure optimal safe performance and shelf life.
M. *Categories of hazards (Objective 4)*—
   1. **Health Hazards**—Acutely toxic, carcinogenic, corrosive, mutagenic, sensitizing agent, irritant, teratogenic
   2. **Physical Hazards**—Flammable, combustible, explosive, oxidizer, water reactive, unstable

VI. Operations with Hazardous Materials (Objective 5)

_____

_____

_____

_____

VII. Procedures to Follow in Case of a Spill (Objective 6)

A. Immediately notify your fellow workers and your supervisor.
B. Clear the area and assist others as required.
C. Implement emergency action plan.
D. Do not clean up spill or fight the fire unless that is your designated responsibility.

VIII. Other Information

_____

_____

_____

Hazard Communication Training Record

Employee name: _____

Employer: _____

I, the undersigned, attended company-provided Hazard Communication Training Phases I and II as indicated below.

Phase I Topics

1. Requirements and overview of the HAZCOM standard
2. Operations and processes where hazardous materials are present
3. Location and availability of the written HAZCOM program
4. Methods and observations used to evaluate hazards and detect the presence of hazardous materials in the work area
5. The following topics were also covered in the company plan:
    - Description of labeling system
    - Material Safety Data Sheet information
    - Information on unlabeled pipes
    - Hazards of non-routine tasks
    - How to obtain and use local hazard information

_____   _____
Employee Signature   *Trainer's Initials/Date*

Phase II Topics

1. The physical and health hazards of chemicals used in the department or work area
2. Measures workers must take to protect themselves from exposure and specific procedures the company has implemented to protect workers such as:
    - Safety rules
    - Engineering controls
    - Emergency response procedures
    - Correct use of personal protective equipment

_____   _____
Employee Signature   *Trainer's Initials/Date*

Sample HAZCOM Quiz

1. Hazardous chemicals are only used by workers in heavy industry.
    a. true
    b. false

2. Detailed information about a chemical's hazardous properties can be found in what document?
    a. Material Safety Data Sheet
    b. Chemical listing
    c. HAZCOM plan

3. A chemical/physical hazard refers to things such as flammability and reactivity.
    a. true
    b. false

4. Chemicals that are strong and affect eyes, skin, or mucous membranes are called:
    a. corrosives
    b. reactives
    c. oxidizers

5. The HAZCOM plan and MSDS file for your department is located _____
    _____.

6. Workers always react to chemical exposure in the same manner.
    a. true
    b. false

7. All hazardous chemicals are toxic to humans.
    a. true
    b. false

8. A flammable liquid has a flash point at or below:
    a. 73°F
    b. 100°F
    c. 200°F

9. Carcinogens are toxic substances capable of causing cancer.
    a. true
    b. false

10. Personal hygiene such as washing your hands after handling chemicals is one of the best protections against exposure.
    a. true
    b. false

APPENDIX L: HEALTHCARE EMERGENCY DRILL EVALUATION

Evaluated area: _____

Type of disaster: _____

Time and date of drill: _____

Person(s) performing critique: _____

Number of physicians responding: _____

Number of nursing personnel responding: _____

Number of other hospital personnel responding: _____

Number of casualties (if any) received: _____

Kind of casualties received: _____

_____

Time first disaster alert was received: _____

Time disaster message authenticated: _____

Time first casualties arrived at emergency department: _____

Time first casualties seen by emergency department physician: _____

Time last casualties arrived: _____

Time last casualties seen by a physician: _____

No.	Subject	Yes	No	Not Observed
	TRIAGE			
1.	Was the triage layout physically arranged to facilitate expedient casualty flow?			
2.	Was only emergency first aid administered in the triage area?			
3.	Were casualties classified and moved quickly?			
4.	Were casualty personal belongings properly secured?			
5.	Was the triage area ready when the first casualties arrived?			
6.	How many physicians were seen working in the triage area?			
7.	How many nursing personnel were working in the triage area?			
	PERSONNEL			
8.	Were personnel calm?			
9.	Did personnel follow their roles as laid out in the emergency preparedness plan?			
10.	Were there any clergymen in the emergency department area?			
11.	Was there a problem with too many personnel showing up on site?			
12.	Were personnel resources adequate?			

No.	Subject	Yes	No	Not Observed
13.	Were those in authority properly identified?			
14.	Were physicians assigned to the following areas:			
	a. Triage			
	b. Major treatment area			
	c. Minor treatment area			
	d. Operating room			
15.	Did personnel seem sincere in their involvement?			
16.	Did observers cause any problems?			
17.	Did command center carry out assigned responsibilities?			
18.	Was the personnel recall system used in this drill?			
19.	Was the physician recall system used during the drill?			
20.	Was the disaster message from the site accurately transmitted/ received?			
21.	Did the vehicle traffic control system operate as planned?			
22.	Was the police department informed of the casualty list?			
23.	Were there any problems with the internal communication system?			
24.	What kind of communication systems were used in this drill?			
	a. Radio			
	b. Telephone			
	c. Walk-around radios			
	d. Messenger			
	e. 911			
	f. Other			
	LOGISTICS			
25.	Were disaster supplies ready and transported to the proper locations?			
26.	Were enough stretchers brought to the disaster areas by nursing personnel?			
27.	Were enough respiratory therapy personnel on site?			
28.	Was a registered pharmacist on site to prepare IV fluids?			
29.	Were laboratory personnel on site?			
30.	Were there any resources lacking? Comment if answer is yes.			
31.	Were observers assigned to key areas?			

No.	Subject	Yes	No	Not Observed
32.	Were all critique personnel and observers in place at the time of the drill?			
33.	Was a critique session held after the drill?			
34.	Have people been assigned the task of identifying deficiencies and getting same corrected?			
35.	Have the results of the drill been reported to the safety committee?			

ADDITIONAL COMMENTS: _____

APPENDIX M: SUGGESTED PRECAUTIONS FOR EXPOSURE TO BLOOD OR BODY FLUIDS

No.	Description of Task	Hand Washing	Gloves	Plastic Apron/Gown	Mask	Eye Protection
1.	Physical assessment	X				
2.	Giving medications:					
	Oral				X	
	Piggybacks		X			
	IM, IV, stopcock	X	X			
	Catheter hub	X	X			
	Suppository, rectal, and/or vaginal	X	X			
3.	Routine bath	X				
4.	Tube feeding	X				
5.	Enema	X	X	+		
6.	Foley irrigation	X	X			
7.	Emptying Foley bag, urine, receptacles, bedpan, emesis basin	X	X			X
8.	Insertion of NG tube	X	X	+		X
9.	Vaginal irrigation	X	X	+		
10.	Sitz bath	X				
11.	Applying pressure to control bleeding	X	X	+	+	+
12.	Feeding patients	X				
13.	Emptying wastebaskets	X	X			
14.	Cleanup of incontinent patient:					
	Urine	X	X			
	Feces	X	X	+		
15.	Decubitus care	X	X	+		
16.	Collecting specimens: sputum, stool, urine, wound	X	X	+	+	+
17.	Keto urine checks	X	X			
18.	Oral suctionings, oral/nasal care	X	X			
19.	Nasotracheal or endotracheal suctioning	X	X	+	X	X
20.	Cleaning up spills of blood/body substance	X	X			

No.	Description of Task	Hand Washing	Gloves	Plastic Apron/Gown	Mask	Eye Protection
21.	Direct contract with patients with frequent forceful coughing	X			X	+
22.	Applying topical ointment to lesion	X	X			
23.	Traction	X				
24.	Vital signs—oral temperature, pulse, respiration, blood pressure	X				
25.	Rectal temperature	X	X			
26.	Shaving	X	X			
27.	Washing hair	X				
28.	Post-mortem care	X	X	+		+
29.	Removal of fecal impaction	X	X	+		
30.	Gastric lavage	X	X	+		+
31.	Changing visibly soiled beds	X	X	+		
32.	Placing oxygen cannula or mask	X				
33.	Cleaning surfaces contaminated by blood/body surfaces	X	X			
34.	Ostomy care, teaching, and irrigation	X	X	+		
35.	Routine dressing changes and wound care	X	X			
36.	Dressing changes for wounds with large amounts of dressing	X	X	+		
37.	Wound irrigation	X	X	+	+	+
38.	Burn dressing changes	X	X	+		
39.	Trach care	X	X			+
40.	Dressing removal	X	X			
41.	Wound packing	X	X	+		
42.	Suture/staple removal—clean, dry wound	X	X			
43.	Suture/staple removal—wound with drainage	X	X			
44.	Wound packing	X	X	+		
45.	I and D of abscess	X	X	+	+	+
46.	Changing pleuravac	X	X			+
47.	Placement of Foley catheter for fecal incontinence and emptying of bag	X	X			

No.	Description of Task	Hand Washing	Gloves	Plastic Apron/Gown	Mask	Eye Protection
48.	Assisting with invasive procedures (lumbar puncture, bone marrow, thoracentesis, paracentesis, liver biopsy):					
	Inside sterile field	X	X	+	+	+
	Outside sterile field	X	X			
	Chest tube insertion	X	X		X	
	Chest tube removal	X	X		X	

Legend: X = use routinely; + = use if soiling or splattering is likely.

APPENDIX N: SUGGESTED PRECAUTIONS FOR RESPIRATORY CARE EXPOSURES

Tasks that routinely involve exposure to blood, body fluids, and tissues

No.	Task	Hand Washing	Gloves	Gown/ Plastic Apron	Mask	Eye Protection
1.	Medical gas administration	●	•			
2.	Ultrasonic nebulization	●	•			
3.	Aerosolized medication	●	•			
4.	Hyperinflation therapy	●	•			
5.	Chest physiotherapy	●	•			
6.	Mechanical ventilation monitoring	●	•			
7.	Weaning parameters	●	•			
8.	Suctioning	●	●	•	•	•
9.	Insertion of airways	●	●	•	•	•
10.	Artificial airway stabilization	●	●			
11.	Artificial airway cuff pressure measurements	●	•			
12.	Artificial airway care	●	●	•	•	•
13.	Cardiopulmonary resuscitation	●	●	●	●	●
14.	Assisting with intubation	●	●	●	●	●
15.	Assisting with bronchoscopy	●	●	●	●	●
16.	Assisting with tracheostomy	●	●	●	●	●
17.	Chest, physical assessment/vital signs	●	•			
18.	Arterial blood gas punctures	●	●			
19.	Pulmonary function tests	●	•			
20.	CO_2 and O_2 monitoring	●	•			
21.	Oximetry	●	•			
22.	Coughing training/incisional support	●	•		•	
23.	Incentive spirometry	●				
24.	Sputum collection	●	●	•	•	
25.	Cleaning equipment/surfaces	●	●	•		

● = required; • = use if soiling or splattering is likely.

APPENDIX O: ACCIDENT INVESTIGATION REPORT

Facility: _____

Date of Accident: _____ Date Reported: _____

Employee(s) Names(s): _____

Time/Location of Accident: _____

Department: _____ Shift: _____ Supervisor: _____

Machines/Tools/Processes/Operations Involved: _____

_____

Brief Description of Injuries: _____

_____

Property Damage: _____

Witnesses: _____

CHECK EACH ITEM THAT CONTRIBUTED TO THE ACCIDENT

☐ Improper Instruction   ☐ Lack of Supervision   ☐ Unsafe Equipment
☐ Lack of Training   ☐ Failure to Use PPE   ☐ Improper Guarding
☐ Horseplay   ☐ Inoperative Safety Device ☐ Physical Impairment
☐ Unsafe Arrangement   ☐ Failure to Lockout   ☐ Improper Clothing
☐ Failure to Secure   ☐ Unsafe Position   ☐ No Authority to Operate
☐ Using Wrong Tool   ☐ Poor Ventilation   ☐ Poor Housekeeping
☐ Improper Maintenance   ☐ Improper Procedure   ☐ Safety Rule Violation
☐ Human Error   ☐ Other _____

(List)

Summary of Investigation: _____

_____

_____

_____

_____

_____

_____

_____

_____

_____

_____   _____
       Signature/Title   *Date*

Recommended Corrective Actions: _____

_____

_____

_____

_____

_____

_____

_____

_____

_____

_____   _____
Position/Signature   *Date*

Review/Comments: _____

_____

_____

_____

_____

_____

_____

_____

_____

Initials: _____   Date Reviewed: _____   Sent to: _____

APPENDIX P: ERGONOMIC (CTD) PAIN/STRAIN REPORT

NAME: _____ DATE: _____

FACILITY: _____ DEPARTMENT: _____ SHIFT: _____

IDENTIFY ALL AFFECTED AREAS:
☐ Forearm   ☐ Wrist   ☐ Knee   ☐ Elbow   ☐ Upper Back
☐ Lower Back   ☐ Hand   ☐ Fingers   ☐ Ankle   ☐ Foot
☐ Shoulder   ☐ Thigh   ☐ Lower Leg ☐ Neck

• What terms *best* describe the problem?
☐ Aching   ☐ Numbness   ☐ Tingling   ☐ Loss of Color
☐ Burning   ☐ Swelling   ☐ Weakness   ☐ Stiffness
☐ Cramping   ☐ Other (describe) _____

• When did the problem first occur?   Month _____   Year _____

• The length of each episode: Days _____   Weeks _____

• How often did the problem occur during the past year? _____

• What do you think caused the problem? _____

• Describe the problem at its worst: _____

• Have you had medical treatment for this problem?   Yes _____   No _____

• What medical treatment helps the problem? _____

• Number of times treated during the past year: _____

• How many work days did you lose because of this problem during the past year?
  _____

• Have you been placed in another job or on "light duty" status because of this
  problem? _____

• What would help the problem? _____

• Do you work another job in addition to this one?   Yes _____   No _____

• What are your off-the-job activities, hobbies or interests? _____

• Other jobs you have done in the last year (for more than two weeks):

  Dept.: _____ Job Name: _____ Time on this job: _____
                                                             Months/Weeks
  Dept.: _____ Job Name: _____ Time on this job: _____
                                                             Months/Weeks
  (If more than two jobs, include those at which you worked the longest.)

  _____
                        Signature

Recommended Corrective Actions: _____

_____

_____

_____

_____

_____

_____

_____

_____

_____

_____   _____
Position/Signature   _Date_

Review/Comments: _____

_____

_____

_____

_____

_____

_____

_____

_____

_____

_____

Initials: _____   Date Reviewed: _____   Sent to: _____

APPENDIX Q: HEARING CONSERVATION ACKNOWLEDGMENT

Employee Name: _____

Employer: _____

I, the undersigned, have been advised that I may be assigned to a job where exposure to noise levels may exceed 85 dB. I have been briefed on the information initialed below.

☐ Noise effects as described in OSHA standards

☐ Purposes, advantages, and disadvantages of the various types of hearing protection

☐ Selection, proper fit, and care of hearing protection

☐ Purpose and procedures of audiometric testing

☐ The company's hearing conservation program

_____   _____
Employee Signature   *Date*

_____   _____
Witness Signature   *Date*

Comments: _____
_____
_____
_____
_____
_____
_____
_____
_____

APPENDIX R: SAFETY AND HEALTH PROGRAM ASSESSMENT WORKSHEET

Instructions—This form is for project use when assessing an employer's safety and health program. Maintain the completed form in the case file. **Column 1**—Essential elements and subelements of effective safety and health programs. **Column 2**—Indicators that elements and subelements are present at the workplace (core requirements which must be present for an exemption to be granted are in **bold type**). **Column 3**—Consultant's judgment on whether the elements and subelements operate in ways which are adequate and appropriate for the workplace. **Column 4**—Use to record how the determination of effectiveness was made and what improvements are needed. Rectangular block may be used to record correction date.

Employer Name (State Option)	Visit Number	Consultant ID	Date
Elements	**Indicators (Core Requirements in bold)**	**In Place? Yes No**	*How were you able to identify and verify their adequacy? What if any improvements are needed?*
Management Commitment and Planning			
Policy and objectives for worker safety and health	**Established policy and objectives communicated to employees**		
Authority and accountability for safety and health protection	**Responsibilities defined and authority assigned**		
	Supervisor performance evaluated and rewarded or corrected		
Commitment of resources to workplace safety and health	**Adequate company resources allocated for safety and health (in staff, equipment, safety promotion, etc.)**		
Management involvement in employee safety and health concerns	Clear lines of communication for safety and health concerns		
	Management sets example of safe and healthful behavior		
Hazard Assessment			
Hazard assessment	**Completion of comprehensive safety and health surveys**		
	Periodic self-inspections conducted by qualified person		

Elements	Indicators (Core Requirements in bold)	In Place? Yes	No	How were you able to identify and verify their adequacy? What if any improvements are needed?
A reliable procedure for employees to report possibly hazardous conditions	**Employees know how and whom to notify, fear no reprisal, and receive timely and appropriate responses**			
Review of accidents or near-miss incidents and injury/illness experience	**Procedures to identify causes and needed corrective action**			
Hazard Correction and Control				
Means of eliminating or controlling hazards	**Engineering and PPE controls in place as appropriate**			
	Administrative controls, including safety and health rules and safe work procedures, established and implemented			
	Procedures for timely correction/ control of newly identified hazards			
Disciplinary action or reorientation for supervisors or employees who break rules or disregard work or emergency procedures	**Systems for rule enforcement and discipline established and carried out**			
Equipment maintenance program	Ongoing monitoring and maintenance of equipment			
Emergency planning and procedures	Preparation, with training and drills as needed, for emergencies requiring PPE, medical care, emergency egress, etc.			
	Emergency telephone numbers, exit routes, etc. visible to all			
Safety and Health Training				
Supervisors understand the hazards associated with a job, the potential effects, and the supervisor's role in ensuring employees follow the rules, procedures, and work practices for controlling exposure	**Supervisors can explain rules, procedures, and work practices for hazard control and how they teach and enforce them**			

Elements	Indicators (Core Requirements in bold)	In Place? Yes	No	How were you able to identify and verify their adequacy? What if any improvements are needed?
Employees are taught hazards and safe work procedures to protect themselves at the same time they are taught to do a job	Employees can explain how and why they do their jobs safely			
	Training is reinforced or repeated as necessary			
Where personal protective equipment is required, employees know when, why, and how to use it and its limitations and maintenance	**Employees use PPE when necessary**			
	Employees know why, how, limits, and maintenance of PPE			

Source: Assistant Secretary for Occupational Safety and Health. OSHA Instruction ADM 1-1.29: The Integrated Management Information System (IMIS) Consultation Forms Manual. Office of Management Data Systems. U.S. Department of Labor, Washington, D.C., November 17, 1992.

APPENDIX S: SAMPLE HEALTHCARE FACILITY SLIP, TRIP, AND FALL REPORT

Occurrence Date: _____　　Report Date: _____　　Time of Occurrence: _____

Person Completing This Report: _____　Charge Nurse/Supervisor: _____

Medical Record Number: _____　Age: _____

Attending Physician: _____　Notified? Y or N

Witness (Name, Address, Phone number):_____

Recipient of Injury/Occurrence				Duty Shift			
☐ Resident	☐ Visitor	☐ Volunteer	☐ Other	☐ 7–3	☐ 3–11	☐ 11–7	☐ Unknown

Nature of Event/Illness/Injury

☐ Abrasion	☐ Sensory Impairment	☐ Electrical Shock
☐ Aggrev. of Condition	☐ Injection Site Injury	☐ Skin Irritation
☐ Allergic Reaction	☐ Inhalation	☐ Toxicity
☐ Burn	☐ Laceration	☐ None Visible
☐ Contusion	☐ Puncture	☐ Not Applicable
☐ Fracture/Dislocation	☐ Sprain/Strain	☐ Other _____

Falls

☐ Found on Floor (Unwitnessed)	☐ From Bed/Stretcher	☐ Transfer Related
☐ Ambulation Related	☐ From Chair/Wheelchair	☐ Other _____

Other Exposures

☐ Blood/Body Fluids	☐ Chemical
☐ Needlestick	☐ Radiation
☐ Other _____	☐ Other _____

Condition Prior to Occurrence

☐ Alert	☐ Medicated	☐ Lethargic/Weak
☐ Disoriented/Confused	☐ Intoxicated	☐ Impaired Senses
☐ Agitated	☐ Unresponsive	☐ Other/NA _____
☐ Uncooperative	☐ Impaired Mobility	

Location

☐ Corridor	☐ Resident Room	☐ Treatment/Exam Room
☐ Elevator	☐ Office	☐ NA
☐ Bathroom	☐ Dining Room	☐ Other _____

Site of Illness/Injury

☐ Abdomen	☐ Chest	☐ Foot L/R	☐ Neck
☐ Arm L/R	☐ Eye R/L	☐ Hand L/R	☐ None Visible
☐ Back	☐ Face	☐ Head	☐ NA
☐ Buttocks	☐ Finger	☐ Leg L/R	☐ Other _____

Treatment Rendered By

☐ Physician	☐ Nurse	☐ None

Note—Use the reverse side of this form for additional comments.

APPENDIX T: SAMPLE EQUIPMENT PROBLEM REPORTING FORM

Affected Department: _____

Date: _____ Time: _____

Equipment Control Number: _____

Equipment Model Number: _____

Equipment Serial Number: _____

Statement of Problem: _____

_____

_____

_____

Immediate Action Taken to Resolve the Problem: _____

_____

_____

_____

Personnel Involved: _____

_____

_____

Was an Incident Report Completed?   Yes _____   No _____

Signature of Hospital Staff Completing Equipment Problem Report:

_____

Signature

DO NOT WRITE BELOW THIS LINE

Follow-up Action: _____

_____

_____

_____

Final Disposition of Equipment: _____

_____

_____

_____

APPENDIX U: SAMPLE HAZARD CORRECTION FORM

Department/Committee: _____   Date: _____

Hazardous Area: _____   Problem No: _____

I.D. Source: _____   Problem Topic: _____

Hazard Description: _____

_____

_____

_____

_____

Review Date	Plan of Action	Next Review Date

Resolution: _____

_____

_____

_____

Initial Resolution Date: _____

Evaluation: _____

_____

_____

_____

Final Resolution Date: _____

APPENDIX V: SAMPLE BOMB THREAT CHECKLIST

Questions to Ask   *Exact Wording of the Threat*

1. When is the bomb going to explode? _____

2. Where is it right now? _____

3. What does it look like? _____

4. What kind of bomb is it? _____

5. What will cause it to explode? _____

6. Did you place the bomb? _____

7. Why? _____

8. What is your address? _____

9. What is your name? _____

10. Gender of caller: _____ Age: _____ Race: _____ Length of call: _____

11. Caller's Voice:

    ☐ Calm   ☐ Laughing   ☐ Lisp   ☐ Disguised
    ☐ Angry   ☐ Crying   ☐ Raspy   ☐ Accent
    ☐ Excited   ☐ Normal   ☐ Deep   ☐ Familiar
    ☐ Slow   ☐ Distinct   ☐ Ragged   If the voice is familiar, who
    ☐ Rapid   ☐ Slurred   ☐ Clearing throat   did it sound like? _____
    ☐ Soft   ☐ Nasal   ☐ Deep breathing   _____
    ☐ Loud   ☐ Stutter   ☐ Cracking voice   _____

12. Background Sounds:

    ☐ Street noises   ☐ Motor   ☐ Local
    ☐ Crockery   ☐ Office machinery   ☐ Long distance
    ☐ Voices   ☐ Factory machinery ☐ Booth
    ☐ PA system   ☐ Animal noises   Other _____
    ☐ Music   ☐ Clear
    ☐ House noises   ☐ Static

13. Threat Language:

    ☐ Well spoken   ☐ Foul   ☐ Incoherent   ☐ Message read by
       (educated)   ☐ Irrational   ☐ Taped   threat maker

14. Remarks: _____

    _____

15. Report call immediately to: _____ Phone Number: _____

16. Fill out completely, immediately after bomb threat.
    Date: _____ Phone Number: _____

17. Name: _____ Position: _____

CHAPTER 13

CHECKLISTS

CHECKLIST 1: NURSING SERVICES

No.	Description	Yes	No	N/A	Comments
1.	All rooms where medications are stored or prepared properly lighted to prevent errors?				
2.	Floors free from breaks, loose tiles or linoleum, or any obstructions that might cause people to stumble or fall?				
3.	Medicine cabinets/carts locked when unattended?				
4.	Employees fully instructed in the proper manner of lifting, handling, and carrying?				
5.	Needles and other sharps discarded only in designated containers?				
6.	Sharps containers emptied in a timely manner?				
7.	Hazardous substances properly labeled and safely stored and handled?				
8.	Staff familiar with MSDS file location?				
9.	Staff familiar with chemical spill response and procedure?				
10.	Staff properly instructed regarding any hazards related to the required work?				
11.	Plastic liners used in all garbage containers?				
12.	Red bags used for medical wastes?				
13.	All liquid containers clearly labeled with type of contents?				
14.	Storerooms well lit?				
15.	Exits and aisles of storerooms clear at all times?				

No.	Description	Yes	No	N/A	Comments
16.	Rubbish, empty cartons, and paper disposed of immediately?				
17.	Spillage items stored below eye level?				
18.	Heavy items stored on the lower shelves?				
19.	Stored materials clear of sprinkler heads (at least 18 inches) as well as other fire-fighting equipment?				
20.	Flammable liquids are:				
	A. Stored in approved containers?				
	B. Stored in safe quantities?				
	C. Stored in approved cabinets?				
21.	Storage shelves adequate for the weight involved?				
22.	Stepladders, rather than "makeshifts" available for use?				
23.	All stepladders in safe condition?				

CHECKLIST 2: BASIC LIFE SAFETY CONSIDERATIONS

Date: _____   Observer: _____

Location: _____   Department: _____

No.	Subjects	Yes	No	N/A	Comments
1.	All fire doors close and latch properly?				
2.	All stairwell doors close and latch properly?				
3.	Patient doors close and latch properly?				
4.	All smoke, fire, and stairwell doors unobstructed and free to close?				
5.	All stairwells free of storage?				
6.	All smoke detectors and sprinkler heads free of paint, dust, etc.?				
7.	All exit lights visible, legible, and properly illuminated (both bulbs burning)?				
8.	All corridors free from obstruction?				
9.	Emergency exits marked?				
10.	Extension cords not used except where approved by the maintenance or biomedical engineering department?				
11.	Electrical cords arranged to prevent trip hazards?				
12.	Electrical distribution rooms and mechanical rooms locked at all times?				
13.	Fire procedures and evacuation routes posted?				
14.	Electrical and mechanical rooms free of stored items?				
15.	All fire extinguishers certified within the last year?				
16.	All floor landings properly identified (per Life Safety Code)?				
17.	Fire extinguishers unobstructed?				
18.	Disaster plans present in the department?				
19.	Fire plan present in the department?				
20.	Staff familiar with fire procedures?				
21.	Staff familiar with disaster plan?				
22.	Staff familiar with evacuation routes?				
23.	Staff familiar with fire-fighting equipment?				
24.	Staff familiar with action to take upon discovery of fire or other disaster?				
25.	Department has conducted and documented annual in-service training on the department's role in the disaster plan (emergency preparedness plan)?				

CHECKLIST 3: HEALTHCARE FACILITY SLIP, TRIP, AND FALL PREVENTION

No.	Description	Yes	No	N/A	Comments
1.	Aisles, halls, and passageways kept clear?				
2.	Aisles, halls, and walkways marked as appropriate?				
3.	Wet surfaces covered with non-slip materials?				
4.	Holes in floors, sidewalks, or other walking surfaces repaired properly, covered, or otherwise made safe?				
5.	Materials or equipment stored in such a way that they will not interfere with the walkway?				
6.	Changes of direction or elevation readily identifiable?				
7.	Adequate headroom provided for the entire length of any aisle or walkway?				
8.	Steps on stairs and stairways designed or provided with a surface that renders them slip-resistant?				
9.	Where stairs or stairways exit directly into any area where vehicles may be operated, adequate barriers and warnings provided to prevent employees from stepping into the path of traffic?				
10.	Surfaces elevated more than 30 inches above the floor or ground provided with guardrails?				
11.	Permanent means of access and egress provided to elevated storage and work surfaces?				
12.	Material on elevated surfaces piled, stacked, or racked in a manner to prevent it from tipping, falling, collapsing, rolling, or spreading?				
13.	Floor openings guarded by a cover, guardrail, or equivalent on all sides (except at entrance to stairways)?				
14.	Grates or similar type of cover over floor openings such as floor drains of such design that foot traffic or equipment will not be affected by the grate spacing?				
15.	WET FLOOR signs utilized when cleaning?				
16.	Hallways mopped one side at a time?				
17.	Warning signs used to identify hazards?				
18.	Sidewalks, curbs, and driveways in good condition?				
19.	Non-skid waxes or polishes used?				
20.	Exits marked with an exit sign and illuminated by a reliable light source?				
21.	Directions to exits, where not immediately apparent, marked with visible signs?				

No.	Description	Yes	No	N/A	Comments
22.	Doors, passageways, or stairways that are neither exits nor access to exits and which could be mistaken for exits appropriately marked NOT AN EXIT, TO BASEMENT, TO STOREROOM, etc.?				
23.	Exit signs marked with the word EXIT in lettering at least 5 inches high and the stroke of the lettering at least $1/_2$-inch wide?				
24.	Special precautions taken to protect employees during construction and repair operations?				
25.	Doors which serve as exits designed and constructed so that the way of exit travel is obvious and direct?				
26.	Windows or glass which could be mistaken for exit doors made inaccessible by means of barriers, railings, or markings?				
27.	Extension cords not placed across pathways?				
28.	Employees instructed not to carry items in a way that impairs vision?				
29.	Spilled materials, trash, and debris cleaned up immediately?				
30.	Employees instructed not to run in hallways or on stairways?				
31.	Workers instructed not to stand on chairs or other makeshift ladders?				
32.	Personnel prohibited from using makeshift platforms?				
33.	Tripping hazards and unsafe conditions reported and corrected immediately?				
34.	Employees wear non-slip shoes correct for their job and work area?				
35.	During winter weather, areas prone to be hazardous (ice, snow, etc.) communicated to all employees?				
36.	Each department trains its personnel on specific fall, slip, and trip hazards found in their work area?				
37.	The facility has identified slip, trip, and fall trends that affect patients and visitors?				
38.	Procedures in place to evaluate patient fall hazards in various departments?				
39.	Parking areas evaluated regularly to identify hazardous conditions?				
40.	Areas prone to slips and falls (i.e., entrance ways, dining rooms, kitchens) inspected at least monthly?				
41.	Janitorial and cleaning process evaluated on a regular basis?				

No.	Description	Yes	No	N/A	Comments
42.	Procedures in place to monitor water accumulating around drinking fountains?				
43.	Housekeeping keeps water cleaned up around doors on rainy days?				
44.	Caution or warning signs used to identify areas with tripping hazards?				
45.	Broken tiles, soiled rugs, and torn carpet repaired/replaced immediately?				
46.	Elevator floors slip resistant and level with landing?				
47.	Wheelchairs inspected regularly?				
48.	Grab bars securely attached and low enough for easy reach around tubs, shower, and toilets?				
49.	Tubs, showers, and floors have non-skid strips or mats?				
50.	Facility personnel understand the basic causes of falls, including:				
	• Equipment or other material in walkway?				
	• Poor housekeeping, type, and condition of floor?				
	• Inadequate illumination?				
	• Human inattention, age, illness, emotional state, fatigue, poor vision?				
51.	The facility uses flush floor-level mats and runners of sufficient length?				
52.	Procedures developed for placing, removing, cleaning, and storing mats?				

CHECKLIST 4: FOOD SERVICE SAFETY

No.	Description	Yes	No	N/A	Comments
1.	Slip-resistant waxes or polishes used to treat floors?				
2.	Portable signs used to indicate wet floors or other temporary hazards?				
3.	Floor coverings checked for holes, tears, loose threads, and other tripping hazards?				
4.	Traffic pattern unobstructed by pans, dishtubs, racks, and other obstacles?				
5.	Floors frequently monitored for excessive water, food spillage, and tripping hazards from worn flooring materials?				
6.	Employees trained in the safe use of cleaning compounds and drying agents to prevent dermatological problems?				
7.	Employees take adequate precautions to prevent burns from hot liquids, hot serving containers, steam, and heat lamps?				
8.	Scheduled inspections of glassware, china, flatware, and plastic equipment for cracks, chips, and defects?				
9.	Counters, steam tables, carts, and serving equipment free from sharp corners and in good condition?				
10.	Microwave ovens in good repair, especially doors and seals? Warnings posted?				
11.	Food properly protected from contamination and maintained at the required temperature?				
12.	Ventilation adequate to remove steam and dampness?				
13.	Ladders used for reaching shelved items?				
14.	Cartons stored away from wet or damp areas?				
15.	Light fixtures operational, guarded, and two feet (minimum) from stored items?				
16.	All fruits and vegetables thoroughly washed?				
17.	Frozen foods properly thawed under refrigeration or under cold running water and cooked from frozen state?				
18.	Raw and cooked/ready-to-serve foods prepared on the same cutting surface without washing and sanitizing between use?				
19.	Food preparation areas clean and free from debris?				
20.	Utensils cleaned, sanitized, and stored to protect them from contamination?				
21.	Grinders, slicers, choppers, and mixers cleaned and sanitized between use?				

No.	Description	Yes	No	N/A	Comments
22.	Cleaning chemicals/pesticides kept in the food preparation and service areas?				
23.	Guards on cutting, chopping, slicing, and grinding machines used?				
24.	Employees familiar with equipment which their duties require them to use?				
25.	Workers know the safeguards and hazards of piece of equipment?				
26.	All electrical equipment properly grounded?				
27.	Employees use pot holders or gloves for handling items?				
28.	Handles of pans in use do not protrude into aisle space?				
29.	Filters free from accumulated grease?				
30.	Grease receptacles emptied regularly?				
31.	Hot-holding equipment maintains food at or above 140°F?				
32.	Cold foods held at 45°F or lower?				
33.	Food-holding cabinets equipped with thermometers?				
34.	Wash and rinse temperatures proper for the type of machine and ware being washed maintained (see manufacturer's specifications on data plate)				
35.	Rinse temperature of at least 170°F maintained for tableware and utensils? Manual dishwashing?				
36.	Detergent concentration maintained at the necessary level for effective washing?				
37.	Cleaned and sanitized ware and utensils stored off the floor and in a clean, dry location?				
38.	Glass and china stored in proper storage facilities and not in temporary storage areas such as counter or table surfaces?				
39.	Employees trained in the proper method of lifting and handling for different types of containers?				
40.	Oily rags kept in closed metal containers?				
41.	Storage of foodstuffs completely separate from storage of cleaning powders, insecticides, and other poisonous substances?				
42.	Emergency exit devices on walk-in refrigerators/freezers operate properly?				
43.	Walk-in refrigerators/freezers have a screen or mesh guard around every blower fan?				
44.	Grease filters, range exhaust hoods, and exhaust hood filters clean?				
45.	Dry chemical or carbon dioxide fire extinguisher located near the cooking area?				

No.	Description	Yes	No	N/A	Comments
46.	Fire extinguishers properly tagged and inspected within the last 30 days?				
47.	Employees trained in how to extinguish a cooking vessel fire (smother it with a lid or use a dry chemical or carbon dioxide fire extinguisher)?				
48.	Food kept covered when in transit to patient care areas?				
49.	Food cart wheels and drawers in good repair?				
50.	Steam, gas, and water pipes clearly marked for identification?				
51.	Knives, saws, and cleavers kept in designated areas when not in use?				
52.	Ice machines clean, free of rust, and not used for food storage?				
53.	Non-skid mats or flooring provided in wet areas?				

CHECKLIST 5: HEALTHCARE FACILITY ELECTRICAL SAFETY

No.	Description	Yes	No	N/A	Comments
1.	Specify compliance with OSHA for all contract electrical work?				
2.	Employees required to report as soon as practicable any obvious hazard to life or property observed in connection with electrical equipment or lines?				
3.	Employees instructed to make preliminary inspections and/or appropriate tests to determine what conditions exist before starting work on electrical equipment or lines?				
4.	When electrical equipment or lines are to be serviced, maintained, or adjusted, necessary switches opened, locked out, and tagged whenever possible?				
5.	Portable electrical tools and equipment grounded double insulated?				
6.	Electrical appliances such as vacuum cleaners, polishers, vending machines, etc. grounded?				
7.	Extension cords that are used have a grounding conductor?				
8.	Multiple-plug adaptors prohibited?				
9.	Ground-fault circuit interrupters installed on each temporary 15- or 20-ampere, 120-volt AC circuit at locations where construction, demolition, modifications, alterations, or excavations are being performed?				
10.	Temporary circuits protected by suitable disconnecting switches or plug connectors at the junction with permanent wiring?				
11.	Electrical installations in hazardous dust or vapor areas meet the National Electrical Code for hazardous locations?				
12.	Exposed wiring and cords with frayed or deteriorated insulation repaired or replaced promptly?				
13.	Flexible cords and cables free of splices or taps?				
14.	Clamps or other securing means provided on flexible cords or cables at plugs, receptacles, tools, equipment, etc. and the jacket securely held in place?				
15.	Cord, cable and raceway connections intact and secure?				
16.	In wet or damp locations, electrical tools and equipment appropriate for the use or location or otherwise protected?				

No.	Description	Yes	No	N/A	Comments
17.	Location of electrical power lines and cables (overhead, underground, under floor, other side of walls, etc.) determined before digging, drilling, or similar work is begun?				
18.	Metal measuring tapes, ropes, hand lines, or similar devices with metallic thread woven into the fabric prohibited where they could come in contact with energized parts of equipment or circuit conductors?				
19.	Use of metal ladders prohibited in areas where the ladder or the person using the ladder could come in contact with energized parts of equipment, fixtures, or circuit conductors?				
20.	Disconnecting switches and circuit breakers labeled to indicate their use or equipment served?				
21.	Disconnecting means always opened before fuses are replaced?				
22.	Interior wiring systems include provisions for grounding metal parts of electrical raceways, equipment, and enclosures?				
23.	Electrical raceways and enclosures securely fastened in place?				
24.	Energized parts of electrical circuits and equipment guarded against accidental contact by approved cabinets or enclosures?				
25.	Sufficient access and working space provided and maintained around all electrical equipment to permit ready and safe operation and maintenance?				
26.	Unused openings (including conduit knockouts) in electrical enclosures and fittings closed with appropriate covers, plugs, or plates?				
27.	Electrical enclosures such as switches, receptacles, junction boxes, etc. provided with tight-fitting covers or plates?				
28.	Electrical cords in good condition and properly used, with no extension cords being substituted for fixed wiring?				
29.	Switches guarded properly to prevent inadvertent or accidental starting?				
30.	Adequate lighting provided throughout the facility, especially at stairs and other hazardous areas?				
31.	Any live but empty light sockets or live but damaged switches?				

No.	Description	Yes	No	N/A	Comments
32.	Fuses (or circuit breakers) on lighting and small appliance circuits of proper capacity, and use of multiple plugs monitored to prevent overloading of circuits?				
33.	Switches for electrical equipment located so that they can be reached easily in the event of an emergency, without having to lean on or against metal equipment?				

CHECKLIST 6: HAZARD COMMUNICATION PROGRAM EVALUATION

No.	Description	Yes	No	N/A	Comments
1.	Hazard Communication Program:				
	A. The program is in writing				
	Our written program provides the following:				
	B. Describes how hazards will be evaluated and described (employers may rely on the chemical manufacturer or importers)				
	C. Tests all hazardous materials in the workplace (employers may rely on the chemical manufacturer or importer)				
	D. Describes our labeling system				
	E. Provides a list of hazardous chemicals referenced on MSDSs for all hazardous materials used in the workplace (see Section B)				
	F. Describes our employee education and training program				
	G. Describes hazards of non-routine tasks				
	H. Describes how hazards of non-labeled pipes will be handled				
	I. Includes procedures for informing on-site contractors of the hazardous substances in the workplace to which their employees may be exposed				
	J. Is available to employees, their designated representatives, assistant secretary of labor for OSHA, and the director of NIOSH				
2.	List of hazardous materials in the workplace—Our list contains all hazardous chemicals including but not limited to:				
	A. Raw materials				
	B. Both isolated and non-isolated intermediates				
	C. Final product				
	D. Cleaning and maintenance chemicals				
	E. Laboratory chemicals for which MSDS information has been received				
	F. Waste products not regulated under the RCRA but which are hazardous under this standard				
	G. Impurities and by-products				
	H. Waste treatment and products				

No.	Description	Yes	No	N/A	Comments
3.	Hazardous materials labeling system:				
	A. All products containing hazardous materials leaving the workplace are labeled (applicable to chemical manufacturers, distributors, and importers only)				
	B. Stationary containers are labeled				
	C. Temporary containers used between work shifts or by different workers are labeled				
	D. A method has been established to ensure that our labels are correct and up to date				
4.	Contents of hazardous material label:				
	A. A chemical name that coincides with name on MSDS				
	B. The identity of hazards with words (in English), pictures, or symbols				
	C. Hazards of immediate and direct consequences of mishandling are included				
	D. Information that does not conflict with DOT regulations				
	E. Other OSHA standards if material is already regulated				
	F. The name and address of a responsible party (or parties)				
5.	In-plant labeling system:				
	A. Containers are labeled with the identity of hazardous chemicals and hazard warnings (unless hazard warning materials are used)				
	B. Hazard warning materials for hazardous chemicals in stationary process containers are readily accessible to the employee in the workplace				
	C. The labels on incoming containers have not been removed or defaced unless immediately replaced with our own labels				
	D. The hazards in pipelines are identified, although they do not have to be labeled under this standard				
	E. Our labels are legible and in English				
6.	Material Safety Data Sheets:				
	A. A MSDS is available for every hazardous chemical which an employee uses				
	B. Our MSDSs are readily accessible to exposed employees in the work area throughout each work shift				

No.	Description	Yes	No	N/A	Comments
7.	Procedures have been established for:				
	A. Updating our MSDSs (or for receiving updated copies from our suppliers)				
	B. Taking appropriate action if a shipment is received without a MSDS				
	C. Getting new and updated MSDSs to employees who handle materials				
	D. Advising employees of any changes in MSDSs				
	E. Documentation of efforts to obtain MSDSs from suppliers (recommended practice but not required by this standard)				
8.	Hazards of non-routine tasks—Procedures have been established to assess the hazards of non-routine tasks as follows:				
	A. All non-routine tasks involving the use of or exposure to hazardous materials are identified				
	B. The hazards involved in the performance of non-routine tasks are described in writing				
	C. A MSDS is prepared or obtained for the hazardous materials involved in these non-routine tasks				
	D. A labeling system or written operating procedure has been established to identify the hazardous substances and the hazards involved in non-routine tasks				
	E. Special training has been established for the performance of non-routine tasks, including written operating procedures				
9.	Employee education and training—Procedures have been established to inform employees of:				
	A. Possible exposure to hazardous materials, including all manufacturing, quality control, plant service, and R&D employees				
	B. Requirements of the hazard communication standard				
	C. Operations where hazardous materials are present				
	D. Location and availability of the written Hazard Communication Program including the hazardous chemical list and MSDSs				
10.	Procedures for training employees include:				
	A. Information about physical and health hazards of chemicals in the work area				

No.	Description	Yes	No	N/A	Comments
B.	Detecting the presence of hazardous materials (monitoring procedures, odors, visibility, etc.)				
C.	Proper use and selection of personal protective equipment				
D.	Emergency procedures in the event of accidental exposure to hazardous materials, including emergency phone numbers and the location of eyewashes and safety showers				
E.	How to determine hazards by reading a label				
F.	The location of MSDSs and the procedure for reviewing them and/or obtaining a copy				
G.	How to obtain the correct MSDS for the hazardous substance used by the employee, such as use of the trade name as a key identifier				
H.	How a MSDS is updated or the procedure for obtaining updated copies from the chemical manufacturer, importer, or distributor				
I.	The significance to the employee of each section of information on the MSDS, how to read it, and what it means				
J.	The measures employees can take to protect themselves from chemical exposure (examples include eyewashes, face shields, respirators, etc.)				
K.	Training conducted prior to the handling of a hazardous chemical, including employees who may only temporarily do such work				
L.	Updated training when an employee transfers jobs or departments				
M.	Updated training is considered when significant changes in chemicals or operations occur				

CHECKLIST 7: ANNUAL SAFETY MANAGEMENT EVALUATION

No.	Description	Yes	No	N/A	Comments
1.	Planned/systematic written safety program?				
2.	Safety committee bylaws/charter approved?				
3.	Committee responsibilities/authority defined?				
4.	Safety program approved by CEO/governing body?				
5.	Medical professionals involved in the safety program?				
6.	Departmental safety policies and procedures reviewed regularly?				
7.	Risk management department monitors and reports on a regular basis?				
8.	Quality assurance department monitors and reports on a regular basis?				
9.	Infection control department monitors and reports on a regular basis?				
10.	Facility-wide hazard surveillance program developed?				
11.	Hazard surveys conducted in each area of the facility semi-annually?				
12.	Patient care equipment inspected and results documented?				
13.	Patient care equipment failures reported on a regular basis?				
14.	Incident and accident reports involving patients, visitors, and guests reported on a regular basis?				
15.	Safety committee meets as required?				
16.	Safety committee membership includes personnel from administration, clinical, and support service areas?				
17.	Safety committee reports to governing board, administration, and department managers as required?				
18.	New employee orientation and training conducted?				
19.	Annual safety training conducted and documented?				
20.	HAZMAT program reviewed annually with all employees?				
21.	Educational/training director reports on education/in-service training on a regular basis?				
22.	Emergency plan implemented at least once each six months?				
23.	Emergency disaster plan updated annually?				
24.	Emergency drills evaluated, critiqued, and documented?				

No.	Description	Yes	No	N/A	Comments
25.	Fire drills conducted quarterly on all shifts?				
26.	Regular environmental testing performed as required?				
27.	Security program reviewed and approved?				
28.	Product recalls monitored, reported, and documented?				
29.	Fire prevention plan reviewed and approved?				
30.	Documentation of inspections by outside regulatory agencies maintained?				
31.	Radiation safety committee reports on a regular basis?				
32.	Laser safety committee reports on a regular basis?				
33.	Interim life safety inspections and corrective actions reported on a regular basis?				
34.	Equipment management issues reported on a regular basis?				
35.	Utilities management issues reported on a regular basis?				
36.	Safe corrective actions documented for utilities management reports?				
37.	Issues concerning ADA requirements reported to risk management?				
38.	Environmental issues handled correctly?				
39.	Safe Medical Device Act safety plan in place?				
40.	Documented corrective actions as a result of Safe Medical Device Act incidents?				

REMARKS:

_____

_____

_____

_____

_____

_____

_____

_____

_____

CHECKLIST 8: SAMPLE OSHA COMPLIANCE FOR HEALTHCARE ORGANIZATIONS

No.	Subject/Standard Reference	Yes	No	N/A	Comments
1.	Occupational Exposure to Bloodborne Pathogens (1910.1030)				
	A. Written exposure control plan describing:				
	1. Exposure determination				
	2. Methods of compliance:				
	• Universal precautions				
	• Engineering and work practice controls				
	• Hepatitis B vaccination; post-exposure evaluation and follow-up				
	• Communication of hazards to employees				
	• Recordkeeping				
	• Evaluation of circumstances surrounding exposure				
	B. Use universal precautions				
	C. Engineering controls and work practices:				
	• Sharps container marked with biohazard label				
	• Handwashing facility				
	• Eyewash facility				
	• No eating, drinking, smoking, applying cosmetics, or handling contact lenses in work area				
	• No food or drink in potentially contaminated areas				
	• Specimens containerized or decontaminated				
	D. Personal protective equipment:				
	• Gloves				
	• Aprons, gowns, other protective body covering				
	• Eye protection—shield, goggles				
	• Shoe covers, surgical caps				
	E. Housekeeping practices:				
	• All work surfaces cleaned and decontaminated after use				
	• Replace protective coverings				
	• Bins, pails, cans, etc. cleaned and disinfected				
	• Procedures for re-usable sharps				

No.	Subject/Standard Reference	Yes	No	N/A	Comments
	F. Regulated waste:				
	• Use approved sharps containers, labeled or color coded				
	• Other regulated waste in approved containers that are labeled or color coded				
	• Approved disposal				
	G. Laundry:				
	• Proper procedures—universal precautions				
	• Proper containers for (wet or dry) linens				
	• Notify laundry service if used				
	H. Hepatitis B vaccinations:				
	• Provided to all exposed workers at no charge within ten days after initial assignment				
	• Documentation				
	I. Post-exposure evaluation and follow-up plan				
	J. Hazard communication:				
	• Biohazard labels				
	• Inform and train employees at the time of initial assignment				
	K. Recordkeeping:				
	• Training: 3 years				
	• Medical: length of employment plus 30 years				
2.	Hazard Communication (1910.1200)				
	A. Written program:				
	• How employer determines what materials are hazardous				
	• MSDS for all hazardous chemicals in facility and how to obtain				
	• List of all hazardous chemicals in facility				
	• Description of labeling system used				
	• Methods to inform employees of the hazard of non-routine tasks and/or unlabeled pipes				
	• Information provided to outside contractors				
	• Hazard communication standard				
	• Training program for employees				
	B. MSDS for each hazardous substance excluding drugs as defined by FDA in pill or tablet form				

No.	Subject/Standard Reference	Yes	No	N/A	Comments
	C. Labels on all hazardous substances excluding drugs as defined by FDA				
	D. Employee training and information				
3.	Radiation safety (1910.96)—Monitoring of exposed workers (X-ray machine)				
4.	Emergency response/evacuation plan first aid (1910.120; 1910.38)				
5.	Emergency exit signs (1910.37)				
6.	Emergency lighting (1910.37)				
7.	Compressed gas storage (1910.101) and nitrous oxide (1910.105)				
8.	Electrical components properly grounded (1910.301–309)				
9.	Abrasive wheels guarded on bench grinder (1910.215)				
10.	OSHA 200—Illness and Injury Log if 11 or more employees (1904.2)				
11.	OSHA Poster (1903.2)				
12.	Fire extinguishers and employee training (1910.157)				
13.	Employee access to medical records (1910.20)				
14.	Miscellaneous toxic substance (i.e., nitrous oxide, mercury, glutaraldehyde) exposures adequately controlled (1910.1000)				
15.	Laboratory standard (1910.1450)				
16.	General duty clause [5(a)(1)]:				
	A. Tuberculosis				
	B. Ergonomic hazards				
17.	Lead (1910.1025)				
18.	Benzene (1910.1028)				

CHECKLIST 9: PERSONAL PROTECTIVE EQUIPMENT

No.	Description	Yes	No	N/A	Comments
1.	Protective goggles or face shields provided and worn where there is any danger of flying particles or corrosive materials?				
2.	Approved safety glasses required to be worn at all times in areas where there is a risk of eye injuries such as punctures, abrasions, contusions, or burns?				
3.	Employees who need corrective lenses (glasses or contacts) in working environments with harmful exposures required to wear *only* approved safety glasses, protective goggles, or use other medically approved precautionary procedures?				
4.	Protective gloves, aprons, shields, or other means provided and required where employees could be cut or where there is reasonably anticipated exposure to corrosive liquids, chemicals, blood, or other potentially infectious materials? (See 29 CFR 1910.1030(b) for the definition of "other potentially infectious materials")				
5.	Approved respirators provided for regular or emergency use where needed?				
6.	Protective equipment maintained in a sanitary condition and ready for use?				
7.	Eyewash facilities and a quick drench shower within the work area where employees are exposed to injurious corrosive materials?				
8.	Food or beverages consumed on the premises are consumed in areas where there is no exposure to toxic material, blood, or other potentially infectious materials?				
9.	Protection against the effects of occupational noise exposure provided when sound levels exceed those of the OSHA noise standard?				
10.	Adequate work procedures and protective clothing and equipment provided and used when cleaning up spilled toxic or otherwise hazardous materials or liquids?				
11.	Appropriate procedures in place for disposing of or decontaminating personal protective equipment contaminated with, or reasonably anticipated to be contaminated with, blood or other potentially infectious materials?				

No.	Description	Yes	No	N/A	Comments
12.	Personal protective clothing or equipment that employees are required to wear or use of a type capable of being cleaned easily and disinfected?				
13.	Employees prohibited from interchanging personal protective clothing or equipment unless it has been properly cleaned?				
14.	Machines and equipment which process, handle, or apply materials that could be injurious to employees cleaned and/or decontaminated before being overhauled or placed in storage?				
15.	Employees prohibited from smoking or eating in any area where contaminates that could be injurious if ingested are present?				
16.	Employees required to change from street clothing into protective clothing in a clean change room, with separate storage facility for street and protective clothing provided?				
17.	Employees required to shower and wash their hair as soon as possible after contact with a known carcinogen?				
18.	Equipment, materials, or other items taken into or removed from a carcinogen-regulated area in a manner that will contaminate non-regulated areas or the external environment?				

CHAPTER 14

AGENCIES

American Association of Health Care
Consultants
11208 Waples Mills Road, Suite 109
Fairfax, VA 22030

American Association of Occupational Health
Nurses
50 Lenox Pointe
Atlanta, GA 30324

American Board of Industrial Hygiene
4600 W. Saginaw, Suite 101
Lansing, MI 48917

American Chemical Society
1155 16th Street, NW
Washington, DC 20036

American College of Emergency Physicians
P.O. Box 619911
Dallas, TX 75261

American College of Occupational and
Environmental Medicine
55 W. Seegers Road
Arlington Heights, IL 60005

American Conference of Governmental
Industrial Hygienists
6500 Glenway Avenue, Building D-5
Cincinnati, OH 45211

American Health Care Association
1201 L Street, NW
Washington, DC 20005

American Hospital Association
840 N. Lake Shore
Chicago, IL 60611

American Industrial Hygiene Association
P.O. Box 8390
345 White Pond Drive
Akron, OH 44320

American National Standards Institute
11 W. 42nd Street
New York, NY 10036

American Public Health Association
1015 15th Street, NW
Washington, DC 20005

American Society for Healthcare Risk
Management
840 N. Lake Shore
Chicago, IL 60611

American Society for Testing and Materials
1916 Race Street
Philadelphia, PA 19103

American Society for Training and
Development
Box 1443, 1640 King Street
Alexandria, VA 22313

American Society of Heating, Refrigerating,
and Air Conditioning Engineers
1791 Tullie Circle, NE
Atlanta, GA 30329

American Society of Safety Engineers
1800 E. Oakton Street
Des Plaines, IL 60018

Asbestos Information Association
1745 Jefferson Davis Highway, Room 509
Arlington, VA 22202

Association of Operating Room Nurses
10170 E. Mississippi Avenue
Denver, CO 80231

Board of Certified Healthcare Safety
Management
8009 Carita Court
Bethesda, MD 20817

Board of Certified Safety Professionals
208 Burwash Avenue
Savoy, IL 61874

Board of Hazard Control Management
8009 Carita Court
Bethesda, MD 20817

Center for Devices and Radiological Health
Food and Drug Administration
5600 Fishers Lane
Rockville, MD 20857

Centers for Disease Control and Prevention
1600 Clifton Road, NE
Atlanta, GA 30333

Compressed Gas Association
1725 Jefferson Davis Highway
Suite 1004
Arlington, VA 22202

Department of Labor
Occupational Safety and Health Administration
200 Constitution Avenue, NW
Washington, DC 20210

Department of Transportation
Office of Hazardous Materials Transportation
400 Seventh Street, NW
Washington, DC 20590

Environmental Protection Agency
401 M Street, SW
Washington, DC 20460

Federal Emergency Management Agency
500 C Street, SW
Washington, DC 20472

Federation of American Hospitals
1405 N. Pierce, Suite 311
Little Rock, AR 72207

Formaldehyde Institute, Inc.
1330 Connecticut Avenue, NW
Washington, DC 20006

Healthcare Financial Management Association
Two Westbrook Corporate Center
Suite 700
Westchester, IL 60154

Insurance Institute of America
720 Providence Road
Malvern, PA 19355

Joint Commission on Accreditation of
Healthcare Organizations
1 Renaissance Boulevard
Oakbrook Terrace, IL 60181

National Council on Radiation Protection and
Measurements
7910 Woodmont Avenue, Suite 800
Bethesda, MD 20814

National Fire Protection Association
P.O. Box 9101
Batterymarch Park
Quincy, MA 02269

National Paint and Coatings Association
1500 Rhode Island Avenue, NW
Washington, DC 20005

National Pest Control Association
8100 Oak Street
Dunn Loring, VA 22027

National Safety Council
1121 Spring Lake Drive
Itasca, IL 60143

National Technical Information Service
Department of Commerce
5285 Port Royal Road
Springfield, VA 22161

Occupation Safety and Health Review
Commission
1825 K Street, NW
Washington, DC 20006

Risk and Insurance Management Society
205 E. 42nd Street
New York, NY 10017

Underwriters Laboratories, Inc.
333 Pfingsten Road
Northbrook, IL 60062

U.S. Government Printing Office
Washington, DC 20402

CHAPTER 15

BIBLIOGRAPHY

American Industrial Hygiene Association. *Do I Work in a Sick Building?* AIHA Essential Source, Fairfax, VA, February 1995.

American National Standards Institute. Safe Use of Lasers in Health Care Facilities, ANSI Standard Z136.3. New York, 1988.

Berry, Michael. *Protecting the Built Environment: Cleaning for Health.* TRICOM 21st Press, Chapel Hill, NC, 1994.

Bird, Frank E. Jr. and Germain, George L. *Practical Loss Control Leadership.* Institute Publishing (ICLI), Loganville, GA, 1985.

Bureau of Labor Statistics, U.S. Department of Labor, Washington, DC:

Occupational Injuries and Illnesses in the United States by Industry, 1990, Bulletin 2399. 1992.

Outlook for Technology and Labor in Hospitals, Bulletin 2404. 1992.

Recordkeeping Guidelines for Occupational Injuries and Illnesses. September 1986.

Survey of Occupational Injuries and Illnesses for 1992. May 1994.

Carson, H. Tom and Cox, Doye B. *Handbook on Hazardous Materials Management,* 4th edition. Institute of Hazardous Materials Management, Rockville, MD, 1992.

Centers for Disease Control and Prevention, U.S. Department of Health and Human Services, Washington, DC:

Biosafety in Microbiological and Biomedical Laboratories, 3rd edition. CDC/NIOSH, May 1993.

CDC Guidelines for the Prevention and Control of Nosocomial Infections and Guidelines for Handwashing and Hospital Environmental Control (Garner, J. and Favero, M.), Publication #99-1117. 1985.

"Essential Components of a Tuberculosis Prevention and Control Program." *Morbidity and Mortality Weekly Report,* Vol. 44, No. RR-11. September 8, 1995.

Guideline for Isolation Precautions in Hospitals, Special Report No. 54, Infection Control and Hospital Epidemiology. January 1996.

Guideline for Preventing the Transmission of Mycobacterium Tuberculosis (TB) in Healthcare Facilities, Federal Register (FR 59:54542). October 28, 1994.

Protect Yourself Against Tuberculosis—A Respiratory Protection Guide for Health Care Workers, DHHS (NIOSH) Publication No. 96-102. December 1995.

Coe, Charles P. *The Elements of Quality in Pharmaceutical Care.* American Society of Hospital Pharmacists, Bethesda, MD, 1992.

Department of Education, Office of Research. Choosing the Right Training Program. Washington, DC, March 1994.

Department of Transportation, Research and Special Programs Administration. Emergency Response Guidebook. Washington, DC, 1993.

Environmental Protection Agency, Washington, DC:

Asbestos Waste Management Guidance Publication. May 1985.

Building Air Quality: A Guide for Building Owners and Facility Managers. December 1991.

EPA Guide for Infectious Waste Management, PB 86-199130. May 1986.

RCRA Orientation Manual. 1990.

Equal Employment Opportunity Commission, Civil Rights Division, U.S. Department of Justice, Washington, DC:

Technical Assistance Manual (Title I), Americans with Disabilities Act. 1992.

The Americans with Disabilities Act, Questions and Answers. 1991.

Federal Emergency Management Agency. Disaster Planning Guide for Business and Industry, FM 141. Washington, DC, 1987.

Food and Drug Administration, Public Health Service, U.S. Department of Health and Human Services, Washington, DC:

FDA/CPSC Public Health Advisory: Hazards Associated with the Use of Electric Heating Pads. December 12, 1995.

FDA Safety Alert: Entrapment Hazards with Hospital Bed Side Rails. August 23, 1995.

Food Service Sanitation Manual. June 1978.

Medical Bulletin, Potential Hazards with Protective Restraint Devices. Vol. 21, No. 3.

General Services Administration, National Archives and Records Administration, Office of the Federal Register. Code of Federal Regulations. Washington, DC:

Title 10: Energy.

Title 29: Labor.

Title 40: Protection of the Environment.

Title 49: Transportation (Parts 100–199).

Gordon, Harold M. *A Management Approach to Hazard Control.* Board of Certified Hazard Control Management, Bethesda, MD, 1994.

Harpster, Linda M. and Veach, Margaret S. (Editors). *Risk Management Handbook for Health Care Facilities.* American Hospital Publishing, Chicago, 1990.

Head, George L. and Horn, Stephen II. *Essentials of Risk Management,* Volume 1, 2nd edition. Insurance Institute of America, Malvern, PA, 1991.

Joint Commission on Accreditation of Healthcare Organizations, Chicago, IL:

1996 Joint Commission Accreditation Manual for Hospitals.

Plant, Technology and Safety Management Handbook. 1989.

Plant, Technology and Safety Management Handbook, Quality of Care. 1994.

National Council on Radiation Protection. Report Number 48: Radiation Protection for Medical and Allied Health Personnel. Washington, DC, 1976.

National Fire Protection Association, Quincy, MA:

Fire Protection Handbook, 17th edition. 1991.

Flammable and Combustible Liquids Code Handbook. 1990.

Healthcare Facilities Handbook. 1993.

Life Safety Code Handbook. 1992.

National Electrical Code (NEC) Handbook. 1993.

NFPA 99: Healthcare.

National Institute of Occupational Safety and Health, U.S. Department of Health and Human Services, Washington, DC:

A Guide to Safety in Confined Spaces, Publication 87-113. 1987.

Guidelines for Protecting the Safety and Health of Health Care Workers. September 1988.

National Restaurant Association, Washington, DC:

A Safety Self-Inspection Program for Foodservice Operators. 1988.

Sanitation Self-Inspection Program. 1983.

National Safety Council, Itasca, IL:

Accident Facts. 1993.

Accident Prevention Manual for Business and Industry, Administration and Programs, 10th edition. 1992.

Accident Prevention Manual for Business and Industry, Engineering and Technology, 10th edition. 1992.

Ergonomics: A Practical Guide. 1991.

Fundamentals of Industrial Hygiene, 3rd edition. 1988

Occupational Health and Safety, 2nd edition, 1994.

Supervisors Safety Manual, 8th edition. 1993.

"Preventing Bed Falls." *Nursing Update,* Vol. 4, No. 1, Spring 1993 (J.T. Posey Company, Arcadia, CA).

"Proper Patient Positioning in Wheelchairs." *Nursing Update,* Vol. 5, No. 1, Winter 1994 (J.T. Posey Company, Arcadia, CA).

Occupational Safety and Health Administration, U.S. Department of Labor, Washington, DC:

All About OSHA, OSHA 2056. 1992.

Concepts and Techniques of Machine Safeguarding, OSHA 3067. 1992.

Controlling Electrical Hazards, OSHA 3075. 1991.

Exposure to Bloodborne Pathogens, OSHA 3127: 1993.

Exposure to Hazardous Chemicals in Laboratories, OSHA 3119. 1989.

Fact Sheets:

    89-32: Control of Hazardous Energy Sources.

    91-17: Ethylene Oxide.

    92-19: Responding to Workplace Emergencies.

    92-27: Occupational Exposure to Formaldehyde.

    93-03: Eye Protection in the Workplace.

    93-06: Better Protection Against Asbestos in the Workplace.

    93-09: Back Injuries—Number One Workplace Safety Problem.

    93-41: Workplace Fire Safety.

Framework for a Comprehensive Health and Safety Program in the Hospital Environment. 1993.

Guidelines for Preventing Workplace Violence for Health Care and Social Service Workers, OSHA 3148. 1996.

Hazardous Waste and Emergency Response, OSHA 3114. 1989.

Industrial Hygiene, OSHA 3143. 1994.

Material Storing and Handling, OSHA 2236. 1992.

OSHA Inspections, OSHA 2098. 1994.

Respiratory Protection, OSHA 3079. 1993.

Stairways and Ladders, OSHA 3124. 1993.

Training Requirements in OSHA Standards, OSHA 2254. 1995.

Working Safely with Video Display Terminals, OSHA 3092. 1991.

Workplace Violence Guidelines, OSHA 3841. 1995.

Work Practice Guidelines for Personnel Dealing with Cytotoxic (Antineoplastic) Drugs, OSHA Instruction Pub. 8.1.1, Appendix A. 1986.

Petersen, Dan. *Safety Management: A Human Approach,* 2nd edition. Aloray, Goshen, NY, 1988.

Reinhardt, Peter A. and Gordon, Judith G. *Infectious Waste Management.* Lewis Publishers, Chelsea, MI, 1991.

Tweedy, James T. Nursing Home Safety and Workers' Compensation Management Guide. 1993.

U.S. Congress. The Occupational Safety and Health Act of 1970, Public Law 91-596 (as amended by Public Law 101-552, November 5, 1990). Washington D.C.

INDEX

A

Abrasive wheels, 176, 485
Absenteeism, 11
Absorption, 217
Accessibility, 105–106
Accident, 43–44
  causal factors, 62
  data, 11, 24–25
  investigation, 9, 12, 35, 60–64
    report, 453–454
  prevention, 5, 9, 44, 332–333
    signs and tags, 178
  report, 12, 47, 48
Accreditation, Joint Commission, 107, 108–109
Acetone, 232–233, 366
ACGIH, see American Conference of Governmental Industrial Hygienists
Acid-fast bacillus, 272, 282
Active surveillance, 48
Acute effects, 52
ADA, see Americans with Disabilities Act
Adhesives, 244
Administrative controls, 51, 459
Administrative Procedure Act, 79
Administrative support, 8
Adult training techniques, 69
Aerosol-delivered drugs, 309–310
Aerosols, 216, 374
Agency for Toxic Substances and Disease Registry, 386
Aggravation, 312
Aggressive behavior, 312
Agreement state, 353
AIDS, 273, 411, see also HIV
Airborne radiation, 353, 357
Airborne substances, hazardous, 216

Airborne transmission, 271
Air conditioning, see Heating, ventilation, and air conditioning system
Air contaminants, 201
Air-line respirators, 59
Air pollutants, 93–94
Air-purifying respirators, 59
Air sampling, 428
Aisles, 165, 468
ALARA, see As low as reasonably achievable
Alarms
  bed, see Bed
  fire, see Fire alarm
Alcohol abuse, 104, see also Substance abuse
Allergens, 217
Alpha particle, 353, 356
American Association of Anatomists, 375–376
American Association of Hospital Pharmacists, 324–326
American Conference of Governmental Industrial Hygienists, 114–115, 385, 489
  exposure limits, 52, 53, 54, 212, 214
  formaldehyde exposure recommendations, 227
American Hospital Association, 54, 489
American National Standards Institute, 54, 55, 57, 112, 386, 489
  cumulative trauma disorders standard, 161
  emergency wash requirements, 60
  ladder standard, 175
  MSDS standard, 220
  warning signs standard, 178
American Society for Testing and Materials, 54, 57, 113, 489
American Society of Healthcare Engineers, 113–114
American Society of Healthcare Risk Managers, 23

American Society of Heating, Refrigerating, and Air Conditioning Engineers, 199, 200, 205–206, 489
American Society of Hospital Pharmacists, 303
American Society of Safety Engineers, 116, 489
Americans with Disabilities Act of 1990, 25, 31, 53–54, 100–106, 385, 482
Ammonia, 224–225
Anemometer, 205, 385
Anesthetic gases, 321, 322
Anesthetic waste gas, 196
Anger, 312
Animal wastes, 250, 251
Announcement codes, 127
Anosmia, 386
ANSI, see American National Standards Institute
Anthropometry, 161
Antibiotics, 96, 201
Antigen, 411
Antimicrobial, 386
Antineoplastic drugs, 201, 229–232, 305
Antineoplastic wastes, 250
Antiseptic, 267, 273
Anti-slip surfaces, 165, 340, 470, 471
Area monitoring, 53
Arm protection, 57
Armrests, 303
Asbestos, 203, 212, 240–244, 259
 OSHA standard, 71, 72
Asbestosis, 242
ASHE, see American Society of Healthcare Engineers
ASHRAE, see American Society of Heating, Refrigerating, and Air Conditioning Engineers
ASHRM, see American Society of Healthcare Risk Managers
As low as reasonably achievable, 351
Asphyxiant, 386
Assessment, in developing written healthcare safety management program, 10
Atomic Energy Act of 1954, 250
Atomic number, 353
Atomic weight, 353
Attending physician, 410
Audiometric testing, 189–191
Authority/responsibility, 10
Autoclave, 255–256, 323, 386
Auto-ignition temperature, 141, 386
Automatic sprinklers, 137 138, 147, 386, 476

B

Bacilli, 267
Background radiation, 353, 386
Back injury, 167–170, 293–297, see also Ergonomics; Lifting
Back support belts, 170
Bacteria, 267–268, see also Infection control
Bactericide, 267
Bacteriostat, 267
Bath lifts, 160
Bathroom, 299, 300, 302
Battery-charging areas, 173
Bed, 160, 295, 296, 302, 306
 alarm, 309
 headboard, 308
 height, 307–308
 mattress, 308
 side rail, 300, 308–309
Behavior, see Human behavior
Benches, 373
Benzene, 212, 233, 485
 OSHA standard, 71
Beta particle, 353, 356, 386
Biohazard, 213, 370
 label/symbol, 276, 287, 379, 381, 408, 484
Biological hazards, 45, see also specific hazards
 control, 265–292, see also specific topics
 bloodborne pathogens, 274–277
 exposure control management, 286–290
 hepatitis, 277–280
 HIV, 280–281
 infection control, 267–272
 infectious waste management, 290
 sharps and needles, 272–274
 tuberculosis, 281–286
Biological monitoring, 32, 53
Biological products, 96
Biological safety cabinets, 229, 230, 326, 378, 380, 381, 382, 415
Biologics, 95, 96
Biomechanics, 161
Biosafety levels, 377–382
Blade guards, 173
Bleach, 267
Blizzard, 131
Blood, 250, 251, 411
 precautions for exposure to, 449–451
Bloodborne pathogens, 89, 271, 274–277, 386, 411
 exposure control plan, 404–413

OSHA standard, 4, 32, 54, 71, 99, 212, 213, 260, 274–276, 288–290, 332, 483–484
Blood products, 274–275, 411
  exposure control management, 286–290
BLS, see Bureau of Labor Statistics
Board of Certified Healthcare Safety Management, 115–116, 490
Board of Hazard Control Management, 115, 490
Body fluids, 250, 251, 274–275
  precautions for exposure to, 449–451
Boiler room, 171
Boiler systems, 194–195
Boiling point, 386
Bomb threats, 133, 464
Bonding, 215, 387
BSCs, see Biological safety cabinets
Budget, 8
Building-related illness, 202–203, 387
Bulletin, safety, 75
Bureau of Labor Statistics, 88
Bureau of Radiological Health, 96
Burns, 179–180, 306, 339, 471

C

Cabinet hoods, 205
Cabinets, 142, 465, 466, see also Biological safety cabinets
Call buttons, 299
Call system, 310
Canopy hoods, 205
Capture velocity, 205
Carbon dioxide, 204, 387
Carbon monoxide, 94, 366, 387
Carcinogens, 212, 326, 376, 387, 419, 427, 487
Carpal tunnel syndrome, 164, 387
Carpets, 201, 299, 301, 302, 470
Cartridges, respirator, 59–60
Carts, 335, 338, 473, 476
Case management, 33
Caution signs, 142, 178–179, 468
CDC, see Centers for Disease Control and Prevention
Ceiling limit, 52, 214, 387, see also specific hazardous materials
Centers for Disease Control and Prevention, 5, 54, 99, 213, 387, 490
  infection control guidelines, 265–266, 274
  tuberculosis guidelines, 283–284, 285–286
Central supply area, 46, 228, 323–324
CERCLA, see Comprehensive Environmental Response, Compensation and Liability Act, 92, 95

Certification, Joint Commission, 106
Certified Hazard Control Manager, 387
Certified Healthcare Safety Professional, 115
Certified Safety Professional, 387
CGA, see Compressed Gas Association
Chain of custody, 38
Chairs, 105, 166, 167, 303, 340, see also Ergonomics
Charting, 310
Chemical dependency, 37, see also Substance abuse
Chemical disinfection, 257
Chemical emergency response training, 136, 137
Chemical hazards, 45, 333, 341, 366, 375–376, see also Hazardous chemicals
Chemical hygiene plan, 367, 388, 414–420
Chemical inventory forms, 93
Chemical inventory list, 400
Chemical Manufacturers Association, 220
Chemical sensitization, 217–218
Chemical storage, 218
Chemistry, 365
Chemotherapeutic agents, 229–232, 305, see also Antineoplastic drugs
Chlorine dioxide, 257
Chlorine gas, 224
Chronic effects, 52
Chrysotile, 240
Circuit breakers, 182, 475, 476
Circuit protection devices, 181
Citations, OSHA, 84–85
Civil disturbances, 133
Claims management, 35
Clean Air Act, 93–94, 95, 135, 206, 387
Cleaning, 331–336, 469
Cleaning agents, 334, 472, see also specific substances
Clean Water Act, 94–95, 247, 388
Cocci, 267
Code of Federal Regulations, 79, 80–81
Combustible, 388
Combustible liquids, 140–142, 215, 223, 427, see also specific substances
Combustible materials, 140, 223, see also specific materials
Combustible wastes, 374
Command center, 129, 139
Commitment, 8, 9, 349, 458
Communication systems, 196
Complaints, 84
Compliance inspections, OSHA, 90

Comprehensive Environmental Response,
  Compensation and Liability Act, 92, 95
Compressed air, 174, 195
Compressed Gas Association, 114, 143, 195, 490
Compressed gases, 143–144, 245, 321, 427, 485
Computer screens, 163, 165–167
Conductive materials, 179
Conductors, 180
Confidentiality, 54, 89, 431
Confined spaces, 185–188
  OSHA standard, 71, 185, 186
Consensus standards, 83
Consent, 33, 54
Construction safety, 177–178, 469
Consumer Product Safety Commission, 307
Contact dermatitis, 218, 224
Contact isolation, 272
Contact transmission, 271
Containers
  handling procedures, 254
  infectious waste, 290, 407, 408
  labeling, 276, 290, 478
  unlabeled, 252
  working safely with, 371–373
Containment, 377
Contaminated, 411
Contamination, 342, 353
Contingency planning, 135
Continuing education, 69–70
Continuous improvement, 28, 29
Continuous quality improvement, 8
Contractors, 402, 434
Control, 8
Controlling, 40
Cooking areas, 155, see also Food service
Coordinating, 40
Cork borers, 371
Corridors, 467
Corrosives, 215, 223, 388, 427
Corrosivity, 211
Cost containment, 35
Countertops, 164
Cross-contamination, 343
CTDs, see Cumulative trauma disorders
Cumulative trauma disorders, 161–162, 164,
  455–456
Curie, 353
Cutting hazards, 339, 341
Cylinders, 323
  compressed gas, 143–144, 321
  flammable gas, 174
  medical gas, 195
Cytology, 365

D

Data processing, 46
Decay, 353
Decontamination, 353, 411, 486
  personal protective equipment, 406
  radioactive, 363
Dental service, 46
Department of Transportation, 98–99, 132, 490
  hazard classes, 223
  hazardous materials markings, 260–261
  hazardous materials regulations, 141
Departmental safety programs, 12, 17
Departmental specialization, 41
DeQuervain's disease, 164
Dermatitis, 217, 218, 224, 227, 334, 366
Desiccators, 373
Desks, 105, 164, see also Ergonomics
Dialysis units, 46, 228, 323
Diesel-powered equipment, 174
Dietary management, 337
Dilution ventilation, 205
Dining room, 342, 469, see also Food service
Dioxane, 233–234
Diphtheria, 272
Directing, 40
Disability, 35–36, 100, see also Americans
  with Disabilities Act
Disaster, 123, see also specific types
Disaster planning
  food service operations, 343–344
  maintenance department, 173
Disaster plans, 17, 467, 481
Disaster preparedness, pharmacy, 327
Discrimination, 100, see also Americans with
  Disabilities Act
Disinfectants, 223–226, 267, 273, 333, 389, see
  also specific substances
Disinfection, 257, 267–268, 331, 487
Disposal, waste, 252–259, 407–408, see also
  Waste; specific type of waste
  radioactive, 361–363, 375
Documentation, 10, 482
  hazardous waste training, 260
  hearing conservation, 190
Doorways, 300, 302
Dose, 353
DOT, see Department of Transportation
Drain-cleaning chemicals, 244–245
Drills, 128–129, 446–448, 467, 481
  fire, see Fire drills
Drive belts, 173
Droplet transmission, 271

Drug-free workplace, 429
Drug-Free Workplace Act of 1988, 37–38
Drug quality, 324
Drug use, 104, see also Substance abuse
    screening for, 38
Dry bulb temperature, 389
Ducts, 205
Dust, 216, 389, 427

E

Earthquakes, 132
Education
    continuing, 69–70
    disaster, 129
    drug use, 430–431
    ergonomics, 161–162
    hazard communication, 479–480
    hazardous materials/waste management, 246
    hearing conservation, 190
    pharmacy safety, 326
    safety, 4, 12
    safety management, 8, 9, 481
    security, 346, 348–349
    topics, 18
    tuberculosis, 286
Electrical appliances, 181, 306, 334, 340, 474
Electrical burns, 179–180, 306
Electrical cords, 165, 332, 434, 439, 467
Electrical distribution systems, 193
Electrical equipment, 474, see also specific
        topics
    in environmental services, 334
    fire safety, 136
    food service, 339–340, 472
    lockout, 434
    in maintenance department, 173
    office safety, 165
    in plant management areas, 197–198
    in surgical/operating room, 321
Electrical fires, 144–145
Electrical hazards, 179
Electrical power distribution, 192
Electrical safety, 17, 179–185, 474–476, see
        also specific topics
    burns, 179–180, 306
    circuits, 181
    conductive materials, 179
    energy sources, 184–185
    equipment, 181
    fire safety, 138
    grounding, 180, 215, 339, 472, 485
    guarding, 173, 180, 339, 439, 472
    hazards, 179

    lockout/tagout, see Lockout/tagout
    National Electrical Code, see National
        Electrical Code
    OSHA standard, 71, 181, 182
    in patient areas, 306
    protecting workers, 57, 180
    requirements, 182–183
    rules, 183–184
    safe work practices, 181
    shock, 179, 340
Electrical shock, 179, 340
Electrical switches, 173, 476
Electric tools, 176
Elevators, 303
Emergency action, OSHA standard, 70
Emergency assistance agencies, 122
Emergency coordinator, 122–123
Emergency drill, 446–448
Emergency escape breathing apparatus, 59
Emergency notification, 93
Emergency plan, 10, 121–122, 481
Emergency planning, 17, 121–157, see also
        specific topics
    assistance agencies, 122
    bomb threats, 133, 464
    categories, 123
    civil disturbances, 133
    compressed gases, 143–145
    coordinating response, 122
    emergency coordinator, 122–123
    evacuation procedures, 139–140
    external disasters, 124–126
    facility management, 124
    fire safety, 136–139, 145–149, 154–155
    food service operations, 343–344
    hazardous materials, 134–136
    ignitable materials, 140–142
    internal disasters, 126–129
    Joint Commission requirements, 123
    Life Safety Code, 149–154
    natural disasters, 130–132
    plan development, 121–122
    process, 123–124
    sabotage, 134
    terrorism, 133
    transportation accidents, 132–133
Emergency Planning and Community Right-to-
        Know Act, 92–93, 134–135
Emergency power, 173, 193
Emergency preparedness drill, 467
Emergency response, 420, 485
Emergency response plans, 93
Emergency response procedures, 127–128

Emergency response standard, 71, 260
Emergency wash requirements, 60
Emotions, 312–315, see also Stress
Employee assistance program, 37, 38, 431
Employee complaints, 84
Employee confidentiality, 54, see also Confidentiality
Employee health, 8, 10
  records, 89
Employee health director, 8
Employee health program, 30–33, see also Occupational health program
Employee risk concerns, 25
Employment trends, 2–3
Energy Reorganization Act of 1974, 350
Engineering controls, 389, 459
  bloodborne pathogens exposure, 404, 412, 483
  formaldehyde exposure, 228
  hazard control, 51
  lab safety, 368
  ventilation, 204–205
Engineering department, 46
Engineering standards, 204
Enteric isolation, 272
Entrances, 105, 335
Entry permits, 187
Environment, 162
Environmental controls, 268–269
Environmental hazards in patient care areas, 301–303
Environmental Protection Agency, 91–95, 135, 389, 490
  asbestos advisory, 244
  infectious waste guidelines, 250
  pesticide regulations, 245
  refrigerant recycling, 206–207
  waste regulation, 248
Environmental services, 46, 217, 331–336
Environment of Care standards, 2, 4, 11–14, 109–111, 311
EPA, see Environmental Protection Agency
Epidemiology, 270
Equal Employment Opportunity Commission, 25
Equipment
  carts, 335
  food service, 338, see also Food service
  form for reporting problem, 462
  grounding, 180, see also Grounding
  inspection, 335
  maintaining, 12, 335
  management, 482
  safety, 17, 306

Equivalency, 153
Ergonomics, 25–26, 45, 159–164, 389, 485, see also Lifting
  cumulative trauma disorders, see Cumulative trauma disorders
  pain/strain report, 455–456
  in patient care areas, 293
  in physical plant, 159–164
  workstation evaluations, 163
Escherichia coli, 343
Essential job functions, 31, 101, 102
Ethyl alcohol, 224, 309
Ethylene oxide, 32, 201, 212, 238–240
  in central supply area, 322, 324
  emergency response procedures, 311
  in medical waste disposal, 256
  OSHA standard, 71, 212, 238
  in surgical/operating room, 321
Evacuation plan, 485
Evacuation procedures, 128, 139–140
Evacuation routes, 467
Evaluating performance, 28
Evaluation guidelines, 10
Evaluation system, 29
Exhaust, 204–205
Exhaust hoods, 325
Exits, 149, 467, 468, 469
Exit signs, 485
Explosives, 99, 215, 223, 427
Exposure control management, infection, 286–290
Exposure control plan, 4
  bloodborne pathogens, 275, 404–413
  tuberculosis, 284–285
Exposure incident, 412
Exposure limits, 52–53, 214, 390, 428, see also specific hazards
Exposure monitoring records, 89
Exposure routes, 216–217
Exposure standards for ionizing radiation, 356
Extension cords, 165, 173, 181, 183, 306, 439, 467, 469, 474
External disaster planning, 124–126
External disasters, 123, 125, see also specific types
Extractions, 372
Eye protection, 56, 333, 365, 421
Eyewash facilities, 486, 483

F

Face protection, 56, 333, 421
Facility security, see Security

Factory Mutual Research Corporation, 113, 145, 390
Fair Packaging and Labeling Act, 96
Falls, 296, 299, 340, 461, see also Slips, trips, and falls
  prevention management, 298–301
Fatalities, 84
  OSHA reporting requirements, 89–90
FDA, see Food and Drug Administration
Federal Emergency Management Agency, 100, 122, 490
Federal Insecticide, Fungicide, and Rodenticide Act, 95, 245
Federal Register, 79, 80, 83, 265, 390
Federal Register Act, 79
FEMA, see Federal Emergency Management Agency
Fibers, 216
Film badge, 353, 360
Filters, 283
Fire, 145–147
  classes of, 145–146, 388
  containment of, 129, 146, 150–152
  controlling smoke, 146–147
  development stages, 146
  electrical, 144–145
  prevention, 17, 70, 159, 482
  protection, 173
    Life Safety Code, see Life Safety Code
  response actions, 138–139
  safety, see Fire safety
Fire alarm, 137, 138, 153–154, 159
Fire drills, 138, 159, 482
Fire exits, 147
Fire extinguishers, 128, 137, 138, 147–149, 339, 438, 467, 473, 485
Fire-extinguishing systems, 154–155
Fire-fighting equipment, 129
Fire inspection, 138, 159
Fire plan, 17
Fire prevention program, 17
Fire response planning, 129
Fire safety, 136–139
  emergency planning, 137–138
  Life Safety Code, see Life Safety Code
  pharmacy, 327
Fire Safety Evaluation System, 152
First Report of Injury Form, 35
First responder, 136, 390
Flammability, numerical ratings, 222
Flammable, 427
Flammable atmospheres, 186
Flammable gas cylinders, 144

Flammable gases, 223
Flammable limits, 141
Flammable liquids, 140–142, 207, 215, 223, 390, 466, see also specific substances
Flammable materials, 140, 174, see also specific materials
Flammable range, 141
Flammable solids, 223, see also specific materials
Flash point, 140–141, 215, 390
Floods, 131–132
Floors, 340, 439, 465, 468, 470
  accessibility, 105
  cleaning, 332, 335
  in food service area, 471
  office safety, 165
  in patient care areas, 299, 301, 302, 309
  slip, trip, and fall prevention, see Slips, trips, and falls
Flywheels, 205
FM, see Factory Mutual Research Corporation
Food and Drug Administration, 95–97, 198, 307
  laser classes, 364
  recommendations for restraints, 312–313
  regulation of medical devices, 314, see also Safe Medical Device Act of 1990
  regulation of radiopharmaceuticals, 358
Foodborne illnesses, 342–343
Food contamination, 342
Food, Drug, and Cosmetic Act, 96
Food service, 114, 337–344, 439, 471–473
  areas, 171
  hazards, 46
  workers, 217
Food service manager, 344
Foot protection, 56–57, 333, 421
Footstools, 300
Formaldehyde, 32, 101, 212, 227–228, 366
  OSHA standard, 71
Formalin, 228–229
Formal investigations, 63
Fuel oils, 223
Fume hoods, 205, 366, 369, 415, 416, 425
Fumes, 216, 427
Fuses, 181, 476

G

Gait belts, 160
Gamma radiation, 354
Gamma ray, 390
Garbage, 247
Gases, 216, 391, 427
  anesthetic, 321, 322

Gas masks, 59
Gasoline, 245
General duty clause, 5, 82, 347, 485
General industry standards, 82
Geri-chairs, 160
Germicide, 267
Glass apparatus, 372
Glass tubing, 371
Glassware, 252, 471, 472
Glove boxes, 415
Glutaraldehyde, 201, 226
Grab bars, 302, 470
Gram-negative bacteria, 224, 268
Gram-positive bacteria, 268
Grinders, 177, 471
Ground-fault circuit interrupters, 181, 474
Grounding, 180, 215, 339, 472, 485
Grounds maintenance, 12
Guarding, 173, 180, 339, 439, 472

H

Half-life, 354
Hallways, 468
Hand grips, 300
Hand protection, 57, 333, 421–422
Handrails, 300, 301
Hand tools, 174
Handwashing, 271, 405, 483
Hazard, 7
Hazard assessment, 55, 458–459
Hazard audits, 48
Hazard communication, 484
Hazard communication program, 218–221,
    399–403
  evaluation checklist, 477–480
  pharmacy, 327
Hazard communication standard, 4, 71, 213,
    218–220, 332, 369, 391, 399, 440–444,
    484
Hazard control, 1–5, 43–78, 459, see also
    Hazards; Healthcare hazard control;
    specific topics
  accidents, 43–44
    investigations, 60–64
  analyzing hazards, 49
  audits, 48
  biological, see Biological hazard control
  categories of hazards, 44–45
  emergency wash requirements, 60
  evaluating hazards, 8, 48–49, 50
  identifying hazards, 8, 44–45, 46
  job hazard analysis, 49–50, 392
  location of hazards, 46

maintenance department, 173–174
personal protective equipment, 54–57
records review, 47
respiratory protection, 57–60
safety
  promoting, 74–76
  training, 65–74
security, 348
surveys, 45, 47–48
types of controls, 50–51
worker exposure, 52–54
worksite analysis, 5, 45, 47, 163, 348
Hazard control coordinator, 8
Hazard control management, 1–2, 45
Hazard control management program, 7
Hazard control manager, 9
Hazard control program, written, 9–10
Hazard correction, form for, 463
Hazard evaluation, 8, 48–49
Hazard identification, 8, 44–45, 46
Hazard Material Identification Guide System,
    222–223
Hazardous chemicals, see also Hazardous
    materials; Laboratory safety; specific
    chemicals
  characteristics of, 211–213
  control measures, 415–416, see also Chemical
    hygiene plan
  hazard communication program, 399–403
  hazard evaluation checklist, 477, 478
  laboratory training requirements, 427–428
  listing, 8
  training, 416–417
  transportation of, 132
Hazardous materials, 211–263, 391, see also
    Hazardous chemicals; specific materials;
    specific topics
  adhesives, 244
  antineoplastic drugs, 229–232
  asbestos, 240–244
  categories, 213–216
  characteristics of, 211–213
  compressed gases, 245
  disinfectants, 223–226
  disposal, 252–259
  drain-cleaning chemicals, 244–245
  exposure to, 216–218
  gasoline, 245
  glutaraldehyde/formaldehyde, 226–229
  hazard evaluation checklist, 477, 478
  HAZCOM training plan, 440–445
  Joint Commission guidelines, 246, 326
  labeling, 221–223

lead, 236–237
management, 17–18
medical, 250–252
mercury, 235–236
methyl methacrylate, 234–235
OSHA hazard communication program,
    218–221
oxides, 237–240
paints, 244
peracetic acid, 235
pesticides, 245
regulation, 248–250
release, emergency response, 134–136
solvents, 232–234
training, 259–261, 401–402
waste management, 246–248
Hazardous Materials Identification System, 223
Hazardous materials list, 375
Hazardous Materials Transportation Uniform
    Safety Act of 1990, 98–99
Hazardous substances, 187, see also specific
    substances
airborne, 216
release, 34–35, 95
safeguards for, 368–369
Hazardous waste, 391
Hazardous waste operation, OSHA standard, 71
Hazard prevention, 5
security, 348
Hazards, 44–51, see also Hazard control;
    specific hazards; specific topics
analyzing, 49
audits, 48
categories of, 44–45
classes of, 391
control of, see Hazard control
evaluating, 8, 48–49, 50
identifying, 8, 44–45, 46
job hazard analysis, 49–50, 392
prevention, 5, 348
records review, 47
surveillance, 12, 48
surveys, 45, 47–48
types of controls, 50–51
worksite analysis, 5, 45, 47, 163, 348
Hazard surveillance program, 12
physical hazards, 299
Hazard surveys, 45, 47–48
Hazard warning systems, 51
HAZCOM plan, 333, 367
HAZCOM standard, 4, 71, 213, 218–220, 332,
    369, 391, 399, 440–444, 484
HAZCOM training plan, 440–445

HAZMAT program, 481
HBV, see Hepatitis B
Headboard, 308
Head protection, 421
Health assessments, 31–32
Healthcare employment trends, 2–3
Health Care Financing Administration, 99–100
Healthcare hazard control, 1–5, 43–78, see also
    Hazard control
benefits of, 1–2
historical perspectives, 2
key program elements, 7
new directions in, 2–3
program management, 4–5
terminology, 7–8
Healthcare safety, 2
orientation and training record, 423–424
Healthcare safety management, 7–42, see also
    Safety management; specific topics
Environment of Care standards, 2, 4, 11–14,
    109–111, 311
healthcare hazard control, 7–8
human resources management, 4, 37–39
information management, 18–21
occupational health program, 30–33
program, 8–11
quality improvement, 26–30
review of management, 39–41
risk management, 21–26
safety committee, 14–18
workers' compensation, 33–37
Health hazards, numerical ratings, 222
Health Resources Services Administration, 99
Hearing conservation, 70, 188–192, 207, 391,
    457, 486
OSHA standard, 70, 191, 486
Heat exhaustion, 172
Heating pads, 307
Heating, ventilation, and air conditioning
    system, 173, 199–207, 425
ASHRAE standards/guidelines, 199, 200,
    205–206
building-related illnesses, 202–203, 387
general requirements, 199
indoor air quality, see Indoor air quality
maintenance, 199–200
refrigerant recycling, 206–207
system design/function, 199
ventilation, 204–205
worker training, 200
Heat-related hazards, 171–172
Heat stroke, 171–172
Heavy work, 33

Helicopter safety, 315–316
Helmets, 55–56, see also Head protection
Hematology, 365
Hepatitis, 273, 277–278, 373, 412
Hepatitis A, 277, 278
Hepatitis B, 54, 260, 271, 272, 274, 275,
    277–280, 409, 410, 412, 484
  exposure control management, 286–290
Hepatitis C, 274, 277, 280
Hepatitis D, 277, 280
Hepatitis E, 277, 280
High-efficiency particulate air filter, 241, 282
HIV, 260, 273, 274, 276, 277, 280–281, 411
  exposure control management, 286–290
Home health services safety, 319–321
Horizontal transport, 193
Hot wires, 180
Housekeeping, 287, 406–407, 438, 469, 470,
    483
Human behavior, 4, 8, 73–74, 312
Human immunodeficiency virus, see HIV
Human resource management, 4, 37–39
Hurricanes, 130
HVAC, see Heating, ventilation, and air
    conditioning system
Hydrogen sulfide, 366
Hygiene factors, 40
Hypochlorite, 333

I

ICES, see Information collection and evalua-
    tion system
Ignitability, 211
Ignitable materials, 140–142
Ignitable solid waste, 249
Ignition source, 215
Ignition temperature, 215, 391
Immediately dangerous to life or health, 214,
    391, see also specific hazardous materials
Imminent danger, 83–84
Immune, 412
Immunization, 271, 272, 273, 412, see also
    Vaccination
Improvement cycle, 29
Improvement process, 28–29
Improving organizational performance, 27–28
Incidence rates, OSHA, 90
Incident report, 9, 310–311
Incineration, 256
Indicator measurement system, 109
Indirect contact transmission, 271
Indoor air quality, 200–204, 205, 392, 425–426
Industrial hygiene surveys, 8

Infection control, 481, see also Biological
    hazard control
  disinfecting, 267–268
  guidelines, 265–266, 269–270
  home care, 320–321
  immunization, 271, 272
  isolation, 271–272
  laboratory, 373–374
  management, 286–290
  principles, 271
  procedures, 268–269
Infection control committee, 269
Infectious materials, 223, see also Bloodborne
    pathogens
  exposure control plan, 404–413
Infectious waste, 250, 392
  disposal, 407–408
  management, 290
Influenza, 271, 272
Information collection and evaluation system,
    19–20, 110
Information management, 18–23
Informed consent, 33, 54
Infrared radiation, 358
Injured workers, ADA and, 104
Injuries
  rates, 3
  reporting, 35
    OSHA requirements, 86–90
Inks, 207
Inspections, 8
  fire, 138, 155
  OSHA, 83–84, 90
Instructional techniques for training programs,
    67
Instruments, 371–373
Insulation, 180
Insurance costs, 1
Interim life safety, 152, 482
Internal disaster, 126
Internal disaster planning, 126–130
Internal emergencies, 123
International Agency on Research on Cancer,
    212
Interviewing witnesses, 63
Investigational drugs, 96
Iodine, 225–226
Iodophors, 225–226, 267, 333
Ionizing radiation, 354, 356–357
  NCRP guidelines, 357–358
  OSHA standard, 70, 357
Irradiation, 257
Irritants, 427

Isolation, 265, 285
  principles, 271
  procedures, 99
  types of, 272
Isopropyl alcohol, 223–224
Isotope, 354 359

J

JCAHO, see Joint Commission on Accreditation of Healthcare Organizations
Job hazard analysis, 49–50, 392
Joint Commission on Accreditation of Healthcare Organizations, 2, 7, 106–111, 490
  accreditation, 107, 108–109
  certification, 106
  electrically powered equipment standards, 197
  emergency planning requirements, 123, 124
  Environment of Care standards, see Environment of Care standards
  equipment management requirements, 197, 198
  fire planning, 129
  hazardous materials and waste management guidelines, 246, 247, 326
  home healthcare infection control requirements, 320–321
  human resources standards, 37
  improvement cycle, 29
  indicator measurement system, 109
  infection control guidelines, 269–270
  information management requirements, 18–20
  Life Safety Code, see Life Safety Code
  quality improvement and, 26–29
  Quality of Care, 26
  radiology/nuclear medicine, 360
  safety education/orientation requirements, 69
  safety standards, 109–111, 344–346
  standards revision, 106–107
  survey process, 107–109
  surveys, 47, 48
  training requirements, 68
  utilities management requirements, 192

K

Kitchen, 171, 469

L

Labeling, 213, 221–223, 253, 400–401, 442, 465, 478, see also specific materials
  FDA requirements, 95
  formaldehyde, 228
  guidelines, 276

ignitable materials, 141
infectious waste, 408
lab safety and, 369
oxides, 240
Laboratory animals, 371
Laboratory hazards, 46
Laboratory safety, 365–382, see also specific topics
  biohazards, 370
  biosafety levels, 377–382
  carcinogens, 376
  chemical hazards, 366, 375–376
  chemical hygiene plan, see Chemical hygiene plan
  fire hazards, 366
  general guidelines, 370
  infection control, 373–374
  instruments and containers, 371–373
  lab animals, 371
  microorganisms, 371
  mutagens, 376
  OSHA standard, 72, 366–369, 414, 427–428, 485
  physical agents, 375
  spills, 376–377
  supervisor responsibilities, 369
  teratogens, 376
  ventilation hoods, 377
  waste storage and disposal, 374–375
  worker responsibilities, 369–370
Laboratory standard, 72, 366–369, 414, 485
  training requirements, 427–428
Ladders, 175, 324, 340, 438, 439, 466, 469, 471, 475
Laminar airflow hoods, 326
Lap belts, 303, 313
Lasers, 363–365
Laser safety committee, 482
Laundry, 171, 276, 336–337, 408, 411, 484
  hazards, 46
  OSHA standard, 71
  workers, 217
Lead, 94, 216, 236–237, 485
  OSHA standard, 71, 236
Leadership, 8
Leading, 40
Leak test, 354
Learning factors, 162
Legionnaires' disease, 201, 202
Life Safety, 106, 467
Life Safety Code, 110, 111, 147, 149–154
Lifting, 167–170, 438, see also Ergonomics
  in food service area, 340–341, 472

hazards, 98
in patient care areas, 293–297, 465
Lighting, 485
in food service area, 471
in parking areas, 346
in patient care areas, 299, 301, 302, 465
workstation, 163, 165, 166
Light work, 32
Limited duty status, 36–37
Linens, 336, see also Laundry
Line organization, 40–41
Liquefied natural gas, 99
Liquid propane gas, 71
Liquids, 140–142
combustible, see Combustible liquids
flammable, see Flammable liquids
Local Emergency Planning Committee, 93
Lockout/tagout, 184–185
OSHA standard, 71, 184, 433, 434
sample policy, 433–437
Long-term care facilities, 100, 268
Loss control, 22, 44
Loss control management, 8
Loss costs, 24–25
Loss trends, 25

M

Maintenance, 12
Maintenance department, 172–177, see also
specific topics
general safety, 173
hazard control measures, 173–174
hazards, 46
HVAC system, 199
ladder safety, 175
lockout/tagout, 433–437
material handling, 174–175
tool safety, 175–177
Management, 349, 458
commitment, 5
review of, 39–41
safety functions, 4
safety involvement, 8–9
Management by committee, 2
Management principles, 40
Manometer, 205
Master fire plan, 129
Material handling
hazards, 174
safe handling, 174–175
Material Safety Data Sheets, 8, 48, 259, 393,
399, 400, 416, 465
ANSI standard, 220–221

in environmental services, 332, 333
in laboratory, 366, 367, 368
OSHA requirements, 220, 366, 367, 368, 440,
442, 478, 479
in pharmacy, 327
in print shop, 207
SARA requirements, 93
storage information, 218
Mattresses, 308
Measles, 271, 272
Measurable quantity, 429
Mechanical lifts, 160, 167–168, 169
Mechanical ventilation, 204
Medicaid, 99
Medical devices, regulation, 96–97, see also
Safe Medical Device Act of 1990
Medical equipment, maintenance, 196–199
Medical evaluations, 8, 53–54, 368, 410,
417–419
ADA and, 103
Medical gas systems, 195
Medical isotopes committee, 359
Medical monitoring, 53
Medical/surgical vacuums, 195–196
Medical surveillance, 8, 31–32, 409
Medical waste, 249, 250–252, 276, 375, 393
disposal/handling, 253–259
worker training, 259–261
Medical waste program, 258–259
Medicare, 99
Medication, 95, 96, 201, 303–305, 309–310,
see also specific types
errors, 304–305
Medium work, 32–33
Mental impairment, 101
Mercury, 235–236, 366
Methicillin, 272
Methyl methacrylate, 234–235
Microbiological wastes, 251
Microbiology, 365
Microorganisms, 371
Microscopy, 365
Microwave radiation, 257, 358
Mists, 216, 393, 427
Mitt restraint, 313
Mixed waste, 249
Mixing equipment, 339
Modified duty categories, 32–33
Monitoring, 53
Monitoring devices, radiation, 361
Motivation, 8, 11, 40
MSDSs, see Material Safety Data Sheets
Mucus, 412

Mucous membrane, 412
Multiple chemical sensitivity, 202
Mumps, 272
Mutagens, 376, 393
Mycobacterium tuberculosis, 281

N

National Ambient Air Quality Standards, 93
National Council of Compensation Insurance, 34
National Council on Radiation Protection, 357–358, 361, 490
National Electrical Code, 145, 181–182, 183, 193, 321, 474
National Fire Protection Association, 111–112, 146, 149, 151, 153–155, 393, 490
  air conditioning/heating ducts, 137
  compressed gases, 321–322
  definitions of flammable/combustible liquids, 140
  fire extinguishers, 148
  fire safety, 136
  hazard rating, 221–222
  healthcare facilities, 154
  HVAC system, 199, 204, 205
  ignitable materials, 140–142
  Life Safety Code, see Life Safety Code
  maintenance department standards, 173
  metal container requirements, 142
  National Electrical Code, see National Electrical Code
  sprinkler system requirements, 155
National Fire Rating System, 221–222
National Institute of Occupational Safety and Health, 54, 97–98, 393
  asbestos exposure, 244
  biosafety levels, 377–382
  electrical shock, 340
  evaluating hazards, 50
  exposure limits, 52, 214
  guidelines for exposure to hot environments, 172
  hazardous substances, 212
  lifting guidelines, 293
  respiratory protection guidelines, 57, 283
  use of back support belts, 170
National Oceanic and Atmospheric Administration, 393
National Paint and Coatings Association, 223, 490
National Restaurant Association, 114
National Safety Council, 116, 490
National Sanitation Foundation, 114

National Toxicology Program, 212
Natural disasters, 130–132
Nature of work, 162
NCRP, see National Council on Radiation Protection
Needles, 272–274, 336, 379, 405, 439, 465
Needlesticks, 32, 272–273
Neutral conductor, 180
Newsletter, safety, 75
NFPA, see National Fire Protection Association
NIOSH, see National Institute of Occupational Safety and Health
Nitric acid, 366
Nitrogen dioxide, 94
Nitrous oxide, 195, 321, 322
Noise, see also Hearing conservation
  exposure, 486
  hazards, 188–192
  level, 207
Non-combustible wastes, 375
Non-intact skin, 412
Non-ionizing radiation, 358–359
Non-skid surfaces, 299, 468, 473
Nosocomial infections, 265, 270
NRC, see Nuclear Regulatory Commission
Nuclear materials, 248
Nuclear medicine, 46
Nuclear Regulatory Commission, 98, 326, 350–352
Nursery, 345–346
Nursing assistants, 217
Nursing homes, 129, 301, 302
Nursing personnel, 217, 299, 446
  back care safety, 295–296
  medication safety, 303
  special concerns, 309–311
Nursing services
  checklist, 465–466
  hazards, 46
Nutrition, 337

O

Occupational exposure, 412, see also specific hazards
  to bloodborne pathogens, 275
  to radiation, 355
Occupational exposure limit, 393
Occupational health program, 18, 30–33, see also specific topics
  case management, 33
  employee health program, 30–33
  health assessments, 31–32
  informed consent, 33, 54

modified duty categories, 32–33
periodic assessment, 32
post-exposure assessment, 32
pre-placement assessment, 31–32, 54
rehabilitation, 33
return-to-work assessment, 32
treatment of occupational injuries, 31
Occupational illness, 87, 162
Occupational injury, 31, 87, 393
Occupational psychology, 162
Occupational Safety and Health Act, 5, 81, 86
Occupational Safety and Health Administration,
31, 32, 81–90, see also specific standards;
specific topics
chemical emergency response training
requirements, 136
emergency wash requirements, 60
ergonomic programs, 161, 162
exposure limits, 52, 214
fire planning, 129
forms, 87–89
Form 101, 88
Form 200, 47, 52, 84, 88, 485
Form 2203, 87, 88
hazard communication program, 218–221
hazardous material regulations, 212, 213
hazardous substance release requirements,
135–136
incidence rates, 90
inspections, 83–84, 90
monitoring requirements, 53
MSDS requirements, 220
PEL, see Permissible exposure limit
poster, 87, 88
recordkeeping, 86–90, 288–290, 349, 483, 484
records disclosure, 84
reporting, 86–90
safety management guidelines, 5
standards, 4–5, 82–83, see also specific
standards
asbestos, 71, 72, 243
benzene, 71
bloodborne pathogens, 4, 32, 54, 71, 99,
212, 213, 260, 274–276, 288–290, 332,
404–413, 483–484
compliance with, checklist for, 483–485
confined spaces, 71
construction, 177–178
electrical safety, 71, 181, 182
emergency action, 70
emergency response, 71, 260
ethylene oxide, 71, 238

fire prevention, 70
formaldehyde, 71
hazard communication, 4, 71, 213, 218–220,
332, 369, 391, 399, 440–444, 484
hazardous waste operation, 71
hearing conservation, 70
ionizing radiation, 70, 357
laboratory, 72, 366–369, 414, 427–428, 485
ladders and stairs, 175
laundry machine, 71
lead, 71, 236
liquid propane gas, 71
lockout/tagout, 71, 184, 433, 434
noise, 191, 486
non-ionizing radiation, 359
personal protective equipment, 54–57
respiratory protection, 57–59, 71, 415
ventilation, 204
working in confined spaces, 71, 185, 186
training requirements, 70–72
tuberculosis regulation, 5
violations and citations, 25, 84–86
Occupational Safety and Health Review
Commission, 99
Off-duty safety, 75–76
Office equipment, 105
Office safety, 164–167
Oil, discharge of, 95
Olfactory warning properties, 217
Omnibus Budget Reconciliation Act, 100
Operating rooms, 46, 321–322
Operational factors, 11
Organic peroxide, 223
Organizational factors, 11
Organizational performance, 27–28, 108
Organizational safety officer, 13, see also
Safety officer
Organizational safety policy, 10
Organizational theory, 40–41
Organizing, 40
Orientation, 8, 12–13, 69
hazardous materials and waste management,
246
record of, 423–424
OSHA, see Occupational Safety and Health
Administration
Over-the-counter drugs, 95
Overtime, 317
Oxides, 238–240, see also specific substances
Oxidizers, 214–215, 222, 223
Oxygen, 142, 186, 195, 223, 306, 321, 322
Ozone, 94

P

Packaging requirements, 253–254, 325
Paints, 174, 244
Parenteral, 412
Parking areas, 105, 346, 469
Parr bombs, 373
Particleboard, 201
Particulate matter, 94, 206, 427
Passageways, 165
Passive surveillance, 48
Pathogen, 412
Pathological wastes, 250, 251–252
Pathology, 46, 365
Patient care areas, 293–329, 481, see also
    specific topics
  beds, 307–309
  call system, 310
  central supply, 323–324
  charting, 310
  dialysis, 323
  electrical safety, 306–307
  equipment, 306
  ethyl alcohol, 309
  ethylene oxide, 311
  home health services, 319–321
  incident report, 310–311
  lifting, 293–297
  medication, 303–305, 309–310
  nursing concerns, 309–311
  patient emotions, 312
  pharmaceutical safety, 324–327
  restraints, 312–314
  Safe Medical Device Act, 314–315
  slips, trips, and falls, 303
  smoking, 311
  special care units, 323
  stress management, 316–319
  surgical/operating room, 321–323
  volunteers, 319
  wheelchairs, 297–298
Patient emotions, 299, 312–315
Patient holding areas, 128
Patient-related risks, 25
Patient security, 346–347, see also Security
Patient transfers, 294, 295–296, see also Lifting
PEL, see Permissible exposure limit
Pentamidine, 309–310
Peracetic acid, 235
Perchloroethylene, 234
Percutaneous, 412
Performance, evaluating, 28

Performance-based standards, 12, 13, 352
Performance improvement, 27
Performance Oriented Packaging Regulation,
    223
Periodic assessment, 32
Permissible exposure limit, 52, 212, 214, 366,
    367, 394, 416, 428, see also specific
    hazardous materials
Permit required spaces, 186
Personal monitoring, 53
Personal protective equipment, 54–57, 287,
    394, 412, 421–422, 428, 443, 459, 460,
    483, 486–487, see also Respirators;
    specific hazards
  for bloodborne pathogens, 405, 406
  decontamination, 406
  environmental services workers, 333
  for formaldehyde, 227
  laboratory workers, 378, 379, 380, 381
  maintenance/operations workers, 200
  periodic worker assessment, 32
  for radiation, 357
Personal safety, 10, see also Security
Pesticides, 95, 245, 472
Pharmaceutical safety, 324–327
Pharmacy, 46, 346
Phenols, 226, 267, 333
Physical hazards, 44, 45, 333, 341
Physical impairment, 100–101
Physical plant safety, 159–209, see also
    specific topics
  confined spaces, 185–188
  construction, 177–178
  electrical, 179–185
  ergonomics, 159–164
  fire prevention, 159
  hearing conservation, 188–192
  HVAC system, 199–207
  heat-related hazards, 171–172
  lifting, 167–170
  maintenance department, 172–177
  office, 164–167
  plant management areas, 192–199
  print shop, 207
  slips, trips, and falls, 170–171
  warning signs, 178–179
Physiological factors, 162
Pipes, 174, 179, 401
Pipetting, 373, 405
Planning, 40
Plant management areas, 192–199
  boiler/steam systems, 194–195

communication systems, 196
medical equipment maintenance, 196–199
medical gas systems, 195
medical/surgical vacuums, 195–196
plumbing, 193–194
utilities management, 192–193
vertical/horizontal transport, 193
Plant services department, see Maintenance
    department
Plumbing, 193–194
Pneumoconiosis, 242
Poisons, 223
Polishing agents, 334
Positive-pressure air-supplied respirator, 243
Positive-pressure personnel suit, 382
Positive-pressure rooms, 199
Post-accident drug screen, 430
Post-exposure assessment, 32
Posture, 166, 168
Powered air-purifying respirator, 59
Power tools, 176
PPE, see Personal protective equipment
Practice drills, see Drills
Pre-employment examination, 103, 430
Pregnant workers, 232, 309, 376, see also
    Reproductive hazards
Pre-placement assessment, 31–32, 54
Prescription drugs, 95, 96
Preventive maintenance, 173, 202
Preventive maintenance program, 9
Print shop, 46, 207
Programmed inspections, 84
Promoting safety, 74–76
Property loss, 10
Proprietary standards, 83
Proton, 356
Psychological factors, 162
Psychological hazards, 45
Public demonstrations, 133
Public Health Service Act, 96
Public Law 99-499, 92
Public Law 102-141, 274
Public Law Update Service, 81
Pulling, 297
Pulmonary edema, 366
Pushing, 297

Q

Qualified individual with a disability, 101
Quality assurance, 11, 26, 481
Quality improvement, 4, 26–30
   continuous improvement process, 29

evaluating performance, 28
implementing, 27
Joint Commission and, 26–27
   improvement cycle, 29
organizational considerations, 29
organizational performance, 27–28
process and performance improvement, 27
process elements, 28–29
risk management and, 26–27
safety and, 30
Quality management, 26–27
Quality of Care, 26
Quarterly report, 20–21
Quaternary ammonium compounds, 224–225,
    333, 395

R

Rad, 354
Radiation, 191, 350–363, see also specific
    topics
   airborne, 353, 357
   background, 353, 386
   control procedures, 360–361
   exposure, 96
   FDA regulations, 358
   general procedures, 361
   ionizing, 70, 354, 356–358
   Joint Commission standards, 360
   measurement, 357
   monitoring devices, 361
   NCRP guidelines, 357–358
   non-ionizing, 358–359
   NRC, 350–352
   occupational exposure to, 355–356
   safety, 18, 485
   sources, 354–355
   terminology, 352–353
   ultraviolet, 359
   waste management, 361–363, 375
   worker safety, 351
   worker training, 351, 359–360
Radiation Control for Health and Safety Act,
    96
Radiation control programs, 98, 351
Radiation safety committee, 351, 359–360, 482
Radiation safety officer, 351
Radiation waste plan, 362
Radioactive materials, 223, 326
Radioactivity, 354
Radioisotope, 354
Radiology hazards, 46
Radiopharmaceuticals, 358

Rage, 312
Ramps, 105
Random testing, 430
Rapid eye movement, 318
Raynaud's syndrome, 164
RCRA, see Resource Conservation and
    Recovery Act
Reaching, 297
Reactives, 214
Reactivity, 211
  numerical ratings, 222
Reasonable accommodation, 101–102
Reasonable suspicion, 430
Recommended exposure limit, 214, 367, see
    also specific hazardous materials
Recordkeeping, 8, 54, 395
  ADA requirements, 102–103
  bloodborne pathogens, 288–290, 410
  lab safety, 369
  medical waste, 258–259
  OSHA, 86–90, 288–290, 349, 483, 484
  radiation, 357
  workplace violence, 349
Records disclosure, 84
Records review, 47
Refrigerant recycling, 206–207
Refrigeration, 425
Refuse, 247
Regulated waste, 404, 484
Regulations, 79–81, see also Safety regulation;
    specific regulations; specific hazards
Regulatory agencies, 81–100, see also specific
    agencies
  inspections, 48
  risk information from, 23
Regulatory standards, 8, see also specific
    standards
Rehabilitation, 33
Relationship-based risks, 25
Rem, 354
Repetitive motion injuries, 161, 163, 296–297
Reporting
  hazardous materials and waste management,
    246
  OSHA, 86–90
Reproductive hazards, 231–232, 309, 376, 419,
    427
Resource Conservation and Recovery Act,
    91–92, 94, 95, 246, 247–248, 249, 395
Respirators, 32, 57–60, 203, 243, 283, 284,
    380, 415, 486
Respirator standard, 415

Respiratory care, precautions, 452
Respiratory effects, 427
Respiratory isolation, 272
Respiratory protection, 57–60, 204, 368, 443,
    see also Heating, ventilation, and air
    conditioning system
  OSHA standard, 57–59, 71, 415
  tuberculosis, 282–283
Restraints, 300, 312–314
Return-to-work assessment, 32
Ribavirin, 309–310
Right-to-know, 221
Risk, 7
Risk assessment program, 12
Risk coordination, 24
Risk management, 4, 21–26, 481, see also
    specific topics
  activity categories, 26
  comparing accident data and costs, 24–25
  employee risk concerns, 25
  functions, 22
  information gathering, 23–24
  loss trends, 25
  patient-related risks, 25
  quality and, 26–27
  relationship-based risks, 25
  role of risk manager, 24
Risk management committee, 22–23
Risk management information system, 23
Risk management program, 8, see also Safety
    management program
Risk manager, 22–26
Roentgen, 354
Rubella, 271, 272

S

Sabotage, 134
Safeguarding, 176
Safe Medical Device Act of 1990, 96–97, 198,
    309, 314–315, 364, 482
Safety
  applying quality improvement principles to, 30
  defined, 7
  fire, see Fire safety
  management, see Healthcare safety manage-
    ment; Safety management
  off-duty, 75–76
  in patient care areas, see Patient care areas
  physical plant, see Physical plant safety
  promoting, 74–76
  regulation, see Safety regulation
  support area, see Support area

Safety action plans, 9
Safety audits, 48
Safety bars, 313
Safety belt, 395
Safety bulletins/newsletters, 75
Safety committee, 2, 9, 10–11, 14–18
  authority, 15
  evaluating program effectiveness, 15
  fall prevention management, 171
  function, 15
  involvement in organizational programs,
    16–18
  membership, 14
  organization, 15–16
  pharmacy safety, 326
  quarterly report, 20
  records review, 47
  responsibility, 14–15
  risk management functions, 22
  role of, 4
Safety director, 171
Safety hat, 395
Safety inspection reports, 8
Safety inspections, 8, 45, see also Inspections
Safety management, 4, see also Healthcare
    safety management
  action plans, 9
  annual evaluation checklist, 481–482
  management involvement, 8–9
  program, see Safety management program
  worker involvement, 9
Safety management performance standards, 13
Safety management plan, 12–13
Safety management program, 12, 109–111
  assessment of, 10
  development steps, 10–11
  elements of, 8, 9–10
  failure of, 11
  success of, 9
Safety management report, 19–20
Safety meetings, 74
Safety officer, 12, 13, 110
  appointment guidelines, 13
  duties of, 14
  involvement in organizational programs,
    17–18
  quarterly report, 20
  risk management functions, 22
Safety orientation, 12–13
Safety policy, 9, 10
Safety policy statement, 438–439
Safety poster, 74, 87, 88

Safety program
  assessment worksheet, 458–460
  improvement 13
Safety regulation, 70–119, see also specific
    agencies/associations; specific regulations;
    specific topics
  ADA, 100–106
  federal regulations, 79–81
  government agencies, 81–100
    EPA, 91–95
    FDA, 95–97
    OSHA, 81–90
  Joint Commission standards, 106–111
  professional associations, 115–116
  voluntary compliance agencies, 111–115
Safety rules, 438–439
Safety standards, see also specific standards
  Joint Commission, 106–111
Safety surveys, 13
Safety training, 65–74, see also Training
  audience for, 67
  budget, 72
  determining needs, 68
  documentation, 70
  evaluation, 70, 72–73
  frequency, 70
  instructional techniques, 67, 69
  Joint Commission requirements, 69–70
  methods, 68–69
  OSHA requirements, 70–72
  preparation checklist, 68
  program development, 65–67
  requirements, 66
  safety meetings, 74
  supervisor responsibility, 72
  topics, 65
  understanding human behavior, 73–74
Safety vest, 313
Safe work practices, 5
Salmonella, 342–343
Sanitary, 331
Sanitation, 338, 342
Sanitizing, 267
SARA, see Superfund Amendments and Re-
    Authorization Act of 1986
Scales, 160
Scavenging, 322
Screening workers, 54
Seating, 303, see also Chairs
Secretion precautions, 272
Security, 312, 344–350, 482
  emergency room, 345

hazards, 46
Joint Commission standards, 344–346
  nursery, 345–346
  parking area, 346
  patient, 346–347
  pharmacy, 346
  workplace violence, 347–350
Security management program, 18
Security program, 10
Sedentary work, 32
Self-contained breathing apparatus, 59, 186, 395
Semi-solid wastes, 255
Senior management, see also Management
  safety involvement, 8–9
Serology, 365
Sewage, 247
Sharps, 395, 411
  disposal of, 250, 252, 255, 290, 322, 324,
      407, 439, 465
  labeling, 276, 483
  in laundry area, 336
  medical wastes and, 250, 252, 255, 290, 407
  working safely with, 272–274, 405
Shift work, 318
Short-term exposure limit, 52, 214, 396, see
      also specific hazardous materials
Shower chairs, 160
Sick building syndrome, 165, 202, 395
Side rails, 308
Skill psychology, 162
Skin disorders, 218
Sleep deprivation, 318
Slicers, 471
Slip-resistant surfaces, 165, 340, 470, 471
Slips, trips, and falls, 187, 298–303, 335
  correcting environmental hazards, 301–303
  prevention checklist, 468–470
  sample report, 461
Sludge, 247
SMDA, see Safe Medical Device Act of 1990
Smoke, 216
  controlling, 146–147
Smoke detectors, 467
Smoking, 138, 144–145, 207, 306, 311, 478
Sodium azide, 174
Sodium hypochlorite, 225
Soft belts, 313
Solid waste, 247–249, 255
Solid Waste Disposal Act, 91, 247
Solvents, 174, 207, 232–234, 396, see also
      specific substances
Somatic factors, 162

Source individual, 412
Span of control, 41
Special care units, 323
Spill control procedures, 231
Spills, 302, 332, 439, 469
  chemical, 465
  containing, 290
  in laboratory, 376–377, 379
  oxides, 239
  in pharmacy, 325
  toxic/hazardous materials, 443, 486
Spirillae, 268
Spores, 267
Sprinklers, 137, 138, 147, 386, 476
Stairs, 300, 302, 468
Stairwell, 467
Standards, see specific standards; specific
      hazards
Standpipe hoses, 147
Staph infections, 272
Staphylococcus, 343
Staphylococcus aureus, 267, 272
State Emergency Response Commission, 92–93
Static pressure, 205
Steam sterilization, 255
Steam systems, 194–195
STEL, see Short-term exposure limit
Stepladders, 325, 466
Steps, 302, 335, 468
Stepstools, 324
Sterile, 331
Sterilization, 255, 256–257, 267, 323, 412
Sterilizers, 239
Storage areas, 142
Storms, 130
Stormwater discharge, 95
Strains, 340–341
Strep infections, 272
Streptococcus, 396
Stress, 316–319
Stretchers, 139, 295
Substance abuse, 37–39, 438
  ADA and, 104
Substance abuse policy, 37, 429–433
Substance abuse testing, 35
Suction flasks, 372
Suctioning, 405
Sulfur dioxide, 94
Superfund Amendments and Re-Authorization
      Act of 1986, 92–93, 134, 135, 395
Supervisor, 458
  laboratory safety responsibilities, 369

role in substance abuse program, 39
training responsibilities, 72
Support area, 331–383, see also specific topics
  environmental services, 331–336
  food service, 337–344
  laboratory, 365–382
  lasers, 363–365
  laundry, 336–337
  radiation, 350–363
  security, 344–350
Support personnel, back care safety, 296–297
Surgical areas, 46, 321–322
Surgical wastes, 252
Surveillance, 48
Survey, 47, 48, 396
  hazard, 45, 47–48, 481
  Joint Commission, 107–109
  safety, 13
Switches, see Electrical switches
Synthetic drugs, 95, 96

T

T$_4$, 412
Tendinitis, 164
Tennis elbow, 164
Tenosynovitis, 164
Teratogen, 376, 396
Terminology, 7–8, 385–397
  bloodborne pathogens, 411–412
  radiation, 353–354
Terrorism, 133
Theory X, 40
Theory Y, 40
Thermal inactivation, 256
Thermoluminescent dosimeter badge, 34
Threshold limit value, 54, 214, 396, see also
    specific hazardous materials
Thunderstorms, 130
Time-weighted average, 214, see also specific
    hazardous materials
TLD badge, see Thermoluminescent dosimeter
    badge
TLV, see Threshold limit value
Toilet chairs, 160
Tool safety, 175–177
Tornadoes, 130–131
Torso protection, 56
Toxic chemical release form, 93
Toxic chemicals, 419, see also specific
    chemicals
Toxicity, 52, 211
Toxic materials, 213–214, see also specific
    materials

Toxic substances, 187, 396, 427, 485, see also
    specific substances
Toxic Substances Control Act, 95
Training, 5, 10, 18, 481, 485
  bloodborne pathogens, 275–277, 409, 411,
    413
  chemical emergency response, 136
  disaster, 129
  environmental services, 333–334
  ergonomics, 161–162
  formaldehyde exposure, 228
  handling hazardous materials, 213
  hazard communication, 219–220, 440–445,
    479–480
  hazardous chemicals, 367–368, 416–417
  hazardous materials, 401–402
  hazardous waste, 259–260
  hearing conservation, 190
  laboratory, 427–428
  lifting practices, 169
  lockout, 435
  personal protective equipment, 55
  pharmacy safety, 326
  radiation, 352
  record of, 423–424
  respiratory protection, 59
  safety, 12, 65–74, see also Safety training
  safety and health, 459–460
  safety management, 8, 9
  security, 348–349
  tuberculosis, 286
Transfers, see Lifting; Patient transfers
Transmission, 271
  precautions, 266
  tuberculosis, 281–282
Transportation
  accidents, 132–133
  DOT hazardous materials markings, 260–261
  untreated waste, 254–255
Trash compactors, 174
Triage, 126, 446
Trigger finger, 164
Trips, 170–171, see also Slips, trips, and falls
Tuberculosis, 201, 224, 267, 268, 281–286
  CDC guidelines, 99, 283- 284, 285–286
  environmental controls, 286
  exposure control, 284–285
  infection control, 284, 373
  isolation, 272
  OSHA guidelines, 5, 89, 282–283, 485
  transmission, 281–282
  worker training and education, 286
Turnover rates, 11

Twisting, 296
Typhoid fever, 267

U

Ultrasound exposure, 191–192
Ultraviolet radiation, 359
Underground storage tanks, 92
Underwriters Laboratories, 112–113, 145, 306, 397, 490
Unit concept, 150–152
Unit dose dispensing system, 325
Unity of command, 41
Universal precautions, 286–287, 397, 404, 412, 439, 483
Untreated waste, 254–255
U.S. Code, 79, 81
USTs, see Underground storage tanks
Utilities management, 192–193, 482

V

Vaccination, 279, 289, 409, 410, 484
Vaccine, 412
Vacuum distillations, 372–373
Vacuum lines, 380
Vapor, 215, 216, 427
Vapor density, 215
Vectorborne transmission, 271
Ventilation, 204–206, 389, 443, see also
    Heating, ventilation, and air conditioning
    system
  asbestos and, 241
  in food service area, 471
  in print shop, 207
Ventilation hoods, 377
Vertical transport, 193
Very heavy work, 33
Video display terminals, 163, 165–167
Video recorders, 63
Violations, OSHA, 84–86
Violence, 347–350
Viral hepatitis, 277–278, see also Hepatitis
Virus, 268
Volunteer safety, 319

W

Warning signs, 142, 178–179, 468
Waste, see also specific types
  animal, 250, 251
  combustible, 374
  containers, 254
  disposal of, 252–259
  gases, 249–250
  handling, 336

hazardous, 71, 248–250, 391
  DOT markings, 260–261
  EPA regulation, 248
  training, 259–260
ignitable solid, 249
infectious, 290, 392, 407–408
laboratory, 374–375
management, see Waste management
medical, see Medical waste
microbiological, 251
mixed, 249
noncombustible, 375
pathological, 250, 251–252
radioactive, 361–363, 375
RCRA and, 246–248
regulated, 404, 484
semi-solid, 255
solid, 247–249, 255
surgical, 252
untreated, 254–255
Waste generator classification, 91
Waste management, 246–261
  disposal, 252–259
  medical wastes, 250–252
  RCRA, 247–248
  regulation, 248–250
  training, 259–261
Wastewater, 94, 247
Water pollutants, 94–95
Water quality standards, 94
WBGT index, 172
Weather-related disasters, 130–132
Wellness program, 75–76
Wheelchairs, 139, 160, 295, 296, 297–298, 300, 470
  safety, 297–298
  types of, 297
Williams-Steiger Act of 1970, 81
Windows, 334–335, 469
Winter storms, 131
Witness interviews, 63
Work areas, 105
Worker complaints, 84
Worker exposure, 52–53, see also Occupational
    exposure
  categories, 52
  effects, 52
  limits, 52–53
  medical evaluations, 53–54
  monitoring, 53
  screening, 54
  toxicity, 52
Worker motivation, 40

Worker safety, 9, 114
Workers' compensation, 25, 33–37
  back injuries, 167
  cost containment, 1, 9, 35
  coverage, 34
  disability categories, 35–36
  falls, 171
  injury data, 48
  limited duty status, 37
  purpose of, 34
  rates, 34–35
  records review, 47
  skin-related claims, 218
Workplace fatalities, 84
Workplace hazards, 97–98

Workplace injuries, treatment of, 31
Workplace violence, 347–350
Work practice controls, 51, 368
Worksite analysis, 5, 45, 47, 163, 348
Worksite maintenance, 407
World Health Organization, ergonomic-related
      definitions, 161–162
Wrist restraint, 313

X

X-radiation, 354
X-ray, 356
X-ray equipment, 355
X-ray table, 296
Xylene, 234, 366